Neurobiology of motor programme selection

Pergamon studies in neuroscience No 4

Series editor Dr William Winlow, *Dept. of Physiology, University of Leeds, LS2 9NQ, UK*

Neuroscience is one of the major growth areas in the biological sciences and draws both techniques and ideas from many other scientific disciplines. *Pergamon studies in neuroscience* presents both monographs and multi-author volumes drawn from the whole range of the subject and brings together the subdisciplines that have arisen from the recent explosive development of the neurosciences.

Pergamon studies in neuroscience includes contributions from molecular and cellular neurobiology, developmental neuroscience (including tissue culture), neural networks and systems research (both clinical and basic) and behavioural neuroscience (including ethology). The series is designed to appeal to research workers in clinical and basic neuroscience, their graduate students and advanced undergraduates with an interest in the subject.

1. **Function and dysfunction in the basal ganglia** *ed.* A. J. Franks, J. W. Ironside, R. H. S. Mindham, R. J. Smith, E. G. S. Spokes *and* W. Winlow
2. **Comparative aspects of neuropeptide function** *ed.* Ernst Florey *and* George B. Stefano
3. **Neuromuscular transmission: basic and applied aspects** *ed.* Angela Vincent *and* Dennis Wray

Previous titles available from Manchester University Press as studies in Neuroscience.

Neurobiology of motor programme selection

New approaches to the study of behavioural choice

edited by Jenny Kien,
Catherine R. McCrohan *and* William Winlow

PERGAMON PRESS

OXFORD · NEW YORK · SEOUL · TOKYO

U.K. Pergamon Press Ltd, Headington Hill Hall, Oxford OX3 0BW, England

U.S.A. Pergamon Press, Inc., 660 White Plains Road, Tarrytown, New York 10591-5153, USA

KOREA Pergamon Press Korea, KPO Box 315, Seoul 110-603, Korea

JAPAN Pergamon Press Japan, Tsunashima Building Annex, 3-20-12 Yushima, Bunkyo-ku, Tokyo 113, Japan

---

First edition 1992

British Library Cataloguing in Publication Data
A catalogue record for this book is available from the British Library.

Library of Congress Cataloging in Publication Data
Neurobiology of motor programme selection: new approaches to the study of behavioural choice/edited by Jenny Kien, Catherine R. McCrohan, and William Winlow.
p. cm.—(Pergamon studies in neuroscience; no. 4)
Includes index.
1. Neuropsychology. 2. Animal behavior. 3. Neural circuitry.
I. Kien, Jenny. II. McCrohan, C. R. (Cathy R.) III. Winlow, W. (William) IV. Series.
[DNLM: 1. Motor Activity—physiology. 2. Nervous System—physiology.
W1 ST927K no. 16/WE 103 N49375]
QP360.L492 1992
591.1′852—dc20
DNLM/DLC 91-21474

ISBN 0 08 041986 0

Printed in Great Britain by B.P.C.C. Wheatons Ltd, Exeter

Contents

Contributors

Garrett E. Alexander Department of Neurology, School of Medicine, Emory University, Atlanta, GA 30322, USA.

Jennifer S. Altman C Fachbereich für Biologie, Universität Regensburg, D-8400 Regensburg, Germany.

David Brody Department of Biology, University of California, San Diego, La Jolla, CA 92093, USA.

Larry B. Cohen Department of Physiology, Yale University School of Medicine, 333 Cedar Street, New Haven, CT 06510, USA.

Michael D. Crutcher Department of Neurology, School of Medicine, Emory University, Atlanta, GA 30322, USA.

Chun Xiao Falk Department of Physiology, Yale University School of Medicine, 333 Cedar Street, New Haven, CT 06510, USA.

Graham P. Ferguson Stazione Zoologica 'Anton Dohrn', Villa Comunale, I-80121 Naples, Italy.

Gary Goldberg Departments of Physiology and Physical Medicine and Rehabilitation, Temple University School of Medicine. *Mailing address*: Moss Rehabilitation Hospital, 1200 W. Tabor Road, Philadelphia, PA 19141-3099, USA.

Paul Grobstein Department of Biology, Bryn Mawr College, Bryn Mawr, PA 19010, USA.

R. Matthias Hennig Abt. Huber, Max Planck Institut für Verhaltensphysiologie, 8130 Seewiesen, Germany.

Hans-Peter Hopp Max Planck Institut für Entwicklungsbiologie, Spemann Str 35, D-7400 Tübingen, Germany.

Rene F. Jansen Department of Biology, Vrije Universiteit Amsterdam, de Boelelaan 1087, 1081 HV Amsterdam, The Netherlands.

Jenny Kien Fachbereich für Biologie, Universität Regensburg, D-8400 Regensburg, Germany.

William B. Kristan Jr Department of Biology, University of California, San Diego, La Jolla, CA 92093, USA.

Michael A. Kyriakides Neuropharm Ltd, 2 Bryngwili Road, Hendy, Dyfed SA4 1XB, UK.

Shawn R. Lockery Computational Neurobiology Laboratory, Salk Institute for Biological Studies, San Diego, CA 92138, USA.

Jill London Department of Biostructure and Function, University of Connecticut Health Science Center, Farmington, CT 06032, USA.

Catherine R. McCrohan Department of Physiological Sciences, University of Manchester, Manchester M13 9PT, UK.

Eve Marder Biology Department, Brandeis University, Waltham, MA 02254, USA.

Leonid L. Moroz Department of Physiology, Laboratory of Bioenergetics, Byelorussian State University, Minsk 220080, USSR.

Willem Mos Department of Experimental Zoology, Biological Centre, University of Amsterdam, Kruislaan 320, 1098 SM Amsterdam, The Netherlands.

Barry L. Roberts Department of Experimental Zoology, Biological Centre, University of Amsterdam, Kruislaan 320, 1098 SM Amsterdam, The Netherlands.

Vadim Roschin Institute of Higher Nervous Activity and Neurophysiology, USSR Academy of Sciences, Butlerova str 5a, 117865, Moscow GSP-7, USSR.

Stephen R. Soffe Department of Zoology, University of Bristol, Woodland Road, Bristol BS8 1UG, UK.

Naweed I. Syed Department of Medical Physiology, Faculty of Medicine, University of Calgary, Calgary, Alberta, Canada T2N 4N1.

Andries Ter Maat Faculty of Biology, Vrije Universiteit Amsterdam, de Boelelaan 1087, 1081 HV Amsterdam, The Netherlands.

James M. Weimann Biology Department, Brandeis University, Waltham, MA 02254, USA.

William Winlow Department of Physiology, Worsley Medical and Dental Building, University of Leeds, Leeds LS2 9NQ, UK.

George Wittenburg Department of Biology, University of California, San Diego, La Jolla, CA 92093, USA.

Jian-Young Wu Department of Physiology, Yale University School of Medicine, 333 Cedar Street, New Haven, CT 06510, USA.

Dejan Zecevic Institute of Biological Research, 11000 Belgrade, Yugoslavia.

Preface
Changing approaches to the neuronal analysis of behavioural choice

Until recently the cellular mechanisms underlying behavioural choice have been largely neglected by neuroethologists. Mechanisms for selection and initiation of behaviour in the vertebrates, for example, were not examined at the neuronal level, because the problem was too complex and no adequate experimental approach was available. The question was first approached in the 1960s when the behaviour of invertebrates began to be analysed in more detail. Their apparently stereotyped behaviour led to the concept of fixed action patterns. This and their simpler nervous systems suggested that behavioural choice might be mediated by relatively simple neural circuitry which might be amenable to analysis. Two characteristics which are advantageous for cellular analysis of invertebrate nervous systems are that they possess fewer nerve cells than those of vertebrates and that these are often constant in position and number. These characteristics could first be exploited during the 1960s owing to major technical advances in microelectrode recording and new dye injection methods for identifying neurones.

One very influential concept that arose in conjunction with the idea of fixed action patterns was that of the command neurone: single neurones capable of initiating or terminating patterned behaviours or behavioural sequences. This was followed by the description of small pattern generating circuits in very reduced preparations. Real behaviour was hardly ever examined in these studies, and so it was never really explained how command neurones and the command neurone concept could account for genuine behaviour. More recent work has demonstrated that invertebrate behaviour is far from stereotyped; indeed it is highly flexible. If variations in behaviour in more intact preparations are analysed, it is hard to imagine a system with one command neurone per output pattern. Furthermore, any elaboration of a command system to fit behavioural flexibility rapidly moves away from a command structure.

Such criticism has made the term 'command neurone' less popular, but, despite this, many workers still base their thinking and experimental design on variations of the command concept. The result is that they approach motor systems as linear command sequences or hierarchies. The increasing dissent from this hierarchical concept, which first arose in the late 1970s, has yet to gain full acceptance, especially in the invertebrate field.

In this book we present alternative approaches and ideas concerning the neural bases of behavioural choice. The authors come from a wide range of backgrounds, studying a variety of animals, from snails to humans. Behavioural choice is treated in terms of selection of motor output. It can mean selection of (a) which motor pattern is to be expressed, (b) which variations of this pattern, and (c) the duration of the pattern and when to stop. We first present a discussion of cellular mechanisms which help to select the motor output pattern in small networks. The next section deals with the neuronal bases of changes in behaviour and the role of distributed networks. Finally, selection of goal-directed and volitional movements are examined.

Jenny Kien, Cathy McCrohan, Bill Winlow

Part I

Cellular bases of motor programme selection

1 *Eve Marder and James M. Weimann*

Modulatory control of multiple task processing in the stomatogastric nervous system

1.1. **Introduction**

Starting with the pioneering work of Don Maynard (1972), the stomatogastric nervous system of decapod crustaceans has provided numerous fundamental insights into the production of motor patterns by rhythmic neural networks. The earliest work focused on attempts to understand the neural mechanisms by which a rhythmic motor pattern can be generated (Maynard, 1972; Mulloney and Selverston, 1974a,b; Eisen and Marder, 1982; Miller and Selverston 1982a,b). Subsequently the study of the stomatogastric nervous system was instrumental in providing an understanding of the mechanisms by which neuromodulators and modulatory neurones can influence the forms of a single motor pattern (Dickinson and Nagy, 1983; Nagy and Dickinson, 1983; Hooper and Marder, 1984, 1987; Marder, 1984; Flamm and Harris-Warrick, 1986a,b; Heinzel and Selverston, 1988; Nagy *et al.*, 1988; Nusbaum and Marder, 1989a,b; Katz and Harris-Warrick, 1990). Most recently, this preparation has allowed us to study interactions among neurones and the neural networks that together produce coordinated movements. In this chapter we will first review the evidence that demonstrates that a single group of neurones can produce multiple forms of a single motor pattern. Then we will summarise recent data which demonstrate that the same neurones can participate in the generation of several different motor patterns, and show how behavioural selection could be achieved by modulating their membrane properties and the connections between them.

The stomatogastric nervous system of decapod crustaceans controls the coordinated movements of the oesophagus and stomach (Fig. 1.1), and consists of four ganglia. There are two commissural ganglia (CGs), each consisting of several hundred neurones, a single oesophageal ganglion (OG) containing 16–18 neurones, and the single stomatogastric ganglion (STG) which consists of approximately 30 neurones. Together the neurones found in these ganglia produce at least four rhythmic motor patterns which govern the movement of food through the foregut region.

Food moves from the mouth into the oesophagus (Fig. 1.1). The oesophageal rhythm has a characteristic period of 10 s, and consists of

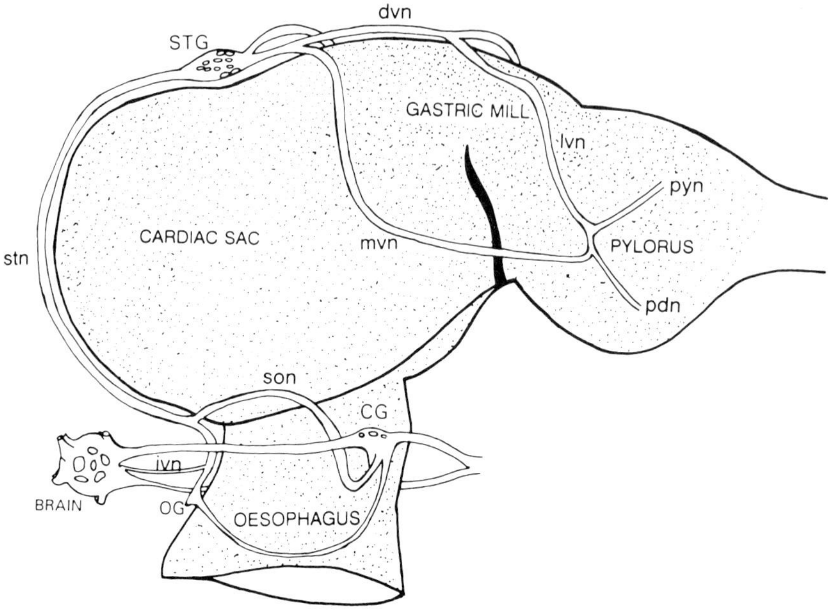

Fig. 1.1. Diagram of the side view of the foregut region and the position of the ganglia and nerves of the stomatogastric nervous system. Abbreviations in this and subsequent figures are as follows: CG, commissural ganglion; OG, oesophageal ganglion; STG, stomatogastric ganglion; dvn, ivn, lvn, mvn, dorsal, inferior, lateral and medial ventricular nerves, respectively; pdn, pyloric dilator nerve; pyn, pyloric nerve; son, superior oesophageal nerve; stn, stomatogastric nerve.

alternating bursts in the oesophageal dilator and constrictor motoneurones. Relatively little is known concerning the mechanisms underlying the generation of the oesophageal rhythm, although it is known that many of the oesophageal motoneurones are found in the CGs and several are found in the OG. From the oesophagus food moves into the cardiac sac region of the stomach (Fig. 1.1). The cardiac sac rhythm has a characteristic period of 30 s to several minutes (Fig. 1.2). It involves motoneurones that are found in both the OG and the STG, and again relatively little is known about the pattern-generating network that underlies it.

The gastric mill receives food from the cardiac sac (Fig. 1.1). The two lateral teeth and the single medial tooth of the gastric mill shred and chew food. The gastric rhythm (Fig. 1.2) has a characteristic period of 5–10 s, and is generated by neurones found in the STG. The gastric mill rhythm has been extensively studied, and is thought to be an example of a rhythmic motor pattern that depends critically for its production on emergent network properties. The pylorus is the last region of the deca-

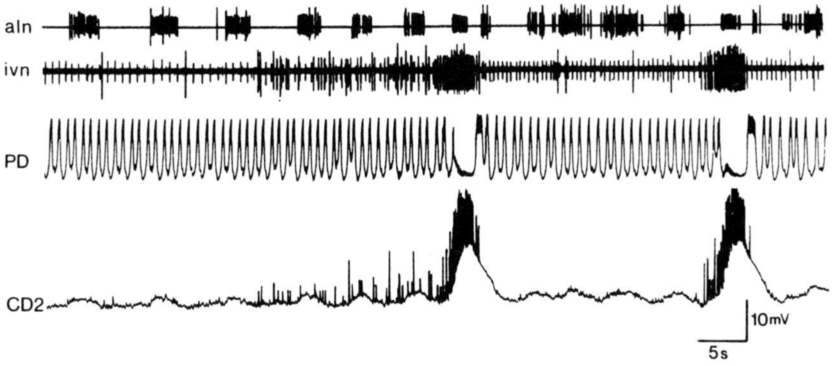

Fig. 1.2. Motor patterns that control the movements of the foregut. Simultaneous extracellular (top two traces) and intracellular (bottom two traces) recordings. The cardiac sac rhythm is seen as slow bursts in the cardiac sac dilator 2 motoneurone (CD2) and bursts of the ivn fibres. The gastric rhythm is shown as bursts of gastric mill (GM) neurone activity recorded on the anterior lateral nerve (aln). The pyloric rhythm is shown as rapid bursts of activity recorded in the pyloric dilator (PD) neurone. (Modified from Dickinson and Marder, 1989.)

pod foregut (Fig. 1.1). The movements of the pyloric region are controlled by the pyloric rhythm (Fig. 1.2), which occurs with a characteristic frequency of approximately 1 Hz. The pyloric network has been extensively studied and depends on the membrane properties of a conditionally bursting neurone for its intrinsic rhythmicity, and on the synaptic interactions among the pattern-generating neurones for the phase relationships of the pyloric rhythm (Marder *et al.*, 1987a).

1.2. **Identification of modulatory inputs and substances**

The earliest studies of the pyloric and gastric rhythms used preparations in which the STG was isolated from the more anterior ganglia of the stomatogastric nervous system by cutting either the stomatogastric nerve (stn) or the inferior and superior oesophageal nerves (ion and son, respectively) (Maynard, 1972; Mulloney and Selverston, 1974a,b). Although these reduced preparations showed rhythmic activity, robust and vigorous pyloric and gastric mill rhythms are most reliably produced by preparations in which the CGs and OG are left attached (Russell, 1976, 1979; Nagy and Moulins, 1987; Hartline *et al.*, 1988). This suggests that neurones of the CGs and OG were likely to function as modulatory inputs to the network in the STG.

The search for these modulatory extrinsic inputs to the STG took two paths. The Moulins laboratory undertook a major attempt to identify specific neurones in the CGs and OG with modulatory actions on the

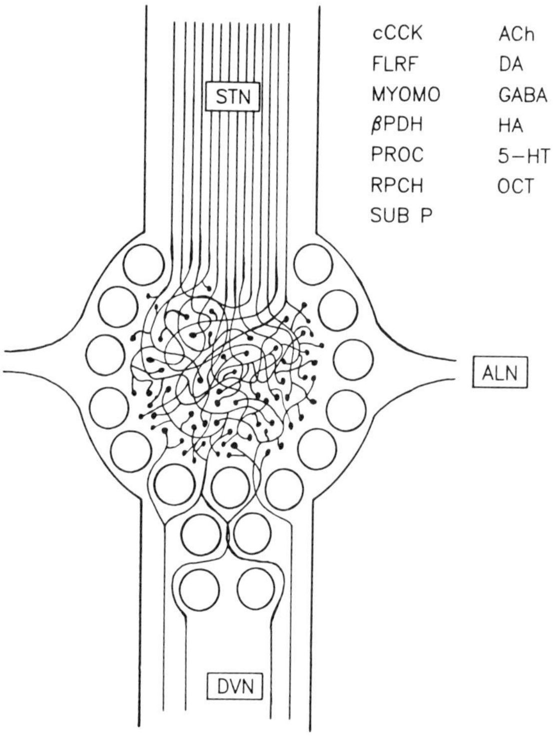

Fig. 1.3. Diagram summarising the neuromodulatory substances found in neuronal projections from the STN into the neuropil of the STG of the generic crustacean. Not all substances are found in all decapod species. Several axons in the DVN also contain GABA, proctolin, FLRFamide and 5-HT. Large clear circles show the somata of the STG. All motoneurones of the STG are either cholinergic or glutamatergic (Marder, 1987). cCCK, Crustacean cholecystokinin, visualised by Turrigiano and Selverston (1989) using antibodies raised against vertebrate CCK; FLRF, FLRFamide-like peptides, visualised with antibodies initially raised against FMRFamide, see Marder *et al.* (1987b); MYOMO, myomodulin-like immunoreactivity, visualised using antimyomodulin, gift of K. Weiss, in unpublished experiments of Lockhart, Oshinsky, Hall and Marder; βPDH, β-pigment dispersing hormone-like immunoreactivity, visualized using anti-βPDH, gift of R. Rao, see Mortin and Marder (1989, 1991); PROC, proctolin, see Marder *et al.* (1986); RPCH, red pigment concentrating hormone-like peptide, see Nusbaum and Marder (1988), Dickinson and Marder (1989); SUB P, substance P-like immunoreactivity, tachykinin-like substance, see Goldberg *et al.* (1988); ACh, acetylcholine, thought to be the neurotransmitter released by the anterior pyloric modulator (APM) neurone described in Dickinson and Nagy (1983) and Nagy and Dickinson (1983); DA, dopamine, demonstrated biochemically and histochemically, see Kushner and Barker (1983), Marder (1987); GABA, γ-aminobutyric acid, see Cazalets *et al.* (1989), Cournil *et al.* (1990), Mulloney and Hall (1990); HA, histamine, demonstrated biochemically in Claiborne and Selverston (1984); 5-HT, serotonin, see Beltz *et al.* (1984); and OCT, octopamine, measured biochemically in Barker *et al.* (1979). STN (stomatogastric nerve), ALN (anterior lateral nerve), DVN (dorsal ventricular nerve).

neurones of the STG. This work has been reviewed by Nagy and Moulins (1987). A complementary approach has been to identify as many as possible of the modulatory substances present in fibres that project into the STG from extrinsic sources (Fig. 1.3). This latter approach has led to the demonstration of a large number of different substances in inputs to the STG, as shown in Fig. 1.3. These include serotonin (Beltz *et al.*, 1984), dopamine and octopamine (Barker *et al.*, 1979; Kushner and Barker, 1983), histamine (Claiborne and Selverston, 1984), GABA (Cazalets *et al.*, 1987; Cournil *et al.*, 1990; Mulloney and Hall, 1990), proctolin (Hooper and Marder, 1984; Marder *et al.*, 1986), several FMRFamide-like peptides that we know now are likely to be extended FLRFamide-like peptides (Hooper and Marder, 1984; Marder *et al.*, 1987b), red pigment concentrating hormone- (RPCH) like peptides (Nusbaum and Marder, 1988; Dickinson and Marder, 1989), crustacean cholecystokinin (CCK)-like peptides (Turrigiano and Selverston, 1989, 1990), a substance-P-like immunoreactivity (Goldberg *et al.*, 1988), a β-pigment dispersing hormone (β-PDH)-like peptide (Mortin and Marder, 1989, 1991), and a myomodulin-like peptide (Lockhart, Hall, Oshinsky and Marder, unpublished results).

The presence of this large number of different substances in the fibres projecting into the STG clearly indicates that neural networks may be modulated by many different substances. Additionally, other substances may reach the neuropil of the STG as hormones, even if they are not found in neuronal projections to the STG. For example, crustacean cardioactive peptide (CCAP), found in neurosecretory structures in crabs and lobsters but not in direct neuropil projections (Dircksen and Keller, 1988; Mortin *et al.*, 1990), has pronounced physiological actions on STG motor patterns in several species (Heinzel, Weimann, Mortin and Marder, unpublished results).

1.3. **Multiple forms of the pyloric rhythm**

Since the pyloric rhythm is the best understood and most stereotyped of the motor patterns produced by the stomatogastric nervous system, the effects of different modulatory substances have been most extensively studied on it (Beltz *et al.*, 1984; Hooper and Marder, 1984, 1987; Marder *et al.*, 1986; Nusbaum and Marder, 1988; Cazalets *et al.*, 1987). Comparison of the effects of several modulatory substances on the pyloric rhythm shows clearly that each substance produces characteristic and different motor patterns (Marder, -1984; Marder and Hooper, 1985; Flamm and Harris-Warrick, 1986a,b; Marder, 1987; Marder *et al.*, 1987a; Harris-Warrick, 1988; Harris-Warrick and Marder, 1991). Figure 1.4 provides a comparison of the effects of eight modulatory substances on the pyloric rhythm of the crab *Cancer borealis*. These records come from

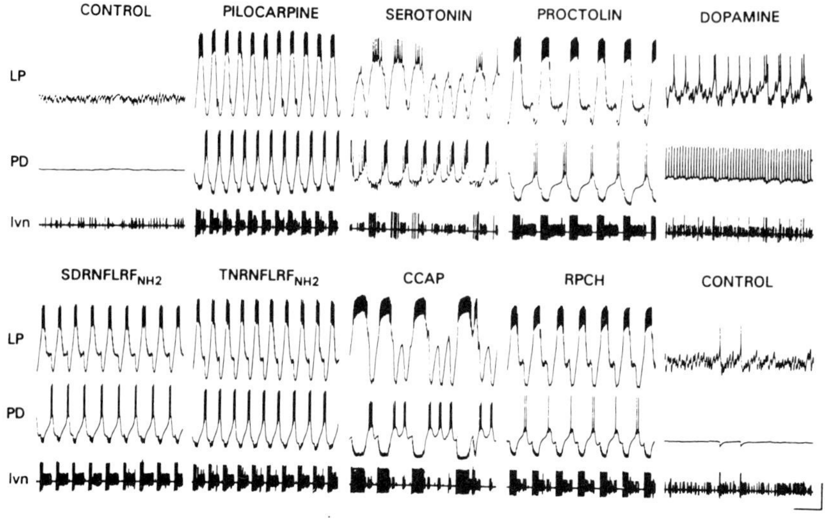

Fig. 1.4. Multiple forms of the pyloric rhythm induced by different neuromodulatory substances. Isolated STG from the crab, *Cancer borealis*. All panels, top trace intracellular recording from the lateral pyloric (LP) neurone; second trace intracellular recording from the pyloric dilator (PD) neurone; third trace, extracellular recording from the lateral ventricular nerve (lvn). The preparation was extensively washed between each modulator application and returned to control levels of activity before new application. Concentrations: pilocarpine, 10^{-5} M; serotonin, 10^{-4} M; proctolin, 10^{-6} M; dopamine, 10^{-6} M; SDRNFLRFamide, 10^{-7} M; TNRNFLRFamide, 10^{-7} M; CCAP, 10^{-6} M; RPCH, 10^{-6} M. Horizontal calibration bar, 2 s. Vertical calibration bar, 10 mV.

a single preparation that was extensively washed between each application. In this experiment, neural inputs from the CGs were removed and the control condition was one in which the pyloric rhythm was quiescent. The intracellular records shown in the top two traces in each panel are from the lateral pyloric (LP) and pyloric dilator (PD) motor neurones that innervate respectively the constrictor and dilator muscles of the pyloric region. Each exogenously applied substance produced a different form of the pyloric motor pattern. When applied to this quiescent preparation all of these substances were 'excitatory' in that they evoked an increase in firing in all cases, and, except for dopamine, induced rhythmic alternations between the PD and LP neurones. However, in each case the exact phase relationships, the intensities of the neuronal bursts, and the intracellularly recorded waveforms of the neurones were different.

Pilocarpine is a muscarinic, cholinergic agonist which is effective in initiating rhythmic pyloric activity (Marder and Paupardin-Tritsch, 1978; Marder *et al.*, 1987a; Marder and Meyrand, 1989). In the presence of pilocarpine the pyloric rhythm was rapid and regular. Pyloric rhythms seen in serotonin, such as that shown in Fig. 1.4, are typically characterised by shallow PD membrane potential oscillations, with long LP neurone bursts that do not occur with each PD neurone burst (Beltz *et al.*, 1984; Katz and Harris-Warrick, 1990). The pyloric rhythm seen in the presence of proctolin is associated with long, high-frequency, LP neurone bursts (Marder *et al.*, 1986). Dopamine has a strong depolarising action on both the PD and LP neurones, which, even at the modest concentration shown here (10^{-6} M), is sufficient to disrupt the burst formation of the pyloric network (see Marder and Meyrand, 1989). The extended FLRFamide-like peptides, SDRNFLRFamide and TNRNFLRFamide, evoked strong pyloric rhythms that are not unlike those seen with pilocarpine (shown just above them for comparison) in that the intensity of the bursts in the PD and LP neurones were similar. CCAP, like serotonin, shown just above it, produced rhythms in which long LP neurone plateaux alternated not with each PD burst, but with every third or fourth PD burst. The pyloric rhythms seen in RPCH most resemble those seen in proctolin (just above it) in that both proctolin and RPCH produce LP neurone activity-dominated pyloric rhythms.

1.4. Multiple forms of the gastric rhythm

Many of the modulatory substances discussed above (Fig. 1.3) also produce dramatic physiological actions on gastric motor patterns. As is the case for the pyloric rhythm, a number of different modulatory substances can activate gastric motor patterns. These include proctolin (Marder *et al.*, 1986; Heinzel and Selverston, 1988), octopamine (Selverston *et al.*, 1983), muscarinic agonists (Weimann, unpublished results), crustacean CCK-like peptides (Turrigiano and Selverston, 1989) and FLRFamide-like peptides (Weimann and Marder, 1989; Weimann *et al.*, 1990). Again, as is the case for the pyloric rhythm, each of the modulatory substances produces different forms of the gastric rhythm. Since the gastric rhythm is less stereotyped than the pyloric, a comparison of the full complexity of the gastric rhythms produced by the many modulatory substances shown in Fig. 1.3 has not yet been completed. However, we anticipate that systematic comparisons will yield much insight into the mechanisms controlling the different forms of gastric patterns produced.

1.5. Neurones that switch between pattern-generating networks

The pyloric, gastric, and cardiac sac rhythms can be simultaneously active, with frequencies that routinely differ by more than an order of magnitude (Fig. 1.2). For example, the pyloric rhythm is commonly found with a period of about 1 s, the gastric rhythm characteristically has a period of 5–10 s, and the cardiac sac rhythm has a characteristic period of 30–120 s. Although these circuits were classically considered to comprise distinct groups of neurones (Selverston and Moulins, 1987), it has been known for many years that there are interactions between them. In preparations from *Panulirus interruptus*, interactions between the gastric and pyloric rhythms are often seen as perturbations of the ongoing pyloric rhythm at distinct phases of the gastric rhythm (Mulloney, 1977). Cardiac sac bursts are associated with disruptions of both the pyloric and gastric rhythms (Nagy and Moulins, 1987; Dickinson and Marder, 1989; Dickinson *et al.*, 1990). In *Homarus gammarus* modulations of robust gastric bursts occur in time with the more rapid pyloric rhythms (Robertson and Moulins, 1984). In a recent study in the lobster *Palinurus vulgaris*, Hooper and Moulins (1989) showed that sensory inputs activated by stretch could switch the ventricular dilator (VD) neurone, which usually functions as part of the pyloric network, into cardiac sac motor patterns. In this case, sensory activation switches the VD neurone between two ongoing motor patterns, and the mechanism underlying the switch is a change in the extent to which the VD neurone expresses plateau properties (Hooper and Moulins, 1989).

In the crab *Cancer borealis*, many neurones switch between ongoing pyloric and gastric motor patterns. In this species neurones can display either gastric-timed or pyloric-timed activity patterns depending on the modulatory environment (Weimann *et al.*, 1990, 1991). In preparations that show only rapid, pyloric rhythms, all of the neurones that are classically considered part of the pyloric circuit participate in the pyloric rhythm. Under these conditions many of the neurones usually classified as gastric neurones also fire in time with the pyloric rhythm. In the top set of recordings shown in Fig. 1.5, the lateral gastric neurone (LG) fires with the pyloric rhythm. However, when the gastric rhythm is activated, as can occur either spontaneously or after the application of the extended FLRFamide-like peptide, SDRNFLRFamide (Weimann *et al.*, 1989, 1990), most of the gastric neurones display gastric rhythm timed activity, and some of the neurones usually considered part of the pyloric circuit show gastric-timed activity (Weimann *et al.*, 1990, 1991). The middle panel in Fig. 1.5 shows a recording from an LG neurone in a hybrid gastric/pyloric pattern associated with weak gastric activity (monitored on the dgn). The bottom trace in Fig. 1.5 shows that the LG neurone can produce a full gastric motor output with little obvious pyloric modulation during robust gastric mill activity.

In many situations, e.g. walking and swimming in man, the same motoneurones are active during different behaviours. In most cases these motoneurones are not part of the pattern-generating circuits but are simply driven by whichever pattern generator is active. However, in the STG, the motoneurones themselves, together with several interneurones, make up the pattern-generating networks. Therefore, the stomatogastric

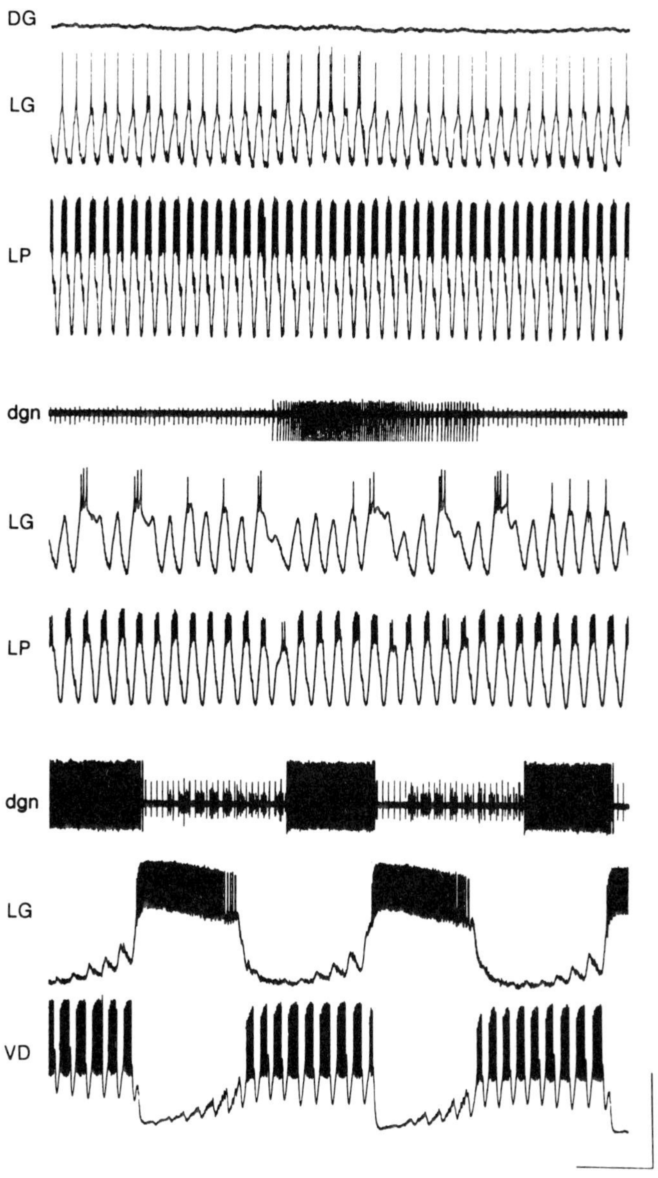

Fig. 1.5. Neurones that switch between gastric and pyloric activity patterns. Top panel: in the absence of an ongoing gastric rhythm (no activity in the dorsal gastric (DG) neurone) the lateral gastric (LG) (one of the 'gastric' neurones) fires in time with the pyloric rhythm (shown as the activity pattern in the LP neurone). Middle panel: during weak gastric activity (see activity of the DG on the dgn), the LG fires in a hybrid pattern, with pyloric timed hyperpolarisations interspersed with longer period short bursts. Bottom panel: during robust gastric activity, seen as the dense DG bursts on the dgn, the LG neurone now fires in intense gastric bursts and shows little pyloric modulation. Horizontal calibration bar is 5 s for the top and bottom panels and 2.5 s for the middle panel. The vertical calibration bar is 16 mV for the LG and 20 mV for the LP neurone in the first two panels; 26 mV for LG and 18 mV for the VD neurone in the bottom panel.

ganglion serves as a model system only for those central networks that directly produce behaviours. As the motoneurones in the crab STG change from gastric-timed to pyloric-timed activity patterns, these changes have important consequences for the operation of the pattern-generating networks in which these neurones function. We have recently shown (Weimann and Marder, unpublished results) that 'gastric' neurones not only can fire in pyloric time, but also can reset the pyloric rhythm. Likewise, 'pyloric' neurones can fire in time with the gastric rhythm, and can also reset the gastric rhythm. This ability to reset the ongoing rhythmic motor pattern is one of the classical experimental demonstrations that identify a neurone as part of the circuit responsible for the generation of the rhythm. Thus, these data argue that all of these neurones have access to the pattern-generating circuitry for both rhythms, and participate at all times to a greater or lesser degree in shaping both rhythms.

The mechanisms responsible for the pyloric/gastric transitions such as those shown in Fig. 1.5 are complex. Preliminary experiments show that some of the peptide-induced changes are associated with inductions of plateau properties in single neurones such as the DG neurone and modifications of defined synaptic interactions (Weimann *et al.*, 1990). The firing patterns of each cell will be uniquely controlled by its membrane properties and the strength of the synaptic contacts at any moment in time. Therefore, it is likely that each 'switch' (i.e. the selection of each motor pattern) is associated with a different pattern of modification in synaptic efficacy and cellular properties.

1.6. Gastric/cardiac sac fusion

Unlike the previous cases in which neurones change their allegiance between two ongoing rhythmic motor patterns, in *Panulirus interruptus* the neuropeptide, RPCH, 'fuses' two motor patterns into one (Dickinson, 1989; Dickinson *et al.*, 1990). Figure 1.2 shows that in control saline the cardiac sac and gastric motor patterns operate simultaneously but with different frequencies. However, in the presence of RPCH, cardiac sac and gastric motor neurones fire in a single alternating motor pattern (Fig. 1.6) which is neither a true cardiac sac pattern nor true gastric pattern. In the example shown in Fig. 1.6, note that cardiac sac CD2 motoneurone and the ivn fibres (also part of the cardiac sac system) fire together in alternation with the gastric mill (GM) and lateral posterior gastric (LPG) motoneurones of the gastric system. Thus, here we see that elements of two initially separate neural circuits have been combined to produce a single conjoint rhythm.

In this case the mechanism responsible for the production of this conjoint rhythm is an RPCH-induced potentiation of the synaptic poten-

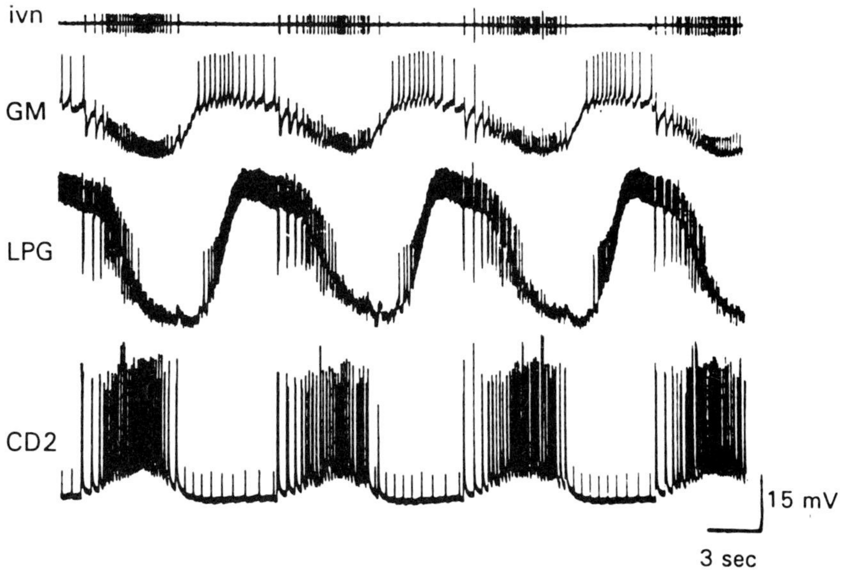

Fig. 1.6. Novel pattern shown by gastric/cardiac sac neurones in the presence of the peptide RPCH. Under control conditions the cardiac sac and gastric motor patterns cycle independently at different frequencies, as shown in Fig. 1.2. In the presence of RPCH the two networks are 'fused' and produce a single rhythmic motor pattern in which CD2 and gastric motoneurones (GM and LPG) fire in alternating bursts. (Modified from Dickinson *et al.*, 1990.)

tials evoked by the ivn fibres in many of the STG motoneurones, including the LPG and GM neurones (Dickinson, 1989; Dickinson *et al.*, 1990). Although the conjoint rhythm occurs because, in the presence of RPCH, the synaptic potentials of the ivn fibres are sufficiently strong to impose a single rhythm on elements that previously fired in separate rhythms, it is important to remember that even here the lateral synaptic interactions among these 'driven' neurones are still shaping the details of the conjoint rhythm, and that these motoneurones are not mere followers.

1.7. **Networks as multiple task processors**

In this paper we have discussed several different ways in which modulatory inputs can reconfigure neural networks. Three of these are shown diagrammatically in Fig. 1.7. First (Fig. 1.7A), modulatory substances can control the specific form of activity produced by an ensemble of neurones, for example by activating or turning off an ongoing rhythm (as illustrated in Fig. 1.4). Additionally, different modulatory inputs can produce rhythms in which the phase relations of the constituent neurones

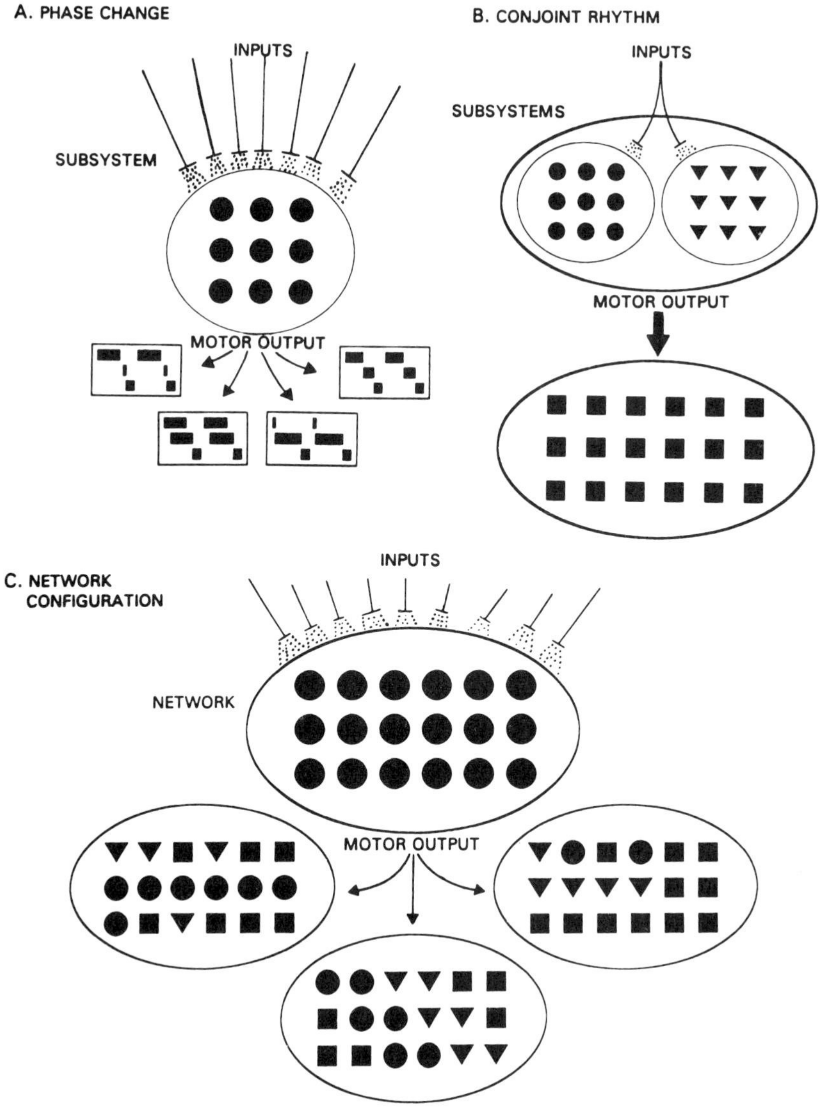

Fig. 1.7. Networks are multiple task processors (MTPs). (A) Different modulatory substances can produce different output patterns from a single subsystem. The block diagrams are a cartoon representation of the phase of the pyloric rhythms seen following bath application of the different modulatory substances. (B) Modulatory substances can 'fuse' two relatively independent subsystems, as represented by circles and triangles, into a single functional unit (squares), which produces a novel conjoint rhythm with a period and element phases different from either of the starting subsystem outputs. (C) Modulatory inputs can select multiple network constellations from a group of neurones (circles), resulting in neurones 'switching' to produce different output patterns. Three of the many types of network configuration are shown with the circles, triangles and squares representing different activity patterns.

vary, burst durations vary, and burst intensities vary (Fig. 1.4, Fig. 1.7A). These kinds of changes might be employed as an animal adapts an ongoing locomotory rhythm to slight modifications of load or terrain. Although many of the changes of this sort can be quite dramatic and might have profound behavioural consequences, we have become accustomed to thinking that they must occur to enable animals to adapt to environmental changes.

The changes illustrated in Figs 1.7B and 1.7C require us to re-evaluate many of our current ideas about the definition of central neural networks. Some neurones may be found in neural subsystems which can be operated either independently or combined with other subsystems to produce different behaviours such as the conjoint rhythms elicited by RPCH (Figs 1.6, 1.7B). Neurones may be found in neural networks that exist in complicated patterns of connectivity where many of the synaptic and cellular properties are subject to modulation. In this case, a modulatory substance might activate neurones and synaptic pathways to form a new functional network. This is diagrammatically shown in Fig. 1.7C, where the individual elements in the unmodulated state are shown as circles. For example, one modulatory substance may elicit pyloric-timed activity in some neurones (squares) and gastric-timed activity in others (triangles). Different modulatory substances might 'select' other neurones and other pathways from the constituent ensemble to form new active circuits (Fig. 1.7C). The organisation shown in Figs 1.7B and 1.7C can lead not only to modifications in frequency and phase, such as illustrated in Fig. 1.7A, but also to neurones switching between two ongoing rhythms to form different operationally defined circuits, rhythmic or otherwise, as illustrated in Figs 1.7B,C. This idea is attractive in that it allows an anatomically defined, synaptically coupled network to perform multiple tasks, some of which can occur simultaneously, others of which may only occur sequentially.

In summary, we now argue that modulatory inputs may allow us to understand how biological 'wetware' can act as a multiple task processor (MTP). Modulatory substances 'program' the MTP by selecting or activating neurones from the ensemble and by changing the gain of the synaptic connections in the ensemble. Although the elements and connections 'selected' serve as the resident 'program', it is important to remember that even non-responsive or silent neurones can have important consequences for circuit operation (Hooper and Marder, 1987; Kepler *et al.*, 1990).

In a nervous system operating as a multiple task processor, a given neurone may operate at different times in entirely different roles in different, functional, neural circuits. The phenotype of such neurones would be that they would 'switch' their activity patterns, but a full understanding of the extent to which they subserve different roles will require far more complete knowledge about the organisation of the neural networks underlying behaviour than is currently available in any preparation. Despite the extensive modulation of networks described in this chapter, network operation in animals seems rarely compromised by 'overmodulation' (Marder and Meyrand, 1989). Thus, an important challenge for the future will be to understand both how network structure constrains the inherent flexibility produced by multiple modulatory processes and how modulation itself is controlled. Research on the changes produced by modulatory substances and neurones in the stomatogastric nervous system should allow us to understand more fully the detailed mechanisms underlying the kinds of modulatory change described in this chapter and so aid our understanding of the selection of motor patterns. We anticipate that similar results will be forthcoming in other nervous systems as the neural networks that produce and control behaviour become better described.

Acknowledgements

We thank Ms Elizabeth Orban, Ms Cassandra Hall and Mr Michael O'Neil for help with figure preparation. Discussions with Drs Patsy Dickinson, Scott Hooper and Pierre Meyrand were critical in the formulation of many of these ideas. Research was supported by NS 17813.

References

Barker, D. L., Kushner, P. D. and Hooper, N. K. (1979) Synthesis of dopamine and octopamine in the crustacean stomatogastric nervous system. *Brain Res.* **161**: 99–113.

Beltz, B., Eisen, J. S., Flamm, R., Harris-Warrick, R., Hooper, S. L. and Marder, E. (1984) Serotonergic innervation and modulation of the stomatogastric ganglion of three decapod crustaceans (*Homarus americanus*, *Cancer irroratus*, and *Panulirus interruptus*). *J. exp. Biol.* **109**: 35–54.

Cazalets, J. R., Cournil, I., Geffard, M. and Moulins, M. (1987) Suppression of oscillatory activity in crustacean pyloric neurons: implication of GABAergic inputs. *J. Neurosci.* **7**: 2884–2893.

Claiborne, B. J. and Selverston, A. I. (1984) Histamine as a neurotransmitter in the stomatogastric nervous system of the spiny lobster. *J. Neurosci.* **4**: 708–721.

Cournil, I., Meyrand, P. and Moulins, M. (1990) Identification of all GABA immunoreactive neurons projecting to the lobster stomatogastric ganglion. *J. Neurocyt.* **19**: 478–493.

Dickinson, P. (1989) Modulation of simple motor patterns. *Seminars in the Neurosciences* **1**: 15–24.

Dickinson, P. S. and Marder, E. (1989) Peptidergic modulation of a multioscillator system in the lobster. I. Activation of the cardiac sac rhythm by the neuropeptides proctolin and red pigment concentrating hormone. *J. Neurophysiol.* **61**: 833–844.

Dickinson, P. S., Mecsas, C. and Marder, E. (1990) Neuropeptide fusion of two motor pattern generator circuits. *Nature* **344**: 155–158.

Dickinson, P. S. and Nagy, F. (1983) Control of a central pattern generator by an identified modulatory interneurone in crustacea. II. Induction and modification of plateau properties in pyloric neurones. *J. exp. Biol.* **105**: 59–82.
Dircksen, H. and Keller, R. (1988) Immunocytochemical localization of CCAP, a novel crustacean cardioactive peptide, in the nervous system of the shore crab, *Carcinus maenas L. Cell Tissue Res.* **254**: 347–360.
Eisen, J. S. and Marder, E. (1982) Mechanisms underlying pattern generation in lobster stomatogastric ganglion as determined by selective inactivation of identified neurons. III. Synaptic connections of electrically coupled pyloric neurons. *J. Neurophysiol.* **48**: 1392–1393.
Flamm, R. E. and Harris-Warrick, R. M. (1986a) Aminergic modulation in lobster stomatogastric ganglion. I. The effects on motor pattern and activity of neurons within the pyloric circuit. *J. Neurophysiol.* **55**: 847–865.
Flamm, R. E. and Harris-Warrick, R. M. (1986b) Aminergic modulation in lobster stomatogastric ganglion. II. Target neurons of dopamine, octopamine, and serotonin within the pyloric circuit. *J. Neurophysiol.* **55**: 866–881.
Goldberg, D., Nusbaum, M. P. and Marder, E. (1988) Substance P-like immunoreactivity in the stomatogastric nervous systems of the crab *Cancer borealis* and the lobsters *Panulirus interruptus* and *Homarus americanus*. *Cell Tissue Res.* **252**: 515–522.
Harris-Warrick, R. M. (1988) Chemical modulation of central pattern generators. In: *Neural Control of Rhythmic Movements in Vertebrates*, A. H. Cohen, S. Rossignol and S. Grillner, eds. John Wiley & Sons, New York, pp. 285–331.
Harris-Warrick, R. M. and Marder, E. (1991) Modulation of neural networks for behavior. *Ann. Rev. Neurosci.* **14**: 39–57.
Hartline, D. K., Russell, D. K., Raper, J. A. and Graubard, K. (1988) Special cellular and synaptic mechanisms in motor pattern generation. *Comp. Biochem. Physiol.* **91C**: 115–131.
Heinzel, H-G. and Selverston, A. I. (1988) Gastric mill activity in the lobster. III. Effects of proctolin on the isolated central pattern generator. *J. Neurophysiol.* **59**: 566–585.
Hooper, S. L. and Marder, E. (1984) Modulation of a central pattern generator by two neuropeptides, proctolin and FMRFamide. *Brain Res.* **305**: 186–191.
Hooper, S. L. and Marder, E. (1987) Modulation of a central pattern generator by the peptide, proctolin. *J. Neurosci.* **7**: 2097–2112.
Hooper, S. L. and Moulins, M. (1989) Switching of a neuron from one network to another by sensory-induced changes in membrane properties. *Science* **244**: 1587–1589.
Katz, P. S. and Harris-Warrick, R. M. (1990) Neuromodulation of the crab pyloric central pattern generator by serotonergic/cholinergic proprioceptive afferents. *J. Neurosci.* **10**: 1495–1512.
Kepler, T. B., Marder, E. and Abbott, L. F. (1990) The effect of electrical coupling on the frequency of model neuronal oscillators. *Science* **248**: 83–85.
Kushner, P. D. and Barker, D. L. (1983) A neurochemical description of the dopaminergic innervation of the stomatogastric ganglion of the spiny lobster. *J. Neurobiol.* **14**: 17–28.
Marder, E. (1984) Modulation of a motor pattern generator. *TINS* **7**: 48–53.
Marder, E. (1987) Neurotransmitters and neuromodulators. In: *The Crustacean Stomatogastric Nervous System: A Model for the Study of Central Nervous Systems*, A. I. Selverston and M. Moulins, eds. Springer-Verlag, New York, pp. 263–300.
Marder, E., Calabrese, R. L., Nusbaum, M. P. and Trimmer, B. (1987b) Distribution and partial characterization of FMRFamide-like peptides in the

stomatogastric nervous systems of the rock crab, *Cancer borealis*, and the spiny lobster, *Panulirus interruptus*. *J. Comp. Neurol.* **259**: 150–163.

Marder, E. and Hooper, S. L. (1985) Neurotransmitter modulation of stomatogastric ganglion of decapod crustaceans. In: *Model Neural Circuits and Behavior*, A. I. Selverston, ed. Plenum Press, New York, pp. 319–337.

Marder, E., Hooper, S. L. and Eisen J. S. (1987a) Multiple neurotransmitters provide a mechanism for the production of multiple outputs from a single neuronal circuit. In: *Synaptic Function*, G. M. Edelman, W. E. Gall and M. W. Cowan, eds. Neuroscience Research Foundation / John Wiley & Sons, New York, pp. 305–327.

Marder, E., Hooper, S. L. and Siwicki, K. K. (1986) Modulatory action and distribution of the neuropeptide proctolin in the crustacean stomatogastric nervous system. *J. Comp. Neurol.* **243**: 454–467.

Marder, E. and Meyrand, P. (1989) Chemical modulation of an oscillatory neural circuit. In: *Neuronal and Cellular Oscillators*, J. W. Jacklet, ed. Marcel Dekker, New York, pp. 317–338.

Marder, E. and Paupardin-Tritsch, D. (1978) The pharmacological properties of some crustacean neuronal acetylcholine, γ-aminobutyric acid and L-glutamate responses. *J. Physiol. London* **280**: 213–236.

Maynard, D. M. (1972) Simpler networks. *Ann. NY Acad. Sci.* **193**: 59–72.

Miller, J. P. and Selverston, A. I. (1982a) Mechanisms underlying pattern generation in lobster stomatogastric nervous system as determined by selective inactivation of identified neurons. II. Oscillatory properties of pyloric neurons. *J. Neurophysiol.* **48**: 1378–1391.

Miller, J. P. and Selverston, A. I. (1982b) Mechanisms underlying pattern generation in lobster stomatogastric nervous system as determined by selective inactivation of identified neurons. IV. Network properties of pyloric system. *J. Neurophysiol.* **48**: 1416–1432.

Mortin, L. I., Lockhart, S. T. and Marder, E. (1990) Crustacea cardioactive peptide (CCAP): a hormone neuromodulator of the stomatogastric nervous system. *Soc. Neuro. Abst.* **16**: 1131.

Mortin, L. I. and Marder, E. (1989) Localization of pigment-dispersing hormone (PDH)-like immunoreactivity in the crustacean stomatogastric nervous system. *Soc. Neuro. Abst.* **15**: 366.

Mortin, L. I. and Marder, E (1991) Differential distribution of β-pigment-dispersing hormone (PDH)-like immunoreactivity in the stomatogastric nervous system of five species of decapod crustaceans. *Cell Tiss. Res.* **265**: 19–33.

Mulloney, B. (1977) Organization of the stomatogastric ganglion of the spiny lobster. V. Coordination of the gastric and pyloric systems. *J. Comp. Physiol.* **122**: 227–240.

Mulloney, B. and Hall, W. M. (1990) GABA-ergic neurons in the crayfish nervous system: an immunocytochemical census of the segmental ganglia and stomatogastric system. *J. Comp. Neurol.* **291**: 383–394.

Mulloney, B. and Selverston, A. I. (1974a) Organization of the stomatogastric ganglion of the spiny lobster. I. Neurons driving the lateral teeth. *J. Comp. Physiol.* **91**: 1–32.

Mulloney, B. and Selverston, A. I. (1974b) Organization of the stomatogastric ganglion of the spiny lobster. III. Coordination of the two subsets of the gastric system. *J. Comp. Physiol.* **91**: 53–78.

Nagy, F. and Dickinson, P. S. (1983) Control of a central pattern generator by an identified modulatory interneurone in crustacea. I. Modulation of the pyloric motor output. *J. exp. Biol.* **105**: 33–58.

Nagy, F., Dickinson, P. S. and Moulins, M. (1988) Control by an identified modulatory neuron of the sequential expression of plateau properties of, and synaptic inputs to, a neuron in a central pattern generator. *J. Neurosci.* **8**: 2875–2886.

Nagy, F. and Moulins, M. (1987) Extrinsic inputs. In: *The Crustacean Stomatogastric System*, A. I. Selverston and M. Moulins, eds. Springer-Verlag, Berlin, pp. 205–242.

Nusbaum, M. P. and Marder, E. (1988) A neuronal role for a peptide similar to crustacean red pigment concentrating hormone: neuromodulation of the pyloric rhythm in the crab, *Cancer borealis*. *J. exp. Biol.* **135**: 165–181.

Nusbaum, M. P. and Marder, E. (1989a) A modulatory proctolin-containing neuron (MPN). I. Identification and characterization. *J. Neurosci.* **9**: 1591–1599.

Nusbaum, M. P. and Marder, E. (1989b) A modulatory proctolin-containing neuron (MPN). II. State-dependent modulation of rhythmic motor activity. *J. Neurosci.* **9**: 1600–1607.

Robertson, R. M. and Moulins, M. (1984) Oscillatory command input to the motor pattern generators of the crustacean stomatogastric ganglion: II. The gastric rhythm. *J. Comp. Physiol.* **154**: 473–491.

Russell, D. F. (1976) Rhythmic excitatory inputs to the lobster stomatogastric ganglion. *Brain Res.* **101**: 582–588.

Russell, D. F. (1979) CNS control of pattern generators in the lobster stomatogastric ganglion. *Brain Res.* **177**: 598–602.

Selverston, A. I., Miller, J. P. and Wadepuhl, M. (1983) Cooperative mechanisms for the production of rhythmic movements. In: *Neural Control of Rhythmic Movements*, A. Roberts and B. L. Roberts, eds. Cambridge University Press, Cambridge, pp. 55–87.

Selverston, A. I. and Moulins, M. (1987) *The Crustacean Stomatogastric System*. Springer-Verlag, Berlin, pp. 1–338.

Turrigiano, G. G. and Selverston, A. I. (1989) Cholecystokinin-like peptide is a modulator of a crustacean central pattern generator. *J. Neurosci.* **9**: 2486–2501.

Turrigiano, G. G. and Selverston, A. I. (1990) A cholecystokinin-like hormone activates a feeding-related neural circuit in lobster. *Nature* **344**: 866–868.

Weimann, J. M. and Marder, E. (1989) Activation of the gastric rhythm of the crab stomatogastric ganglion by SDRNFLRFamide. *Soc. Neuro. Abst. 15*: 1047.

Weimann, J. M., Meyrand, P. and Marder, E. (1990) Neurons that participate in several behaviors. In: *Frontiers in Crustacean Neurobiology*, K. Wiese, W-D. Krenz, J. Tautz, H. Reichert and B. Mulloney, eds. Birkhäuser Verlag, Basel, pp. 424–430.

Weimann, J. M., Meyrand, P. and Marder, E. (1991) Neurons that form multiple pattern generators: identification and multiple activity patterns of gastric/pyloric neurons in the crab stomatogastric system. *J. Neurophysiol.* **65**: 111–122.

2 *Andries Ter Maat, Graham P. Ferguson and Rene F. Jansen*

Control of egg laying behaviour patterns in *Lymnaea stagnalis*

2.1. Introduction

Significant progress towards understanding how animal behaviour is produced and modulated by the nervous system has been made by utilising the simple systems approach (see Selverston, 1985); the analysis of the neuronal and hormonal mechanisms underlying behaviour is facilitated by using invertebrates with simple behavioural repertoires and nervous systems containing relatively small numbers of cells, many of which are individually identifiable. One such model system is the pond snail *Lymnaea stagnalis*, in which the neural circuits controlling several behaviours have been identified (e.g. locomotion: Haydon and Winlow, 1986; Winlow and Haydon, 1986; respiration: van der Wilt *et al.*, 1987; whole-body withdrawal: Benjamin *et al.*, 1985; Ferguson, 1985; Ferguson and Benjamin, 1985). Two behavioural systems are now understood especially well, the neural control of feeding (reviewed by Benjamin, 1983) and the control of egg laying behaviours (reviewed by Geraerts *et al.*, 1987). In *Lymnaea* egg laying behaviour, a group of specialised neurosecretory cells (the caudodorsal cells, CDCs) produces the ovulatory hormone. Sensory information converges on the CDC system to trigger electrical activity and the release of ovulation hormone as well as other CDC peptides. These are released both systemically to regulate target organs, and within the brain to influence neuronal activity and behaviour.

Our interest is in determining which aspects of *Lymnaea* egg laying behaviour are controlled by the CDC peptides and in identifying other inputs important for the production and coordinated timing of egg laying behaviour patterns. In this analysis it is important to remember that peptides can have multiple and complex effects on the nervous system and act as hormones, neurotransmitters and neuromodulators. Therefore to describe the functional role of a given peptide it is necessary to know its identity, to identify the conditions that normally trigger its release and to understand the organisation of the neural circuits and effectors through which the resulting behaviour is expressed (Dismukes and Leibeskind, 1978). The objective of this chapter is to summarise our present understanding of the CDC peptides, how they are released and which aspects of the resulting egg laying behaviour pattern are due to their actions. We will also demonstrate that, in addition to the CDC peptides, input from the ovulated egg cells is important for the full expression of the egg laying behaviour patterns. We show that many inputs converge and are necessary for the release and execution of egg laying behaviour.

2.2. **The CDC peptides**

Two genes encoding the presumptive precursors for multiple CDC peptides have been isolated, cloned and sequenced (Vreugdenhil *et al.*, 1988). These genes encode two different biologically active CDC peptides that have been purified and sequenced: CDCH (Ebberink *et al.*, 1985) and calfluxin (Dictus and Ebberink, 1988). They also encode several additional small peptides, viz. $\alpha CDCP_{1\text{-}11}$ and three bCDCPs (Table 2.1). It has been hypothesised that four forms of αCDCP (1-11; 3-11; 3-10 and 3-9) may be cleaved from the precursor. This hypothesis is based on (i) the presence of potential dibasic cleavage sites on the CDCH precursors, and (ii) partial structural homology with similar peptides derived from the egg laying hormone precursor in the bag cells of *Aplysia* (Rothman *et al.*, 1983; Scheller *et al.*, 1983). During the CDC discharge there is a massive release of multiple neuropeptides, including the egg laying hormone (CDCH; Geraerts and Hogenes, 1983, 1985). Thus it is likely that the CDCs release up to 11 peptides which could have functional roles in egg laying.

Table 2.1. Primary structures of candidate peptides. CDCH and calfluxin have been purified and sequenced. βCDCPs and $\alpha CDCP_{1\text{-}11}$ have been synthesised based on known dibasic cleavage sites on both sides. The structures of the shorter αCDCPs are based on potential cleavage sites and their partial structural homology with similar peptides (αBCPs) derived from the ELH precursor of the neuroendocrine bag cells of *Aplysia*. Classical nomenclature is used for sequence indication.

$\alpha CDCP_{1\text{-}11}$	Ala-Arg-Glu-Pro-Arg-Leu-Arg-Phe-His-Asp-Val
$\alpha CDCP_{3\text{-}11}$	Glu-Pro-Arg-Leu-Arg-Phe-His-Asp-Val
$\alpha CDCP_{3\text{-}10}$	Glu-Pro-Arg-Leu-Arg-Phe-His-Asp
$\alpha CDCP_{3\text{-}9}$	Glu-Pro-Arg-Leu-Arg-Phe-His
$\beta_1 CDCP$	Arg-Leu-Arg-Phe-His
$\beta_2 CDCP$	Arg-Leu-Arg-Ala-Ser
$\beta_3 CDCP$	Arg-Leu-Arg-Phe-Asn
Calfluxin	Arg-Val-Asp-Ser-Ala-Asp-Glu-Ser-Asn-Asp-Asp-Gly-Phe-Asp
CDCH	Leu-Ser-Ile-Thr-Asn-Asp-Leu-Arg-Ala-Ile-Ala-Asp-Ser-Tyr-Leu-Tyr-Asp-Gln-His-Lys-Leu-Arg-Glu-Arg-Gln-Glu-Glu-Asn-Leu-Arg-Arg-Arg-Phe-Leu-Glu-Leu-NH_2

2.3. The CDC discharge is the command for egg laying

The CDC network consists of 100 electrotonically coupled neurones divided into two clusters, one in each cerebral ganglion (Vlieger *et al.*, 1980). Normally these cells are silent. In the isolated central ganglia, CDC discharges that last about 40 min can be elicited by repetitive electrical stimulation using suprathreshold depolarising current pulses (Vlieger *et al.*, 1980). Following a discharge, the CDCs cannot be made to generate another discharge for several hours (Kits, 1980).

In intact *Lymnaea*, a long-lasting discharge of simultaneous spiking activity occurs in all CDCs in response to the presentation of water conditions favourable for egg laying (Ter Maat *et al.*, 1983, 1989). Similar CDC discharges invariably precede spontaneous egg laying (Ter Maat *et al.*, 1986). Conversely, the triggering of CDC discharges in intact animals by electrical stimulation of the CDCs is always followed by egg laying (Ter Maat *et al.*, 1989). It had been shown earlier that removing the CDCs abolishes egg laying (Geraerts and Bohlken, 1976). These results show that the CDC discharge is necessary for egg laying.

2.4. CDC peptides act as autotransmitters

In isolated brains of *Lymnaea*, repeated intracellular stimulation (2 Hz usually) of one CDC with depolarising current pulses of sufficient strength to trigger action potentials depolarises the whole network by 15–20 mV and induces spikes in all the other CDCs. Once the rate of spiking in both the stimulated and the follower CDCs reaches 0.5 Hz, electrical stimulation can be stopped and the CDCs generate an afterdischarge (Fig. 2.1; Kits, 1980; Vlieger *et al.*, 1980). To explore the possibility that the CDCs release a self-excitatory substance we performed the following experiments. Two brains were pinned close to each other with their intercerebral commissures, the neurohaemal area of the CDCs, less than 200 μm apart. An electrically triggered afterdischarge of the CDCs in one brain always induces a similar discharge in the other brain (Fig. 2.1). This result indicates that, contingent upon the afterdischarge, a CDC-depolarising substance is released from the stimulated brain.

In a further study designed to identify the CDCs as the source of the excitatory substance, different combinations of predicted cleavage products of the CDCH precursor were tested for their discharge-inducing effect (Brussaard *et al.*, 1990). It appeared that none of the CDC peptides were sufficient by themselves to induce a discharge, but that a combination, consisting of CDCH and two αCDCPs, was the minimum requirement for CDC discharge induction. Antibodies against CDCH (monoclonal) and αCDCPs (polyclonal) caused inexcitability of the CDCs and stopped CDC discharges prematurely.

Thus, some of the CDC-peptides may act as excitatory autotransmitters

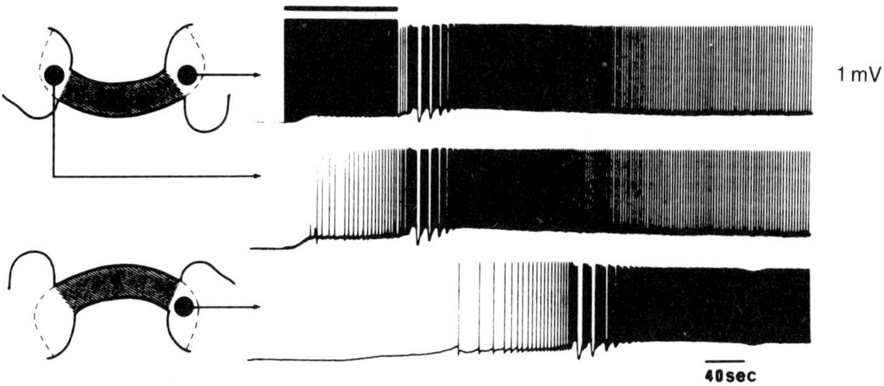

Fig. 2.1. Chemically mediated autoexcitation in the CDCs. Two central ganglia rings were pinned down in close proximity. Two CDCs were impaled in one brain and one in the other. Repetitive stimulation (bar over top trace) of one single CDC depolarised both the stimulated and the unstimulated CDC of the same brain (middle trace), giving rise to action potentials in the unstimulated CDC. After stimulation is stopped, the CDC discharge developed synchronously in both recorded CDCs and probably in the whole cluster. As the discharge developed in the stimulated brain, the CDC impaled in the other brain (bottom trace) slowly depolarised to generate action potentials and finally a high-frequency discharge. CDCs in unstimulated brains never show any spontaneous spiking activity.

amplifying the (unknown) external excitatory input to cause the CDCs to fire at their maximum rate. These findings clearly demonstrate that the CDC peptides are important for the generation of the all-or-none firing pattern of the CDCs.

2.5. **Egg laying behaviours**

As described above, the CDC discharge marks the start of egg laying behaviour. During the CDC discharge, peptides are released both into the blood and locally, within the CNS. It is to be expected that the actions of the peptides released locally will have behavioural effects not observed after peptide injections. Therefore, the study of the role of CDC peptides in behaviour ideally begins with the description of the behaviour initiated by CDC discharges, rather than by injection of homogenates or synthetic CDC peptides. We have analysed the egg laying behaviour of *Lymnaea* under five different conditions: (1) spontaneous egg laying of unoperated animals; (2) induced egg laying after clean water stimulus (CWS); (3) spontaneous egg laying of animals where the CDCs were monitored using fine wire electrodes (cf. Parsons *et al.*, 1983); (4) egg laying elicited by electrical stimulation of the CDCs through fine wire

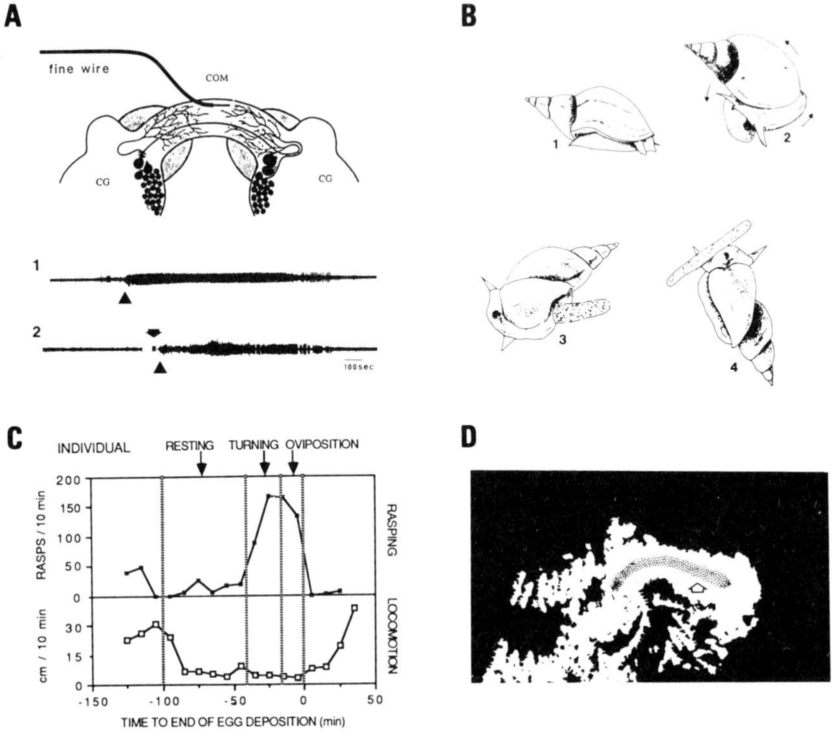

Fig. 2.2. The egg laying behavior of *Lymnaea* is initiated by the discharge of the neurosecretory caudodorsal cells (CDCs). These cells are electrotonically coupled and have their somata in the cerebral ganglia (CG). Their neurites form a neurohaemal area in the periphery of the intercerebral commissure (COM). (A) Precise timing of the start of egg laying can be achieved by using fine wire electrodes to record from the neurohaemal area of the CDCs *in vivo*. CDC discharges can be monitored as they occur spontaneously (A1), following CWS (not shown), and following electrical stimulation (arrow) of their neurites through the fine wire electrode (A2). The triangles denote the onset of the CDC discharge. After the CDCs have fired, the animals express three phases of egg laying behaviour: resting (B1), turning (B2) and oviposition (B3), each of which is characterised by specific postures and action patterns. Once oviposition is complete, animals typically inspect the egg mass with their lips and tentacles (B4). Characteristic patterns of buccal rasping activity and locomotion are also associated with egg laying (C). During resting, levels of rasping and locomotion are low. During turning, locomotion remains low, but rasping gradually increases, reaching a maximum level just before the start of oviposition. Levels of rasping drop during oviposition. Once the egg mass is deposited, locomotion increases, but rasps stay low for the remainder of the observation period. During turning, the animal moves slowly along a tortuous path while rasping against the substrate at a high rate. When egg laying was induced with the animals on glass plates covered with starch, the animals removed the starch from the area in which the egg mass was deposited (D). This suggests that, during egg laying, the function of the rasps is to clean the area of substrate where the egg mass is deposited.

electrodes (Fig. 2.2A); and (5) induced egg laying following injection of purified and synthetic CDCH.

The postures (Fig. 2.2B) and patterns of buccal and locomotory activity (Fig. 2.2C) characteristic of egg laying have been identified. After the CDC discharge is started, three distinct phases of egg laying occur in sequence. The first of these, resting, has a duration of about 50 min and is characterised by a reduction in both locomotion and the number of rasping movements of the buccal mass, as well as the adoption of a posture where the foot is longitudinally contracted, the shell is pulled down over the head-foot and the tentacles are curved downwards (Fig. 2.2B1). Resting is followed by turning (Fig. 2.2B2), which lasts approximately 60 min and is the most active phase. During turning, the animal starts to make buccal rasping movements against the substrate (Fig. 2.2C) and moves around, in a winding path, within a small area of the tank. The animal makes frequent turning movements with its shell. The frequency of rasping movements increases gradually throughout the turning phase, reaching a maximum at about the time when the egg mass starts to be deposited on the substrate. This marks the beginning of oviposition. During this phase the right caudal quadrant of the foot is folded and the egg mass exits the reproductive tract through the vagina (Fig. 2.2B3). The animal moves forward slowly, with the shell tilted until the edge is against the substrate, and the egg mass is pressed on to the substrate by the swollen pneumostome, right mantle edge and the shell. Oviposition takes about 10–20 min (depending on the size of the egg mass). There is a gradual reduction in the number of rasping movements until they stop at about the time oviposition is complete (Fig. 2.2C). The animal then turns around and inspects the egg mass by slowly moving along its length and brushing it with the lips and tentacles (Fig. 2.2B4). After inspection, rasps remain low, but locomotory activity resumes (Fig. 2.2C).

Under all the conditions where egg laying follows normal CDC discharges (1–4) animals show virtually identical patterns of egg laying behaviours. Following injections of CDCH, however, the speed of locomotion is not reduced immediately but continues unchanged for about 50 min before it drops to low levels (see below).

The following sections will describe how the different phases of egg laying behaviour are initiated and timed. There, we will differentiate between two parts of egg laying behaviour: Part I, consisting of resting, and Part II, consisting of turning and oviposition.

2.6. **Part I of egg laying may involve local effects on an identified interneurone**

The egg laying behaviour following injections of CDC extracts and synthetic CDCH is different from the behaviour that follows a CDC

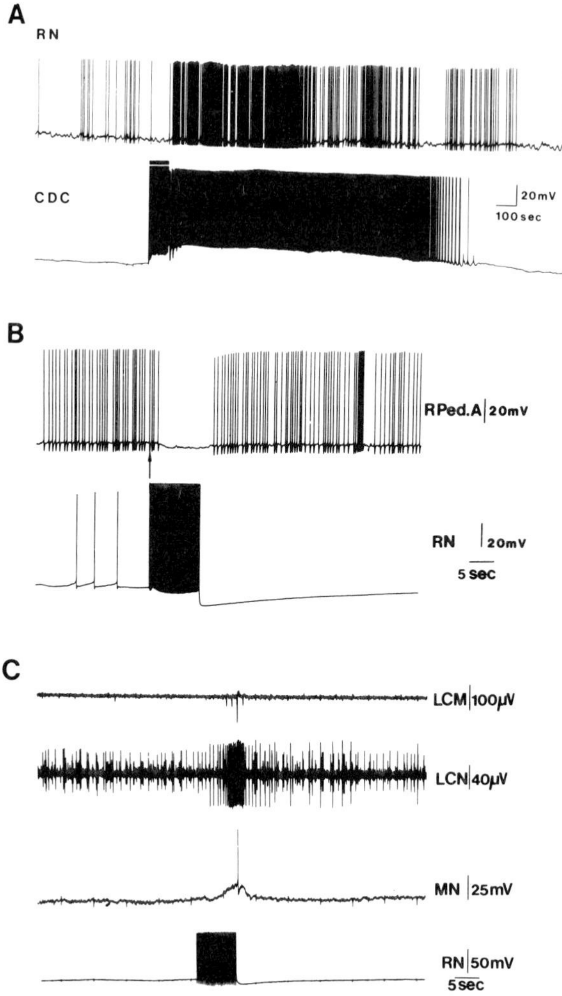

Fig. 2.3. Possible role of the ring neurone in Part I of egg laying behaviour. A: Excitatory effect of the CDC discharge on the ring neurone. One of the CDCs and the ring neurone were impaled. The CDC discharge was initiated by repeated intracellular stimulation as in Fig. 2.2 (bar). Immediately after stimulation was stopped, the CDC discharge developed causing a substantial increase in the firing rate of the ring neurone. This excitation lasted approximately as long as the CDC discharge. B: Inhibition of pedal 'A' cluster neurones (R.Ped.A) by the ring neurone. The pedal 'A' cluster neurones are putative ciliomotoneurones. The results presented here might, therefore, explain the inhibition of locomotion which immediately follows the CDC discharge. C: Excitation of a columellar motoneurone in the pedal ganglion. LCM, Left columellar muscle; LCN, left columellar nerve; MN, motoneurone; RN, ring neurone. The nerve recording shows that, along with the impaled motor neurone, other neurones are excited after RN stimulation. The extracellular muscle recording shows that a slow potential, probably a junction potential, follows the motoneurone action potential. These results demonstrate the existence of a connection between the CDC system and motoneurones.

discharge, where locomotion stops immediately after the start of a discharge. Following an injection, locomotion continues until the start of turning when it is suppressed. Apart from the absence of an initial effect on locomotion, there is no effect on the high frequency of rasping and the suppression of locomotion during turning and oviposition. This suggests that Part I of egg laying behaviour (resting) is caused either by local effects of CDC peptides, or hormonal effects of CDC peptides other than CDCH.

In the right cerebral ganglion an unpaired neurone (ring neurone) is excited during a CDC discharge (Fig. 2.3A; Jansen and Bos, 1984). The ring neuron projects bilaterally to the pedal ganglia. Action potentials in the ring neurone inhibit pedal 'A' cluster neurones (Fig. 2.3B) that are motoneurones of the cilia on the foot (Winlow and Haydon, 1986; Syed and Winlow, 1989; Winlow *et al.*, Chapter 4, this volume), and have excitatory and inhibitory effects on motoneurones of the columellar muscle that moves the shell (Fig. 2.3C; Jansen and Ter Maat, 1985). Thus, the ring neurone seems to be part of the pathway from the CDCs to motoneurones involved in egg laying behaviour. Because the effect starts very soon after the start of the CDC discharge, it is likely that the first part of egg laying behaviour is at least partly due to local influences on the circuitry responsible for locomotion and shell position.

2.7. **Egg mass size determines duration of Part II of egg laying behaviours**

The total duration of egg laying behaviour is positively correlated with the size of the egg mass laid (Fig. 2.4A). The slopes of the lines describing this relationship are similar under all conditions that allow the determination of onset and end of egg laying behaviour (fine wire monitor of CDC activity, CWS, and injection of hormone). However, in the CWS-stimulated animals the correlation is weaker, and the time elapsed between CWS and the end of oviposition is about 30 min longer than under the other conditions. This suggests that the CDCs in these animals were not fired immediately after CWS. Indeed, in animals that are opened at different intervals following CWS and where the CDCs are recorded extracellularly (cf. Kits, 1980), the CDCs of the majority are not active until over 15 min after CWS (Ter Maat *et al.*, 1988).

Next, we examined the effect of egg mass size on rasping behaviour by examining the duration of the period of high rasping in animals that laid different numbers of eggs (Fig. 2.4B). This demonstrated that there is a clear correlation between the total duration of the period of high rasping and the number of eggs laid. This correlation cannot be explained by variations in the length of the oviposition phase because this only varies by 10 min in the different animals, whereas the difference in the duration of rasping is in the order of 60 min depending on the size of the egg mass.

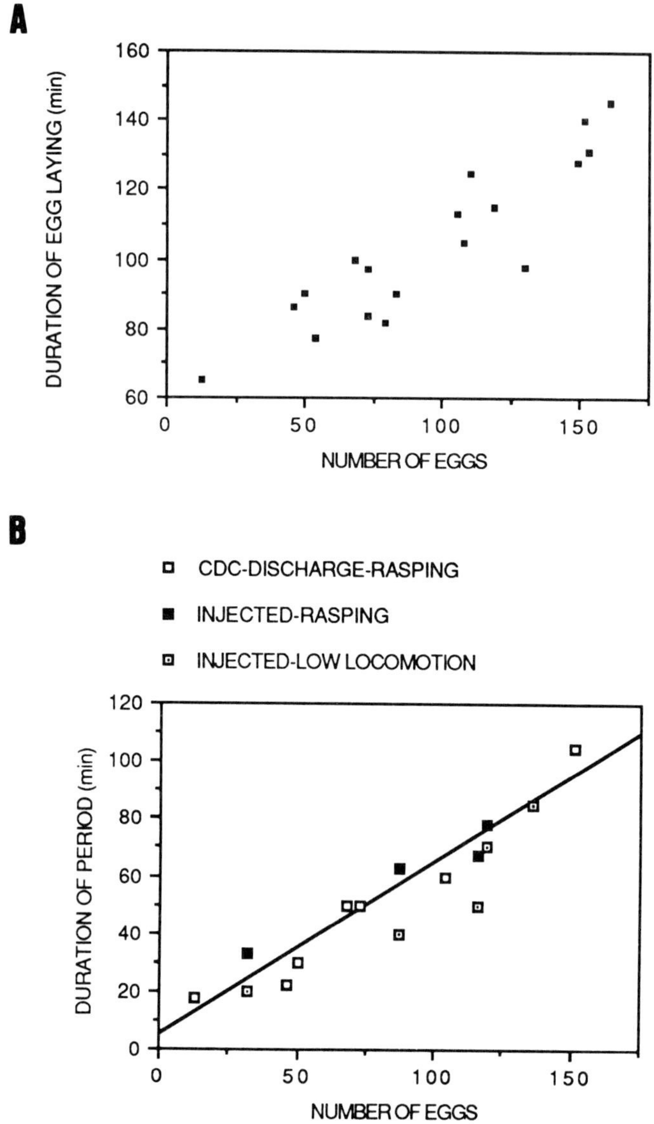

Fig. 2.4. A: The relationship between the size of the egg mass and the total duration of egg laying behaviour. The start of egg laying behaviour is marked by the onset of the CDC discharge as monitored by chronically implanted electrodes on the CDCs. Larger egg masses are associated with longer egg laying behaviour. B: The durations of the periods of high rasping (min) plotted against the size of the egg mass. Following injection of CDCH, spontaneous CDC discharges, and electrically stimulated CDC discharges, there was a clear positive correlation (regression line shown) between the duration of rasping behaviour and the size of the egg mass. The durations of the periods of low locomotion are also shown for injected animals. They also show a positive correlation with egg mass size. Together A and B suggest the feedback from the covert processes occurring in the reproductive tract play a role in maintaining overt egg laying behaviour.

The duration of the period of low rasping following both injection and CDC discharges (Part I) is not correlated with egg mass size. These results suggest that input from the eggs determines the duration of Part II of egg laying, encompassing turning and oviposition, which is characterised by high rasping and low locomotion.

2.8. Activation of Part II of egg laying is due to sensory input from the female duct to the motor circuitry

The hypothesis that sensory input from the covert processes occurring in the reproductive tract is important for the expression of rasping behaviour during egg laying was further tested in lesion experiments. We investigated the timing of ovulation and egg packaging after injections of extracts containing CDCH. At about 50 min after injection, eggs are present in the oothecal gland where they are packaged to form the egg mass (Plesch *et al.*, 1971). The onset of turning is also about 50 min after injection. It may, therefore, be determined by the animal perceiving that eggs are present in this gland. To examine the role of sensory input from the female duct we lesioned the pleuroparietal connectives bilaterally (Ferguson *et al.*, 1988; Ferguson, Pieneman, Jansen and Ter Maat, unpublished observations). These connectives provide the major pathway (Fig. 2.5A) connecting the female reproductive organs with the cerebral, pleural, pedal and buccal ganglia where the neurones of the buccal system are located. The rasping activity of five animals following injections of CDCH is shown in Fig. 2.5B. The effect of the lesion is dramatic, and rasping movements are abolished almost completely. Rasps only occur after the egg mass has started to exit the vagina and to be deposited on the substrate, but even then their frequency is very low. The lesion also affects the locomotory behaviour as well as shell turning. The animals continue to locomote around the tank, keeping their shells in the normal position, until the egg mass is out of the vagina. Only then do they slow down until oviposition is complete, but they do not use their shells to press the egg mass on to the the substrate. Lesioning of the cerebrobuccal connectives (the major pathway between the buccal ganglia and the rest of the CNS) causes a massive reduction in the number of rasps during CDCH-induced egg laying (Fig. 2.5C). The animals still make some rasping movements that continue at the same low rates after CDCH injection. An interesting feature of the behaviour of these animals is that although the pattern of rasps is reduced greatly, the locomotory behaviour during turning remains normal. Cutting all the visceral nerves causes a severe reduction in the number of rasps during egg laying but does not interfere with normal feeding. In addition, the pattern of locomotion does not show the characteristic curves of normal turning. These results demonstrate that sensory input might play a very important

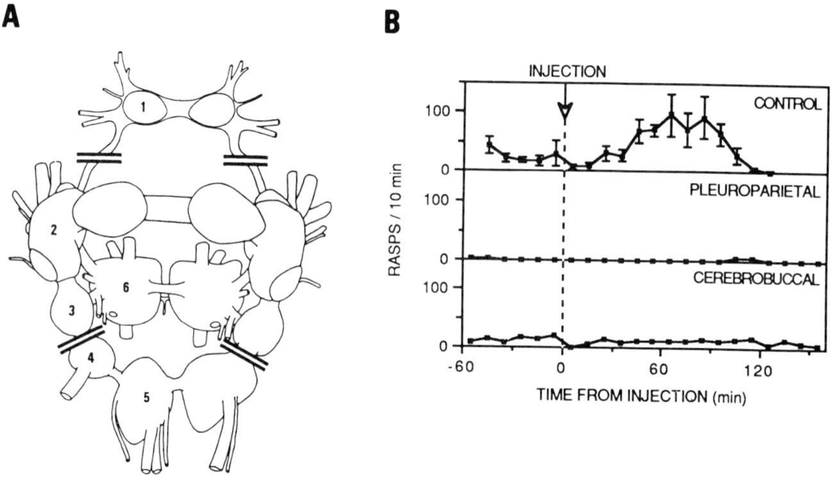

Fig. 2.5. The effects of bilateral lesions of the pleuroparietal and the cerebrobuccal connectives. A: Schematic drawing of the CNS of *Lymnaea* to show the location of the lesions as indicated by the double lines. The bilateral pleuroparietal lesions severed the central connections of the buccal (1), cerebral (2), pleural (3) and pedal (6) ganglia on the one hand with the parietal (4) and the unpaired visceral (5) ganglia on the other. The nerves innervating the female sex organs arise from the visceral ganglion. The bilateral cerebrobuccal lesions isolated the buccal ganglia, where the main components of the buccal system are located, from the rest of the CNS. B: The patterns of activity before and after injections of synthetic CDCH. The normal pattern of unoperated animals is shown in the upper panel. Bars show ±SD (n = 5). Both lesions (middle panel: pleuroparietal; bottom panel: cerebrobuccal) virtually abolished all rasping activity during egg laying, even though the animals laid eggs.

role in the transition from Part I to Part II of egg laying behaviour. In addition, the results of lesioning the cerebrobuccal connectives show that CDCH is not sufficient to initiate rasping in Part II of egg laying.

Similarly, in *Aplysia* egg laying behaviour, initiated by the bag cells (Kupfermann, 1967; Pinsker and Dudek, 1977), there are also two phases, one termed 'appetitive', which precedes the extrusion of the egg cordon from the genital tract and one 'consummatory', which accompanies it (Ferguson *et al.*, 1989a,b). The appetitive phase comprises two patterned behaviours (waves and undulations), as does the consummatory phase (weaves and tamps). The consummatory behaviours are clearly dependent on the presence of eggs since they do not occur in gonadectomised animals that continue to generate bag cell discharges (Ferguson *et al.*, 1986). This indicates that, unlike the CDCs in *Lymnaea*, the *Aplysia* bag cells may act directly on neural circuits to generate rhythmic behaviours. However, the feedback from the genital tract is essential for the expression of the full complement of egg laying behaviours in both *Lymnaea* and *Aplysia*.

2.9. **Buccal muscle activity during egg laying and feeding**

Although the behavioural analysis allows us to quantify rasping behaviour during egg laying (Fig. 2.2), it does not provide qualitative information about whether the buccal mass motor programmes are the same, or different, during egg laying and feeding. This information is important for future work, especially as it has been shown that in *Pleurobranchaea* the motor activity of the buccal mass is only slightly different during several feeding-related behaviours (McClennan, 1982). To characterise the different motor programmes we implanted fine wire electrodes (Parsons *et al.*, 1983) on to the anterior jugalis and posterior jugalis/tensor muscles of the buccal mass and recorded their activity during egg laying and feeding. These muscles were selected because previous experiments with semi-intact preparations have shown that the posterior jugalis/tensor muscle shows activity during protraction of the radula, and the anterior jugalis shows activity during both phases of retraction (Rose and Benjamin, 1979). Thus, by monitoring both of these muscles, it should be possible to obtain a correlate of all three phases of the feeding cycle. Egg laying was induced in four animals by the CWS, and the buccal EMGs underlying rasping during egg laying and rasping during feeding were recorded (Fig. 2.6). In slow displays (Fig. 2.6A) the durations of the bursts of excitatory junction potentials (EJPs) (which were related to complete cycles of buccal activity in a one-for-one manner) were longer and the frequency of bursts was slower during egg laying than during rasping against the substrate, eating fish food, or eating lettuce. These differences were more obvious in fast displays (Fig. 2.6B). During egg laying the total duration of each burst was approximately 8 s as compared with 2–3 s when the animal was rasping against the substrate or eating fish food and 4 s when it was eating lettuce. To date we have not been able to separate the EMG activity underlying the two separate phases of retraction. However, muscle activity during radular protraction was distinct from that underlying retraction.

A major difference in buccal activity was that the protraction phase lasted for about 4 s during egg laying whereas its duration was approximately 1.5 s during feeding and rasping against the substrate. Also during egg laying the muscle activity underlying protraction and retraction was separated by a period when the muscle showed very little activity (Fig. 2.6B1). In contrast, muscle activity was continuous during the other motor programmes (Fig. 2.6B2–4). Thus, the motor programme underlying egg laying is clearly different from the feeding motor programme, suggesting that CDC peptides may be modulating the activity of the

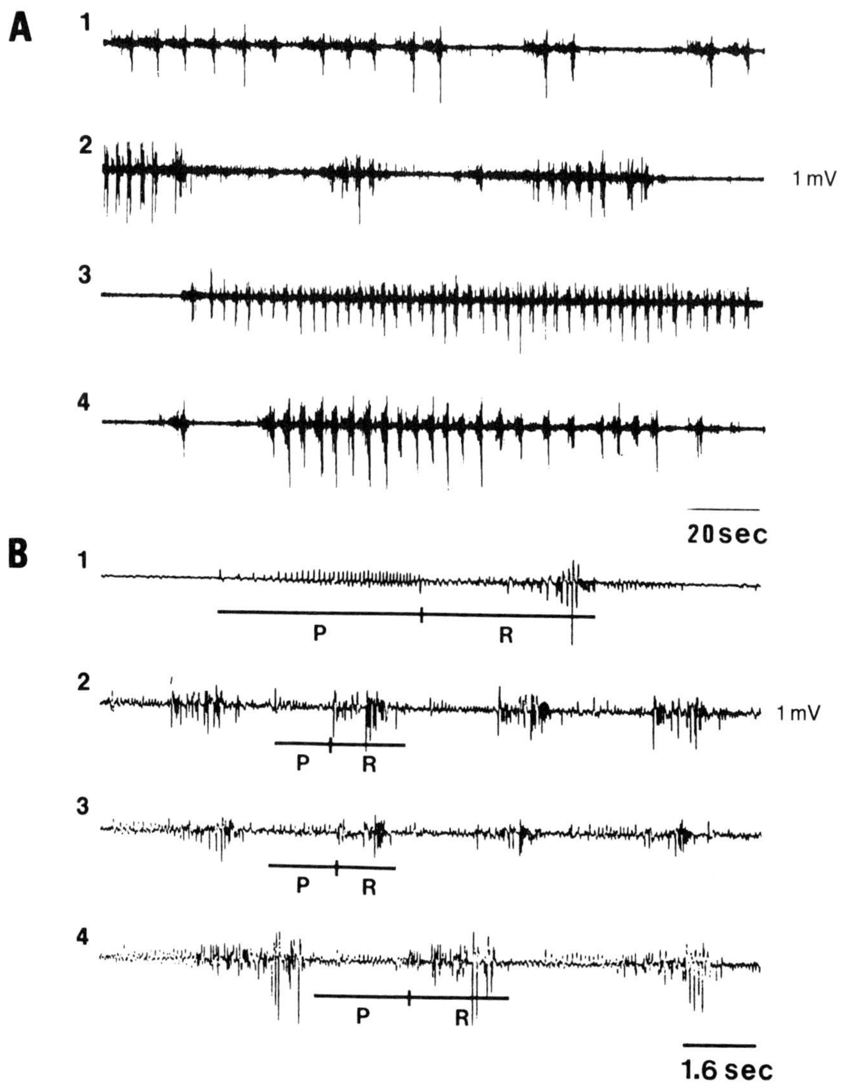

Fig. 2.6. Buccal motor programmes were recorded in freely behaving animals by implanting a fine wire electrode on the anterior jugalis and posterior jugalis/tensor muscles. Egg laying was then induced by the water change stimulus, and the buccal motor programme underlying rasping was recorded (A1, B1). Once egg laying was finished, food was presented to the animal and motor programmes were recorded during rasping against the side of the tank (A2, B2), eating fish food (A3, B3), and eating lettuce (A4, B4). During all behaviours the bursts of excitatory junction potentials (EJPs) recorded from the muscles were correlated with a complete protraction/retraction (P/R) cycle of the radula. The slow displays (A) show that the durations of the individual bursts were longer during egg laying rasps, and occurred at a lower frequency, than during the other behaviours. Faster displays (B) showed that the total duration of the muscle activity during egg laying rasps was more than twice as long as during the other behaviours and was characterised by a different pattern of EJPs.

buccal system. It should also be noted that the amplitude of the muscle activity was larger during ingestion of lettuce (which is tough) than during ingestion of soft fish food. This suggests that sensory input may be causing load compensation, as has previously been described in *Limax* (Reingold and Gelperin, 1980).

2.10. **Conclusions**

The most important conclusion to be drawn from the study of *Lymnaea* egg laying behaviour is that the CDC discharge is not sufficient to trigger any externally observable patterned motor output unless there is input from the genital tract, i.e. many inputs are required for the normal expression of this behaviour. The multiplicity of CDC peptides cannot readily be interpreted as the programme producing and coordinating egg laying behaviour. In fact, the only effect that we know to be mediated by more than one CDC peptide is the peptidergic positive feedback involved in discharge generation. This is a simple effect, even though the underlying mechanisms may be complex.

The behavioural experiments suggest local effects of the CDC peptides, especially during Part I of egg laying. These observations are borne out by studies on the isolated central ganglia ring of *Lymnaea*. There is a local effect of the CDC discharge on the ring neurone as well as on other neurones (Jansen, 1984). Although we do not know whether or not these effects are direct, i.e. monosynaptic or 'mono-nonsynaptic', they might explain why *Lymnaea* stops ongoing locomotion and adopts a clearly defined posture when the CDCs discharge.

The neuroendocrine system controlling egg laying behaviour in *Lymnaea* is similar to systems controlling the reproductive behaviour of a wide range of animals, which suggests that similar mechanisms may be used by different species. CDC discharges are similar to those of the neuroendocrine bag cells of the intact sea hare *Aplysia* (Pinsker and Dudek, 1977). The CDCs also resemble the *Aplysia* bag cells in that both release multiple neuropeptides locally as well as into the blood, some of which have modulatory effects on identified neurones (Jansen and Mayeri, 1988; Mayeri *et al.*, 1979a,b). These molluscan neuroendocrine systems resemble magnocellular oxytocinergic and vasopressinergic cells in the supraoptic and paraventricular nuclei of rats which show afterdischarge-like activity (Andrew and Dudek, 1983). In the case of rat oxytocinergic cells, discharge-like neuronal activity can be induced by suckling (Lincoln and Wakerley, 1974; Freund-Mercier and Richard, 1984).

In summary, egg laying behaviour involves both the neuroendocrine system and a number of motor routines 'borrowed' from other behaviours including locomotion, feeding and shell twisting (see Winlow *et al.*,

Chapter 4, this volume). These routines are clearly modified during egg laying behaviour, and this raises the question of how these motor routines are reorganised depending on the behaviour in which they occur.

References

Andrew, R. D. and Dudek, F. E. (1983) Burst discharge in mammalian neuro-endocrine cells involves an intrinsic regenerative mechanism. *Science* **221**: 1050–1052.

Benjamin, P. R. (1983) Gastropod feeding: behavioural and neural analysis of a complex multicomponent system. In: *Neural Origin of Rhythmic Movements*, A. Roberts and B. Roberts, eds. Cambridge University Press, Cambridge, pp. 159–193.

Benjamin, P. R., Elliott, C. J. H. and Ferguson, G. P. (1985) Neural network analysis in the snail brain. In: *Model Neural Networks and Behavior*, A. I. Selverston, ed. Plenum, New York, pp. 87–108.

Brussaard, A. B., Schluter, N. C. M., Ebberink, R. H. M., Kits, K. S. and Ter Maat, A. (1990) Discharge induction in molluscan peptidergic cells requires a specific set of autoexcitatory neuropeptides. *Neuroscience* **39**: 479–491.

Dictus, W. J. A. G. and Ebberink, R. H. M. (1988) Structure of one of the neuropeptides of the egg-laying hormone precursor of *Lymnaea. Mol. Cell. Endocrinol.* **60**: 23–29.

Dismukes, R. K. and Leibeskind, J. C. (1978) How much can psychopharmacology tell us about the role of neuropeptides in behavior? *Neurosci. Res. Prog. Bull.* **16**: 493–497.

Ebberink, R. H. M., Van Loenhout, H., Geraerts, W. P. M. and Joosse, J. (1985) Purification and amino acid sequence of the ovulation neurohormone in *Lymnaea stagnalis. Proc. Natl Acad. Sci. USA* **82**: 7767–7771.

Ferguson, G. P. (1985) Neurophysiological analysis of whole-body withdrawal in the snail, *Lymnaea stagnalis* (L). D. Phil. Thesis, University of Sussex.

Ferguson, G. P. and Benjamin, P. R. (1985) Whole-body withdrawal of the pond snail *Lymnaea stagnalis. Soc. Neurosci. Abstr.* **11**: 513.

Ferguson, G. P., Parsons, D. W., Ter Maat, A. and Pinsker, H. M. (1986) Spontaneous and elicited bag cell discharges in gonadectomized *Aplysia. J. exp. Biol.* **123**: 159–173.

Ferguson, G. P., Pieneman A. W. and Ter Maat, A. (1988) Organization of egg laying behaviors in the pond snail, *Lymnaea stagnalis. Soc. Neurosci. Abstr.* **14**: 690.

Ferguson, G. P., Ter Maat, A., Parsons, D. W. and Pinsker, H. M. (1989a) Egg laying in *Aplysia* I. Behavioral patterns and muscle activity of freely behaving animals after selectively elicited bag cell discharges. *J. Comp. Physiol. A* **164**: 835–847.

Ferguson, G. P., Ter Maat, A. and Pinsker, H. M. (1989b) Egg laying in *Aplysia* II. Organization of central and peripheral pathways for initiating neuro-secretory activity and behavioral patterns. *J. Comp. Physiol. A* **164**: 849–857.

Freund-Mercier, M.-J. and Richard, Ph. (1984) Electrophysiological evidence for facilitatory control of oxytocin neurons during suckling in the rat. *J. Physiol.* **352**: 447–466.

Geraerts, W. P. M. and Bohlken, S. (1976) The control of ovulation in the hermaphrodite freshwater snail *Lymnaea stagnalis* by the neurohormone of the caudodorsal cells. *Gen. Comp. Endocrinol.* **28**: 350–357.

Geraerts, W. P. M. and Hogenes, Th. M. (1983) Multiple release of peptides by electrical active neurosecretory caudo-dorsal cells of *Lymnaea stagnalis. Neurosci. Lett.* **41**: 151–155.

Geraerts, W. P. M. and Hogenes, Th. M. (1985) Heterogeneity of peptides released by electrical active neurosecretory caudodorsal cells of *Lymnaea stagnalis. Brain Res.* **33**: 51–61.

Geraerts, W. P. M., Ter Maat, A. and Vreugdenhil, E. (1987) The peptidergic neuroendocrine control of egg laying behaviour in *Aplysia* and *Lymnaea.* In: *Invertebrate Endocrinology, Vol. 2, Endocrinology of Selected Invertebrate Types*, H. Laufer and R. Downer eds. Alan Liss, New York, pp. 144–231.

Haydon, P.G. and Winlow, W. (1986) Shell movements associated with locomotion of *Lymnaea* are driven by a central pattern generator. *Comp. Biochem. Physiol.* **83A**: 23–25.

Jansen, R. F. (1984) Neuronal and hormonal control of the egg laying behavior in the pond snail *Lymnaea stagnalis.* PhD Thesis, Vrije Universiteit.

Jansen, R. F. and Bos, N. P. A. (1984) An identified neuron modulating the activity of the ovulation hormone producing caudo-dorsal cells of the pond snail *Lymnaea stagnalis. J Neurobiol.* **15**: 161–167.

Jansen, R. F. and Mayeri, E. (1988) The neuropeptide egg-laying hormone modulates multiple ionic currents in single target neurons of the abdominal ganglion of *Aplysia. J. Neurosci.* **8**: 3074–3084.

Jansen, R. F. and Ter Maat, A. (1985) Ring neuron control of columellar motor neurons during egg-laying behavior in the pond snail. *J. Neurobiol.* **16**: 1–14.

Kits, K. S. (1980) States of excitability in ovulation hormone producing cells of *Lymnaea stagnalis* (Gastropoda) and their relation to the egg-laying cycle. *J. Neurobiol.* **11**: 397–410.

Kupfermann, I. (1967) Stimulation of egg laying: possible neuroendocrine functions of bag cells of abdominal ganglion of *Aplysia californica. Nature* **216**: 814–815.

Lincoln, D. W. and Wakerley, J. B. (1974) Electrophysiological evidence for the activation of supraoptic neurons during the release of oxytocin. *J. Physiol.* **242**: 533–554.

McClennan, A. D. (1982) Movements and motor patterns of the buccal mass of *Pleurobranchaea* during feeding, regurgitation and rejection. *J. exp. Biol.* **98**: 195–211.

Mayeri, E., Brownell, P. H. and Branton, W. D. (1979) Multiple, prolonged actions of neuroendocrine ‘bag cells’ in *Aplysia.* II. Effects on beating pacemaker and silent neurons. *J. Neurophysiol.* **42**: 1184–1197.

Mayeri, E., Brownell, P. H., Branton, W. D. and Simon, S. B. (1979) Multiple, prolonged actions of neuroendocrine ‘bag cells’ on neurons in *Aplysia.* I. Effects on bursting pacemaker neurons. *J. Neurophysiol.* **42**: 1165–1184.

Parsons, D. W., Ter Maat, A. and Pinsker, H. M. (1983) Selective recording and stimulation of individual identified neurons in freely-behaving *Aplysia. Science* **221**: 1203–1206.

Pinsker, H. M. and Dudek, F. E. (1977) Bag cell control of egg laying in freely behaving *Aplysia. Science* **197**: 490–493.

Plesch, B., Jong-Brink, M. de and Boer, H. H. (1971) Histological and histochemical observations on the reproductive tract of the hermaphrodite pond snail *Lymnaea stagnalis. Neth. J. Zool.* **21**: 180–201.

Reingold, S. C. and Gelperin, A. (1980) Feeding motor programme in *Limax.* II. Modulation by sensory inputs in intact animals and isolated central nervous systems. *J. exp. Biol.* **85**: 1–19.

Rose, R. M. and Benjamin, P. R. (1979) The relationship of the central motor pattern to the feeding cycle of *Lymnaea stagnalis*. *J. exp. Biol.* **80**: 137–163.

Rothman, B. S., Mayeri, E., Brown, R. O., Yuan, P.-M. and Shively, J. E. (1983) Primary structure and neuronal effects of a-bag cell-peptide, a second candidate neurotransmitter encoded by a single gene in bag cell neurones of *Aplysia*. *Proc. Natl Acad. Sci. USA* **80**: 5753–5757.

Scheller, R. H., Rothman, B. S. and Mayeri, E. (1983) A single gene encodes multiple peptide neurotransmitter candidates involved in a stereotyped behavior. *TINS* **6**: 340–345.

Selverston, A. I. (1985) (ed.) *Model Neural Networks and Behavior*, Plenum, New York.

Syed, N. I. and Winlow, W. (1989) Morphology and electrophysiology of neurons innervating the ciliated locomotor epithelium in *Lymnaea Stagnalis* (L). *Comp. Biochem. Physiol.* **93A**: 633–644.

Ter Maat, A., Dijcks, F. A. and Bos, N. P. A. (1986) In vivo recordings of neuroendocrine cells (caudo-dorsal cells) in the pond snail. *J. Comp. Physiol. A* **158**: 853–859.

Ter Maat, A., Lodder, J. C. and Wilbrink, M. (1983) Induction of egg laying in the pond snail *Lymnaea stagnalis* by environmental stimulation of the release of ovulation hormone from the caudo-dorsal cells. *Int. J. Invert. Reprod.* **6**: 239–247.

Ter Maat, A., Roubos, E. W., Lodder, J. C. and Buma, P. (1983) Integration of biphasic synaptic input by electrotonically coupled neuroendocrine caudo-dorsal cells in the pond snail. *J. Neurophysiol.* **49**: 1392–1409.

van der Wilt, G. J., Roest, M. van der and Janse, C. (1987) Neuronal substrates of respiratory behaviour and related functions in *Lymnaea stagnalis*. In: *Neurobiology Molluscan Models*, H. H. Boer, W. P. M. Geraerts, and J. Joosse, eds. North Holland Publishing Co, Amsterdam, pp. 292–296.

Vlieger, T. A. de, Kits, K. S., Ter Maat, A. and Lodder, J. D. (1980) Morphology and electrophysiology of the ovulation hormone producing neuro-endocrine cells of the freshwater snail *Lymnaea stagnalis* (L.). *J. exp. Biol.* **84**: 259–271.

Vreugdenhil, E., Bogerd, J., Mahon, A., Bouwmeester, T., Klootwijk, C., Van Heerikhuizen, M. and Joosse, J. (1988) The genomic organization of the ovulation hormone gene family of the snail *Lymnaea stagnalis*. *J. Neurosci.* **8**: 4184–4191.

Winlow, W. and Haydon, P. G. (1986) A behavioural and neuronal analysis of the locomotory system of *Lymnaea stagnalis*. *Comp. Biochem. Physiol.* **83A**: 13–21.

Winlow, W., Moroz, L. L. and Syed, N. I. (1991) Mechanisms of behavioural selection in *Lymnaea stagnalis*. In: *Neurobiology of Motor Programme Selection: New Approaches to the Study of Behavioural Choice*, J. Kien, C. R. McCrohan and W. Winlow, eds. Manchester University Press, Manchester, pp. 52–72.

3 *Catherine R. McCrohan and Michael A. Kyriakides*

Motor programme selection and the control of feeding in the snail

3.1. Introduction

One of the ultimate aims of neurobiology is to explain how behaviour is generated and controlled by the central nervous system (CNS). The behavioural repertoire of even the simplest animal involves a variety of activities, which may or may not be mutually compatible. In its normal environment an animal must choose which behaviour to perform at a given time, depending on external conditions and on its own internal state. In order to understand how such decisions are made, many invertebrate neurobiologists have concentrated on an approach known as 'circuit breaking'. This involves detailed analysis of the cellular and synaptic properties of neuronal networks which generate motor output for specific behaviour patterns. Once these networks are understood, it then becomes possible to study their interactions and to examine the role of such interactions in behavioural decision making in the whole animal.

The pond snail *Lymnaea stagnalis* provides an excellent subject for detailed analysis of neuronal interactions involved in the control of behaviour. The behavioural repertoire of *Lymnaea* is relatively simple and the CNS is easily accessible for electrophysiology. The CNS comprises 11 major ganglia (Fig. 3.1). Neuronal somata lie on the surface of the ganglia, and many are individually identifiable owing to their large size, constant location, and electrophysiological and morphological characteristics. There is some localisation of function to particular ganglia; for example, much of the neural circuitry underlying rhythmic radula movements is located in the buccal ganglia (Benjamin, 1983). However, most motor patterns involve the concerted action of neurones in several ganglia (McCrohan and Winlow, 1985). Considerable progress has already been made in unravelling the neuronal circuitry underlying a number of behavioural activities in *Lymnaea* including rhythmic feeding (radula) movements, locomotory behaviour, whole body withdrawal, breathing, and behaviour associated with egg laying. Probably the best described in terms of its interneuronal and motoneuronal components is the neural network controlling the rhythmic radula and swallowing movements which accompany ingestion of food. In this chapter we review the properties of interneurones that regulate the occurrence and nature of rhythmic feeding motor output, and describe what is known about central

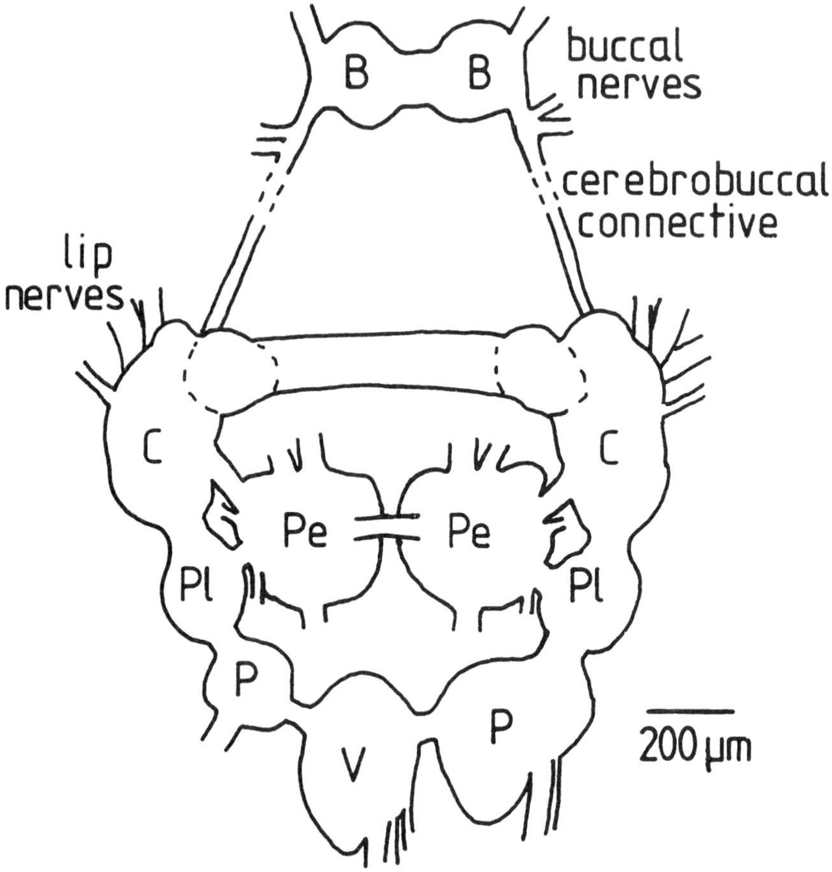

Fig. 3.1. Dorsal view of the CNS of *Lymnaea stagnalis*, showing major ganglia, and some nerves. Ganglia labelled are: B, buccal; C, cerebral; P, parietal; Pe, pedal; Pl, pleural; V, visceral.

interactions between feeding circuitry and neurones concerned with control of other behaviours such as locomotion.

3.2. Generation of feeding rhythm in *Lymnaea*

Neuronal activity underlying rhythmic movements of the radula during feeding is generated by a central pattern generator (CPG) in the buccal ganglia (Benjamin, 1983). This comprises three subnetworks of interneurone types, N1, N2 and N3 (Rose and Benjamin, 1981b; Elliott and Benjamin, 1985a). These are connected synaptically in such a way that they fire successively during each phase of the feeding cycle; N1 interneurones fire during protraction of the radula, N2 during retraction (i.e. radula rasping), and N3 during swallowing. Rhythmic activity generated by the CPG is imposed on motoneurones in the buccal and cerebral ganglia via direct synaptic connections (Rose and Benjamin, 1981b). Movements of the radula are controlled by the musculature of the buccal mass, which is innervated by buccal motoneurones (Rose and Benjamin, 1979). Buccal motoneurones also innervate the pro-oesophagus and salivary glands. The motoneurones in the cerebral ganglia innervate the lips, probably to control opening and closing of the mouth during feeding (McCrohan, 1984a).

The CPG can generate rhythmic feeding motor output in the isolated CNS (i.e. in the absence of sensory input) although the rhythm is usually slower than that observed in the whole animal. Activity in the N1, N2 and N3 interneurones may be monitored by recording from identified buccal motoneurones, in which the occurrence and polarity of synaptic inputs from each of the three interneuronal subnetworks is known. So, for example, the B4 retractor motoneurone receives inhibitory input from both N1 and N2 interneurone types, and then fires a burst of spikes by postinhibitory rebound. During this burst B4 also receives inhibitory inputs from N3 interneurones, leading to fractionation of the burst (see Fig. 3.2). Another buccal motoneurone, B3, receives inhibitory input from N1 and excitation from N2, followed by a series of brief excitatory inputs from N3, during each feeding cycle (see Fig. 3.4). Most work to date, however, has been carried out on the isolated CNS or on semi-

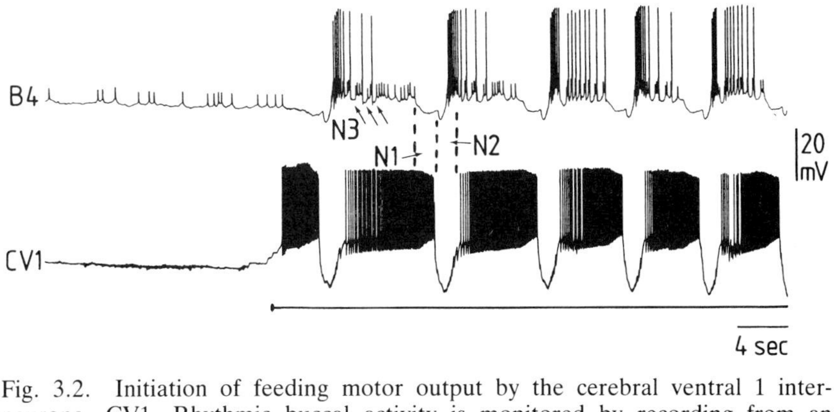

Fig. 3.2. Initiation of feeding motor output by the cerebral ventral 1 interneurone, CV1. Rhythmic buccal activity is monitored by recording from an identified buccal retractor motoneurone, B4. Steady depolarisation of CV1 by current injection (indicated by bar) leads to onset of rhythm, i.e. discharges. Phases of activity in N1, N2 and N3 interneurones of the feeding CPG are clearly seen as identified synaptic inputs (all inhibitory) in B4. CV1 also receives rhythmic feedback from the CPG.

intact preparations in which peripheral structures such as the buccal mass or lips and tentacles are retained, attached to the CNS by the nerves that innervate them.

Because the CPG underlying feeding motor output in *Lymnaea* is so well understood, it becomes possible to ask questions about the precise mechanism of action of cells which may have a role in switching feeding behaviour on or in modulating its frequency or intensity. A knowledge of the actions of these cells could increase our understanding of the cellular basis of behavioural choice or decision making.

3.3. **Initiation and modulation of feeding rhythm in *Lymnaea***

A current view of 'command' function within neural networks controlling behaviour is that it is an emergent property of the network rather than being confined to individual command neurones (e.g. Getting and Dekin, 1985). This view arose following the discovery of extensive feedback between lower and higher levels within so-called motor control hierarchies for behavioural output. Thus the concept of command arising from a specific neuronal location breaks down, since elements at lower levels in the hierarchy could theoretically take on a command function by activating higher order elements via feedback loops. However, individual neurones can still be identified which are particularly effective in initiating and modulating motor output, and several have been described in the feeding system of *Lymnaea*. These neurones have marked effects on motor output underlying feeding, at least in reduced preparations.

Two types of higher order interneurone have been described in the *Lymnaea* feeding system, which can initiate rhythmic feeding motor output in isolated and semi-isolated CNS preparations. These are the buccal slow oscillator (SO) neurone (Rose and Benjamin, 1981a; Elliott and Benjamin, 1985b) and the cerebral ventral 1 (CV1) cell (McCrohan, 1984b; McCrohan and Kyriakides, 1989). At first sight the actions of these two neurone types appear the same. Steady depolarisation of either SO or CV1 leads to initiation and maintenance of rhythmic feeding motor output from the buccal ganglia. This is illustrated for CV1 in Fig. 3.2. The SO and CV1 in turn receive rhythmic feedback from the CPG, in particular excitation during the N1 phase and inhibition during the N2 phase (Fig. 3.2). The SO and CV1 are not mutually connected so they must act independently, and we have observed one major difference in their functional outputs. When CV1 is depolarised tonically, the evoked buccal rhythm is maintained indefinitely (for longer than 7 min in our experiments). In contrast, steady depolarisation of SO leads to a rhythm which is maintained for only a limited time (between 0.5 and 2 min) before cycles start to be missed or the rhythm is disrupted. This is shown in Figure 3.3; the rhythm evoked by CV1 alone is more regular than that

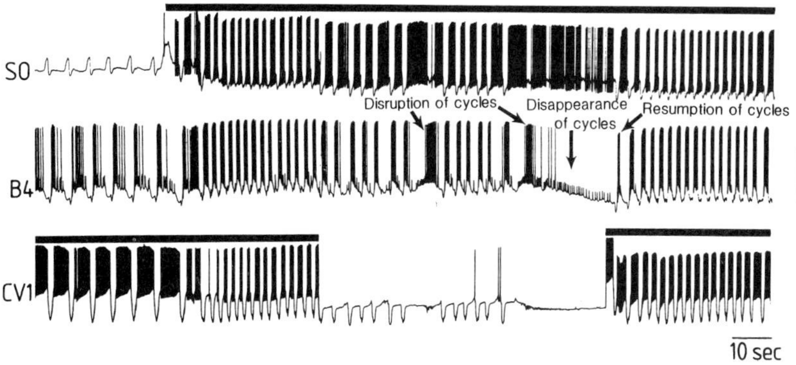

Fig. 3.3. Effects of CV1 and the slow oscillator interneurone, SO, on rhythmic buccal activity, monitored in B4 motoneurone. Bars indicate periods of steady depolarisation of CV1 and SO. Depolarisation of CV1 evokes regular rhythm. Depolarisation of CV1 and SO together increases the frequency of the rhythm. When SO is activated alone, the rhythm is less regular; cycles are missed and eventually the rhythm breaks down. Rhythmic activity is resumed immediately on depolarisation of CV1.

seen during stimulation of SO alone. Fig. 3.3 also shows that the effects of CV1 and SO on feeding cycle frequency are additive.

Other differences between SO and CV1 neurones have been found. The SO occurs as a single unpaired neurone in either left or right buccal ganglion (Elliott and Benjamin, 1985b). CV1 cells comprise a population of at least three individuals in each cerebral ganglion, which are of two subtypes ($CV1_a$ and $CV1_b$) based on the strengths of their synaptic inputs (McCrohan and Kyriakides, 1989). Both SO and CV1 make monosynaptic, facilitating, excitatory connections with N1 interneurones of the CPG (Elliott and Benjamin, 1985b; McCrohan and Kyriakides, 1989). SO also inhibits N2 interneurones monosynaptically (Elliott and Benjamin, 1985b) but no connection from CV1 to N2 has been demonstrated. The location of SO and CV1 in different ganglia may reflect fundamental differences in their roles. Since SO is less effective than CV1 in maintaining a feeding rhythm for long periods of time, it is possible that the two cell types have different functions, CV1 acting to gate the rhythm while SO is more important for cycle-by-cycle modulation of its frequency and intensity.

The location of CV1 interneurones in the cerebral ganglia means that they could have access both to sensory input related to the presence of food (via the lip nerves near whose roots the CV1 cells lie), and to information about current activity in centres in the CNS that control other behaviours such as breathing and locomotion. Putative homologues of the CV1 cells are found in at least two other gastropod species, *Limax*

(Delaney and Gelperin, 1990) and *Pleurobranchaea* (Davis *et al.*, 1983). In the latter, synaptic inputs in response to food stimuli are altered following aversive conditioning of the whole animal, and in satiated animals. This leads to a reduction in the cells' ability to initiate rhythmic buccal motor activity. It is possible therefore that the CV1 neurones, and their equivalents in other species, provide a site at which changes in behavioural responsiveness in the intact animal are reflected by changes in synaptic responses in identified neurones. Our present work is directed towards exploring this possibility.

Other cells are also involved in control of the feeding rhythm. A pair of giant serotonin-containing neurones, the cerebral giant cells (CGCs), is located in the anterior lobes of the cerebral ganglia. These neurones have axonal projections to the buccal ganglia, via the cerebrobuccal connectives, and to the periphery in nerves which innervate the lips and buccal musculature (McCrohan and Benjamin, 1980a). The CGCs have post-synaptic actions on CPG interneurones and on buccal motoneurones (McCrohan and Benjamin, 1980b; Benjamin *et al.*, 1981). They probably also have modulatory effects on muscles of the buccal mass and lips (McCrohan *et al.*, 1989). Their precise role in modulating feeding motor output has been difficult to determine and is still unclear. In the isolated CNS, the CGCs usually fire steadily at a rate of 0.5 to 2.0 spikes per second. An increased firing frequency in the CGCs leads to an increase in the intensity of bursting in buccal motoneurones (McCrohan and Benjamin, 1980b), due mainly to mono- and polysynaptic connections between CGCs and motoneurones. In some preparations, tonic depolarisation of a CGC to increase firing rate to about 5 spikes per second can lead, after a delay, to initiation of rhythmic buccal activity in a previously non-rhythmic preparation (McCrohan and Audesirk, 1987). However, this action is not reliable, and the rhythm is not maintained for more than 20 cycles. In a preparation already showing rhythmic activity, an increase in CGC firing rate leads, paradoxically, to a decrease in the frequency of the buccal rhythm. A similar response is seen following superfusion of the preparation with serotonin (Tuersley and McCrohan, 1988). In fact, intense evoked CGC activity or prolonged application of serotonin can result in complete cessation of rhythmic feeding motor outpur (Tuersley and McCrohan, 1988; Kyriakides and McCrohan, 1989), usually accompanied by tonic activation of N3 interneurones. The CGCs have been shown to excite N3 interneurones, and also to weakly inhibit N1 and N2 interneurones (Benjamin *et al.*, 1981), whilst the N3 interneurones themselves feed back to inhibit N1 interneurones (Elliott and Benjamin, 1985a). We hypothesise that increased activity in CGCs acts to reduce the activity of the N1 and N2 interneurones in the CPG and enhance the activity of N3 interneurones (McCrohan *et al.*, 1989); this would have the effect of slowing down the rhythm by delaying the onset of the N1 activity

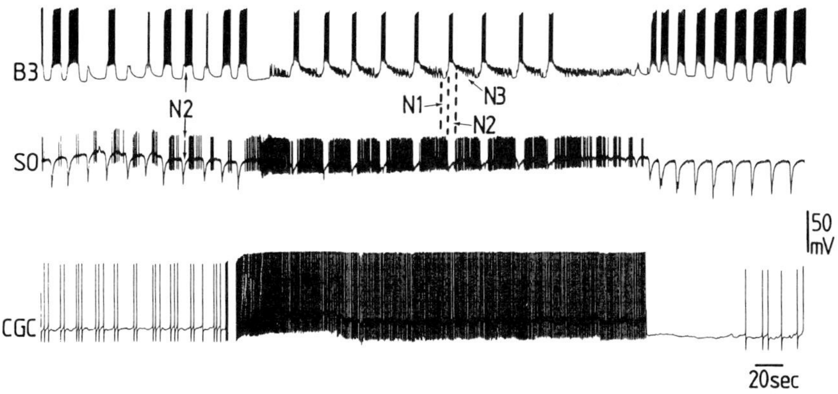

Fig. 3.4. Effect of increased firing rate of CGC on rhythmic buccal motor output driven by SO. SO was steadily depolarised throughout the recording. Rhythmic buccal activity is monitored from a buccal motoneurone, B3. At low CGC firing rate (0.4 spikes per second), the rhythm is dominated by N2 interneuronal activity, seen as strong depolarising input to B3 and hyperpolarisation of SO. When CGC is driven to fire at a higher frequency (2.0 spikes per second), the rhythm changes; N2 interneuronal inputs to B3 and SO become weaker, and N1 and N3 activity becomes more apparent. After the end of CGC activation, the rhythm reverts abruptly to its original form.

which starts each cycle, and could lead ultimately to continuous activity in N3 neurones at the expense of N1 and N2.

One of the more interesting actions of the CGCs which we have observed in the isolated CNS, and which can be interpreted according to the hypothesis outlined above, is illustrated in Fig. 3.4. Prolonged depolarisation of a CGC leads to an abrupt and maintained change in the nature of rhythmic motor output driven by the SO. In the experiment illustrated the SO was excited tonically by current injection in order to drive a buccal rhythm, which was monitored in a B3 motoneurone. At the start of the record the rhythm was dominated by N2 activity, leading to strong rhythmic excitation and bursting in B3 and marked N2 inhibitory inputs to SO (arrows). We have called this the ‘N2 rhythm’, and suggested that it may underlie rasping in the absence of ingestion of food such as might occur during exploratory activity (Kyriakides and McCrohan, 1989). The CGC was then depolarised tonically, to increase its firing rate from 0.4 to 2.0 spikes per second. This led to an immediate change to a feeding rhythm in which inputs from all three interneuronal components of the CPG (N1, N2, N3) are clearly seen. This rhythm is the typical feeding rhythm as described by Benjamin and his colleagues, and is thought to underlie true feeding in the intact animal. The most obvious changes in the rhythmic activity in Fig. 3.4 following activation of the

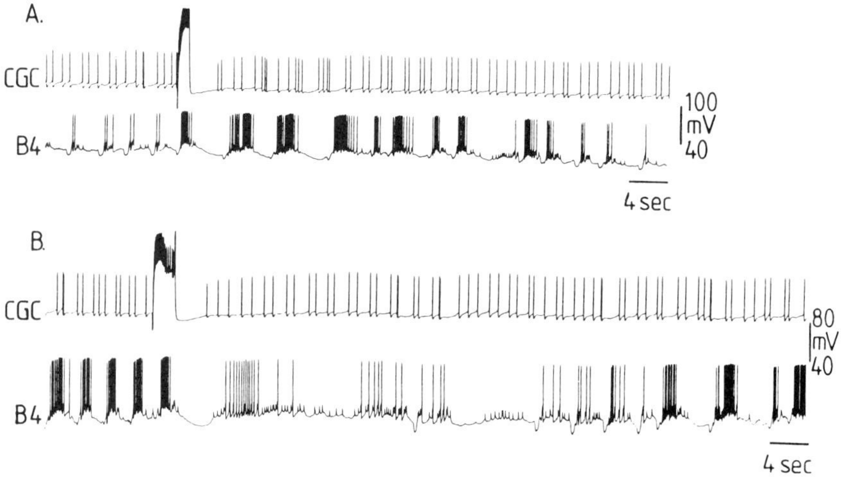

Fig. 3.5. Brief evoked bursts of activity in CGC lead to long-lasting changes in rhythmic buccal motor output, monitored from buccal motoneurones, B4. (A) Following a 2-s burst in CGC, the period and intensity of the buccal rhythm increase; this response lasts for about 45 s. (B) In another preparation an evoked burst in CGC leads to disruption of rhythmic activity in B4, which resumes after about 40 s.

CGC are a reduction in the amplitude and duration of N2 inputs to B3 and SO, and the onset of N3 activity, seen as a series of brief depolarising potentials in B3. This is consistent with our hypothesis that increased activity in CGCs enhances N3 activity whilst reducing activity in N2 interneurones. It seems therefore that the CGCs may function to modify the type of rhythmic activity generated by the CPG by altering the balance of activities in the different types of CPG interneurone.

The effects of the CGCs on the buccal rhythm are often prolonged, outlasting the increased activity in the CGCs themselves. This is illustrated in Fig. 3.5. In a preparation showing weak bursts of activity in a B4 retractor motoneurone a brief (~2 s) evoked burst in a CGC leads to a marked increase in intensity and a decrease in frequency of retractor bursts, lasting up to 40 s (Fig. 3.5A). In contrast, in Fig. 3.5B, the rhythm at the start of the recording is strong; an evoked CGC burst leads to complete disruption of rhythmic activity which is only restored after about 40 s. These prolonged effects of the CGCs may involve production of second messengers (e.g. cyclic AMP) in postsynaptic neurones (McCrohan *et al.*, 1989). What seems clear is that the effects of the CGCs on feeding motor output depend on the initial state of activity in the buccal network. Brief bursts of activity in CGCs, similar to those evoked in Fig. 3.5, occur spontaneously and in response to food stimuli applied to the lips (Kemenes *et al.*, 1986; McCrohan and Audesirk 1987). Prolonged high frequency activity in CGCs (cf. Fig. 3.4) may not be physiological. Indeed, recordings taken from CGCs in freely moving intact snails (M. S. Yeoman, personal communication) suggest that their activity in the isolated CNS does not reflect that in the whole animal, and that the relatively high spike frequencies that we have used in experiments to investigate the function of the CGCs may never actually occur. Clearly further work on more intact preparations is required.

What is also needed if we are to understand more about the neural mechanisms underlying the snail's decision to feed is information about the role of sensory cues and the pathways by which sensory information is fed into the system. Some work in this direction has already been carried out. Primary oesophageal mechanoreceptive neurones that signal gut distension have been identified in the buccal ganglia. These receptors appear to be involved in switching off feeding, possibly in response to 'satiety' cues (Elliott and Benjamin, 1989), and could therefore be said to have a role in decision making. With regard to sensory triggering of feeding behaviour via food stimuli, the conclusion so far is that multimodal sensory inputs are used and that these act at several levels (i.e. interneurones, motoneurones) in the system (Kemenes *et al.*, 1986).

3.4. **Central interactions between neural networks**

So far, we have described ways in which the higher order interneurones, SO, CV1 and CGC, initiate and modulate rhythmic motor outputs underlying feeding in *Lymnaea*, and we have suggested that access to appropriate sensory inputs and to information about the animal's internal state could allow these neurones to play some role in mediating decisions about whether or not to feed. In the freely behaving animal, however, more complex decisions are also required. For example, when presented simultaneously with sensory triggers for two or more incompatible behaviours, what are the mechanisms that ensure that only one of these behaviours is generated at any one time? For mutually compatible behaviours (i.e. those that can occur together), how are their activities coordinated so that each occurs to best effect? Because much is known about the cellular organisation and function of the CNS of *Lymnaea*, we can at least begin to answer these questions.

Two behaviours that are mutually compatible in *Lymnaea* are feeding and locomotion. Snails feed whilst moving forward across the substrate and there is clear coordination of these activities. For example, it has been shown that rasps never occur at the same time as changes in the direction of head movement (Dawkins, 1974). Indeed, a rasp can only be efficiently completed if movement ceases briefly; forward movement can then resume during the other phases of the cycle (i.e. radula protraction and swallowing). We examined the relationship between activity in the buccal feeding circuitry and in neurones in the pedal ganglia which are thought to be involved in the generation of locomotory and body movements. It was already known that phasic information about ongoing feeding activity in the buccal ganglia was distributed as far as the cerebral ganglia. The CV1 neurones receive cyclical inputs in phase with N1, N2 and N3 interneuronal activity, as do lip motoneurones in the cerebral ganglia (McCrohan, 1984a,b). Benjamin and Elliott (1989) described a pair of buccal neurones, called Nc, which have axonal projections to the cerebral ganglia and which may carry an 'efference copy' of the buccal rhythm. Furthermore, N1 interneurones also project to the cerebral ganglia and convey phasic information (Elliott and Benjamin, 1985a). An identified interneurone with its cell body located in the ventral cerebral ganglion and an axonal projection into the ipsilateral pedal ganglion has been described (McCrohan, 1982). This neurone, CV2, bursts rhythmically in phase with N1 activity in the buccal ganglia, and could therefore provide pedal neurones with phasic information about activity in the buccal CPG.

A considerable amount is known about the electrophysiological and morphological properties of neurones in the pedal ganglia (Kyriakides *et al.*, 1989). The ganglia are largely symmetrical. Neurones occur either as paired clusters of similar cells (each cluster assigned a letter; e.g. R.Pe.D.F = right pedal dorsal F cluster), or as individually identifiable 'giant' cells (assigned a number; e.g. R.Pe.D.1 = right pedal dorsal 1 cell) most of which are also paired. By paired recordings we have shown that many pedal neurones receive rhythmical synaptic inputs in phase with the feeding rhythm recorded from buccal motoneurones (Kyriakides and McCrohan, 1988). These inputs, which were largely inhibitory and most often occurred during the N2 phase of the feeding cycle (i.e. at the time of radula rasping in the intact animal), were recorded in neurones of three pedal cell clusters, E, F and M, and in five types of giant neurone. Fig. 3.6A shows an example of these inputs in an F-cluster neurone. The rhythmic inputs to pedal neurones appear to originate from the buccal ganglia because they are abolished when the cerebrobuccal connectives are cut. Furthermore, steady depolarisation of SO or CV1 to initiate rhythmic buccal activity in a previously non-rhythmic preparation also leads to rhythmic inputs to pedal neurones (Fig. 3.6B). The fact that these inputs are inhibitory and occur mainly during the radula retraction phase suggests that they may act to provide widespread inhibition of at least some pedal networks during rasping. Pedal F-cluster neurones innervate the body wall, column and foot (Winlow and Haydon, 1986; Kyriakides *et al.*, 1989), and receive inputs in phase with rhythmic locomotory movements (Winlow and Haydon, 1986). They may also be involved in controlling whole body withdrawal. Other pedal neurones

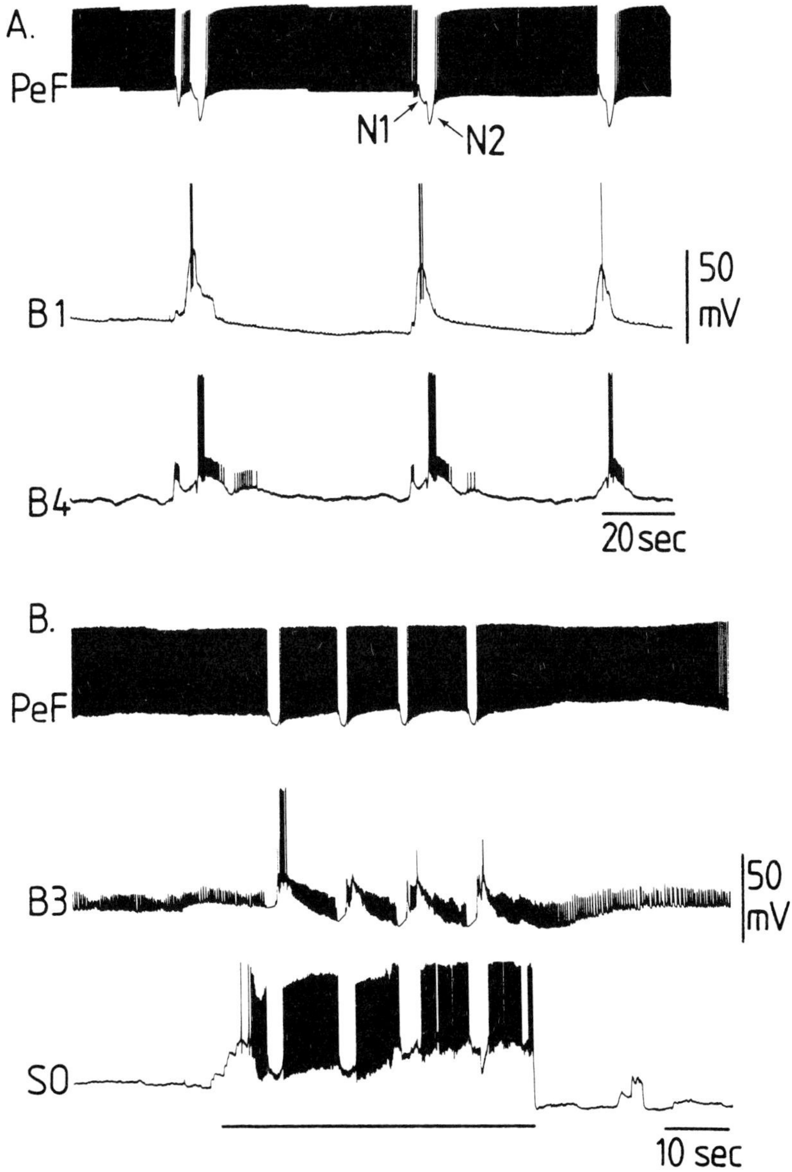

Fig. 3.6. (A) Cyclical inhibitory inputs to a pedal F-cluster neurone (PeF) occur at the same time as N1 and N2 inputs to buccal motoneurones, B1 and B4. B. Depolarisation of SO (indicated by bar) activates the buccal CPG leading to rhythmic inputs to a buccal motoneurone (B3) and simultaneous inhibitory inputs to PeF.

which receive cyclical inputs related to the buccal rhythm are thought to be involved in generation of a variety of behavioural responses including locomotion, withdrawal, and opening and closing of the pneumostome during ventilation (Benjamin *et al.*, 1985; Ferguson and Benjamin, 1985; Winlow and Haydon, 1986; and see Winlow *et al.*, Chapter 4). We suggest that inhibitory input to pedal neurones at the time of radula rasping acts to prevent inappropriate movements at the time that the radula is in contact with the substrate. The pedal neurones are then released from inhibition, allowing the animal to move to a fresh site for further sampling of food.

Feeding and locomotory behaviours are coordinated and can occur at the same time, but some behaviours are not compatible. For example, all rasping activity ceases immediately after egg laying (Ter Maat *et al.*, 1989). This prevents the animal from devouring its own eggs. We know little about the interrelationships of neuronal circuitry controlling these two behaviours. However, it is appropriate to summarise some of what we do know about the overlapping neural networks that generate and control overt behaviour in *Lymnaea*.

It is clear that the control of most activities is multiganglionic. So, for example, generation of feeding behaviour involves buccal, cerebral and probably also pedal ganglia. Whole body withdrawal is controlled by motoneurones in at least four different ganglia, cerebral, pleural, parietal and pedal (Benjamin *et al.*, 1985). Egg laying is triggered by neurones in

Table 3.1. Involvement of pedal neurones in neuronal networks controlling different behavioural activities in *Lymnaea*. See text for description. L, left; R, right.

	Feeding	*Locomotion and body movements*	*Ciliary locomotion*	*Withdrawal*	*Egg laying behaviour*	*Ventilation*
R/LPeD1	√					√
R/LPeD2	√	√ (L)				√
R/LPeD3	√					√
R/LPeD5	√					
R/LPeD10	√					√
R/LPeA		√	√		√	√
R/LPeD		√		√	√	√
R/LPeE	√	√				√
R/LPeF	√	√		√		√
R/LPeG		√		√	√	√
R/LPeH						√
R/LPeM	√					

Compiled from the following: Benjamin and Winlow, 1981; Benjamin *et al.*, 1985; Jansen and Ter Maat, 1985; Winlow and Haydon, 1986; Kyriakides, 1988; Kyriakides and McCrohan, 1988; Syed and Winlow, 1988; Syed *et al.*, 1988; Kyriakides *et al.*, 1989; Winlow *et al.*, Chapter 4, this volume.

the cerebral ganglia, but associated body movements are controlled by the pedal ganglia (Jansen and Bos, 1984; Jansen and Ter Maat, 1985; and see Ter Maat *et al.*, Chapter 2, this volume). Moreover, many individual neurone types appear to be multifunctional, being incorporated into neuronal networks controlling a number of different behaviours. To illustrate this point, Table 3.1 summarises what we now know about the involvement of some of the pedal neurones in the control of five different behaviours, feeding, locomotion, withdrawal, egg laying and ventilation. Neurones said to be involved in the network controlling a particular activity are those that have been demonstrated to have a direct role in the motor output, and also those whose activity is correlated in some way with the behaviour, usually owing to the receipt of synaptic input. Even for the limited number of neurone types included in Table 3.1 it is clear that much overlap exists, which must be important for the selection, coordination and integration of these behaviours. Some of the behaviours listed in Table 3.1 are considered in more detail in the following chapter (Winlow *et al.*, 1991).

3.5. **Conclusion**

What emerges from studies of the CNS of *Lymnaea* is that neuronal networks controlling different behavioural activities are highly distributed and interwoven. We can no longer confine our analysis to discrete motor networks in single ganglia, but must move on to higher levels of complexity. The scope for interaction between behavioural outputs at the central level is enormous, presumably providing for the high degree of behavioural flexibility that is required of the intact animal.

Acknowledgements

Some of the work described here was supported by a grant from the SERC to C.R.M.

References

Benjamin, P. R. (1983) Gastropod feeding: behavioural and neural analysis of a complex multicomponent system. In: *Neural Origin of Rhythmic Movements*, A. Roberts and B. Roberts, eds. SEB Symposium XXXVII, Cambridge University Press, Cambridge, pp. 159–193.

Benjamin, P. R. and Elliott, C. J. H. (1989) Snail feeding oscillator: the central pattern generator and its control by modulatory interneurons. In: *Cellular and Neuronal Oscillators*, J. Jacklett, ed. Marcel Dekker, New York, pp. 173–214.

Benjamin, P. R., Elliott, C. J. H. and Ferguson, G. P. (1985) Neural network analysis in the snail brain. In: *Model Neural Networks and Behavior*, A. I. Selverston, ed., Plenum, New York, pp. 87–103.

Benjamin, P. R., McCrohan, C. R. and Rose, R. M. (1981) Higher order interneurones which initiate and modulate feeding in the pond snail *Lymnaea stagnalis*. *Adv. Physiol. Sci.* **23**: 171–200.

Benjamin, P. R. and Winlow, W. (1981) The distribution of three wide-acting synaptic inputs to identified neurones in the isolated brain of *Lymnaea stagnalis* (L.). *Comp. Biochem. Physiol.* **70A**: 293–307.

Davis, W. J., Gillette, R., Kovac, M. P., Croll, R. P. and Matera, E. M. (1983) Organization of synaptic inputs to paracerebral feeding command interneurons of *Pleurobranchaea californica*. III. Modifications induced by experience. *J. Neurophysiol.* **49**: 1557–1572.

Dawkins, M. (1974) Behavioural analysis of coordinated feeding movements in the gastropod *Lymnaea stagnalis*. *J. Comp. Physiol.* **92**: 255–271.

Delaney, K. and Gelperin, A. (1990) Cerebral interneurons controlling fictive feeding in *Limax maximus*. II. Initiation and modulation of fictive feeding. *J. Comp. Physiol. A* **166**: 311–326.

Elliott, C. J. H. and Benjamin, P. R. (1985a) Interactions of pattern-generating interneurons controlling feeding in *Lymnaea stagnalis*. *J. Neurophysiol.* **54**: 1396–1411.

Elliott, C. J. H. and Benjamin, P. R. (1985b) Interactions of the slow oscillator interneuron with feeding pattern-generating interneurones in *Lymnaea stagnalis*. *J. Neurophysiol.* **54**: 1412–1421.

Elliott, C. J. H. and Benjamin, P. R. (1989) Esophageal mechanoreceptors in the feeding system of the pond snail, *Lymnaea stagnalis*. *J. Neurophysiol.* **61**: 727–736.

Ferguson, G. P. and Benjamin, P. R. (1985) Whole-body withdrawal of the pond snail *Lymnaea stagnalis*. *Soc. Neurosci. Abstr.* **11**: 513.

Getting, P. A. and Dekin, M. S. (1985) *Tritonia* swimming. A model system for integration within rhythmic motor systems. In: *Model Neural Networks and Behavior*, A. I. Selverston, ed. Plenum, New York, pp. 3–20.

Jansen, R. F. and Bos, N. P. A. (1984) An identified neuron modulating the activity of the ovulation hormone producing caudo-dorsal cells of the pond snail *Lymnaea stagnalis*. *J. Neurobiol.* **15**: 161–167.

Jansen, R. F. and Ter Maat, A. (1985) Ring neuron control of columellar motor neurons during egg-laying behavior in the pond snail. *J. Neurobiol.* **16**: 1–14.

Kemenes, G., Elliott, C. J. H. and Benjamin, P. R. (1986) Chemical and tactile inputs to the *Lymnaea* feeding system: effects on behaviour and neural circuitry. *J. exp. Biol.* **122**: 113–137.

Kyriakides, M. A. (1988) Neural mechanisms underlying modulation and coordination of motor output in *Lymnaea stagnalis*. PhD Thesis, University of Manchester.

Kyriakides, M. A. and McCrohan, C. R. (1988) Central coordination of buccal and pedal neuronal activity in the pond snail *Lymnaea stagnalis*. *J. exp. Biol.* **136**: 103–123.

Kyriakides, M. A. and McCrohan, C. R. (1989) Effect of putative neuromodulators on rhythmic buccal motor output in *Lymnaea stagnalis*. *J. Neurobiol.* **20**: 635–650.

Kyriakides, M. A., McCrohan, C. R., Slade, C. T., Syed, N. I. and Winlow, W. (1989) The morphology and electrophysiology of the neurones of the paired pedal ganglia of *Lymnaea stagnalis* (L.). *Comp. Biochem. Physiol.* **93A**: 861–876.

McCrohan, C. R. (1982) Properties of two cerebral neurones whose activity is related to feeding motor output in *Lymnaea stagnalis* (L.). *J. Moll. Stud.* **48**: 225–226.

McCrohan, C. R. (1984a) Properties of ventral cerebral neurones involved in the

feeding system of the snail *Lymnaea stagnalis*. *J. exp. Biol.* **108**: 257–272.
McCrohan, C. R. (1984b) Initiation of feeding motor output by an identified interneurone in the snail *Lymnaea stagnalis*. *J. exp. Biol.* **113**: 351–366.
McCrohan, C. R. and Audesirk, T. E. (1987) Initiation, maintenance and modification of patterned buccal motor output by the cerebral giant cells of *Lymnaea stagnalis*. *Comp. Biochem. Physiol.* **87A**: 969–977.
McCrohan, C. R. and Benjamin, P. R. (1980a) Patterns of activity and axonal projections of the cerebral giant cells of the snail *Lymnaea stagnalis*. *J. exp. Biol.* **85**: 149–168.
McCrohan, C. R. and Benjamin, P. R. (1980b) Synaptic relationships of the cerebral giant cells with motoneurones of the feeding system of *Lymnaea stagnalis*. *J. exp. Biol.* **85**: 169–186.
McCrohan, C. R. and Kyriakides, M. A. (1989) Cerebral interneurones controlling feeding motor output in the snail *Lymnaea stagnalis*. *J. exp. Biol.* **147**: 361–374.
McCrohan, C. R., Kyriakides, M. A. and Tuersley, M. D. (1989) Initiation and modification of rhythmic buccal motor output in the isolated central nervous system of *Lymnaea stagnalis*. *J. Moll. Stud.* **55**: 183–192.
McCrohan, C. R. and Winlow, W. (1985) Interganglionic coordination and bilateral symmetry in the nervous systems of gastropod molluscs. In: *Coordination of Motor Behaviour*, B. M. H. Bush and F. Clarac, eds. SEB Seminar Series, 24, Cambridge University Press, Cambridge, pp. 33–62.
Rose, R. M. and Benjamin, P. R. (1979) The relationship of the central motor pattern to the feeding cycle of *Lymnaea stagnalis*. *J. exp. Biol.* **80**: 137–163.
Rose, R. M. and Benjamin, P. R. (1981a) Interneuronal control of feeding in the pond snail *Lymnaea stagnalis*. I. Initiation of feeding cycles by a single buccal interneurone. *J. exp. Biol.* **92**: 187–201.
Rose, R. M. and Benjamin, P. R. (1981b) Interneuronal control of feeding in the pond snail *Lymnaea stagnalis*. II. The interneuronal mechanism generating feeding cycles. *J. exp. Biol.* **92**: 203–228.
Syed, N. I., Harrison, D. and Winlow, W. (1988) Locomotion in *Lymnaea* — role of serotonergic motoneurones controlling the pedal cilia. *Symp. Biol. Hung.* **36**: 387–402.
Syed, N. I. and Winlow, W. (1988) The role of central neurones in respiratory behaviour in *Lymnaea*. *J. Physiol.* **403**: 62P.
Ter Maat, A., Ferguson, G. P. and Jansen, R. F. (1991) Control of egg laying behaviour patterns in *Lymnaea stagnalis*. In: *Neurobiology of Motor Programme Selection: New Approaches to the Study of Behavioural Choice*, J. Kien, C. R. McCrohan and W. Winlow, eds. Manchester University Press, Manchester, pp. 20–36.
Ter Maat, A., Pieneman, A. W., Goldschmeding, J. T., Smelik, W. F. E. and Ferguson, G. P. (1989) Spontaneous and induced egg laying behavior of the pond snail, *Lymnaea stagnalis*. *J. Comp. Physiol. A* **164**: 673–683.
Tuersley, M. D. and McCrohan, C. R. (1988) Serotonergic modulation of patterned motor output in *Lymnaea stagnalis*. *J. exp. Biol.* **135**: 473–486.
Winlow, W. and Haydon, P. G. (1986) A behavioural and neuronal analysis of the locomotory system of *Lymnaea stagnalis* (L.). *Comp. Biochem. Physiol.* **83A**: 13–21.
Winlow, W., Moroz, L. L. and Syed, N. I. (1991) Mechanisms of behavioural selection in *Lymnaea stagnalis*. In: *Neurobiology of Motor Programme Selection: New Approaches to the Study of Behavioural Choice*, J. Kien, C. R. McCrohan and W. Winlow, eds. Manchester University Press, Manchester, pp. 52–72.

4 *William Winlow, Leonid L. Moroz and Naweed I. Syed*

Mechanisms of behavioural selection in *Lymnaea stagnalis*

4.1. Introduction

In an intact animal the functions of individual neurones may vary from time to time as circumstances change and different behaviour patterns emerge out of highly complex neuronal ensembles within the nervous system. Although a great deal is known about the neural circuitry and central pattern generators underlying specific behaviours (see Selverston, 1985, for a series of reviews), there is little understanding about the ways in which animals select between behaviours or coordinate synchronously occurring behaviours. In 1985, Getting and Dekin proposed that behavioural selection and coordination in *Tritonia* was underlain by polymorphic neuronal networks, and evidence for such networks is now accruing in insects (Altman and Kien, 1989; Kien and Altman, Chapter 9, this volume) and the pond snail, *Lymnaea stagnalis* (Kyriakides and McCrohan, 1988; McCrohan and Kyriakides, Chapter 3, this volume; Syed, 1988; Syed and Winlow, 1989b, 1991).

Since its introduction as a model system for the study of neurosecretory phenomena by Prof J. Lever at the Free University of Amsterdam in 1956, *Lymnaea stagnalis* has attracted the attention of many neurobiologists (see, for example, Winlow and Benjamin, 1976; Lever and Boer 1983; Benjamin *et al.*, 1985; Boer *et al.*, 1987; Ter Maat *et al.*, 1987; Roubos *et al.*, 1988; Moroz, 1989, 1991; Kyriakides *et al.*, 1989; Winlow and Syed, 1991). It is an advantageous preparation for studies in neuroethology since it has an interesting and varied behavioural repertoire and distinct central ganglia lying in close proximity to one another. In this review we consider a neural ensemble in *Lymnaea* that underlies several different behaviours, including respiration, locomotion and withdrawal. In addition transmitter control of respiratory behaviour is discussed in detail.

4.2. Behaviour in Lymnaea is multiganglionic in origin

Experiments on isolated brains and minimally dissected semi-intact preparations of *Lymnaea* demonstrate that behaviour is organised on a multiganglionic basis (McCrohan and Winlow, 1985) and is influenced by a series of wide-acting synaptic inputs (Benjamin and Winlow, 1981), some of which have been localised to particular cells. Recent work (Syed *et al.*, 1988; Syed and Winlow, 1989b, Winlow and Syed, 1991) indicates that a neuronal network spans the major central ganglia and acts as a substrate out of which varied and complex behaviours may emerge. The neurones of this multiganglionic ensemble are believed to interact to organise the ciliary (Syed and Winlow, 1989a) and muscular (Winlow and Haydon, 1986; Syed and Winlow, 1988c; Winlow and Syed 1991) aspects of locomotion, whole animal withdrawal (Ferguson, 1984; Syed and Winlow, 1989c), respiratory movements of the pneumostome (Syed and Winlow, 1988b; Moroz, 1989, 1991; Syed *et al.*, 1990) and also to influence neurones innervating the heart. Synaptic inputs associated with the feeding system are also detectable on some of these neurones (Kyriakides and McCrohan, 1988). Activity in some parts of the network appears to be modifiable so that specific synaptic inputs may be temporarily abolished during particular modes of behaviour (see Fig. 4.5).

It is probable that various wide-acting synaptic inputs (Benjamin and Winlow, 1981; Syed *et al.*, 1988; Syed and Winlow, 1989b, and see Table 3.1 in the chapter by McCrohan and Kyriakides, Chapter 3, this volume), and a variety of sensory inputs act upon this neuronal ensemble to generate the various behaviours. This implies that behavioural selection is a product of the activity of the whole neuronal ensemble and that individual neurones and groups of neurones may be reorganised into several different circuit configurations according to the behaviour required and according to the balance of synaptic inputs impinging upon them. The wiring diagram of the multiganglionic network is presented in Fig. 4.1. By itself the wiring diagram tells us little, and indeed we are already aware that many of the cells in the network participate in, or are active during, several different behaviour patterns. However, once we begin more fully to comprehend the subcircuits and overlapping contributions of the neurones of the network, the weighting of the individual synaptic and sensory inputs to it and the role of the transmitters and humoral agents acting upon it, we ought to be able to model its actions and gradually increase our understanding of how it initiates and controls behaviour.

4.3. **Whole animal withdrawal**

The whole animal withdrawal response of *Lymnaea* is a protective reflex which takes precedence over all other behaviours in the behavioural hierarchy of the animal (Fig. 4.7). It is mediated by the columellar muscle which, as shown below, is also used in driving the shell movements associated with locomotion and also during egg-laying behaviour (see Ter Maat *et al.*, Chapter 2, this volume). The role of central neurones is of prime importance in whole animal withdrawal behaviour, since the peripheral nervous system of *Lymnaea*, and probably of other basom-

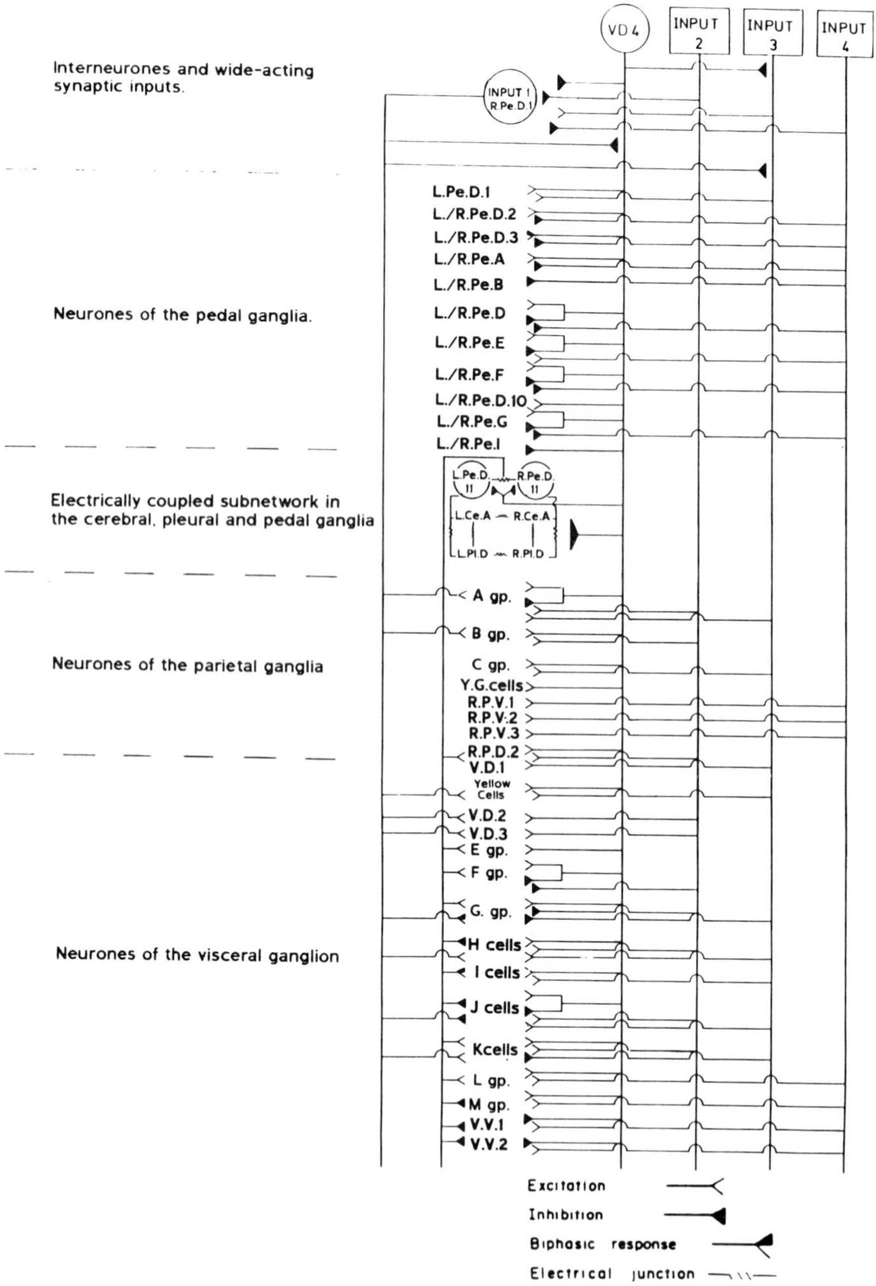

VD 4
INPUT 2
INPUT 3
INPUT 4
Interneurones and wide-acting synaptic inputs.
INPUT 1 R.Pe.D.1
Neurones of the pedal ganglia.
L.Pe.D.1
L./R.Pe.D.2
L./R.Pe.D.3
L./R.Pe.A
L./R.Pe.B
L./R.Pe.D
L./R.Pe.E
L./R.Pe.F
L./R.Pe.D.10
L./R.Pe.G
L./R.Pe.I
Electrically coupled subnetwork in the cerebral, pleural and pedal ganglia
L.Pe.D. 11
R.Pe.D. 11
L.Ce.A
R.Ce.A
L.Pl.D
R.Pl.D
Neurones of the parietal ganglia
A gp.
B gp.
C gp.
Y.G.cells
R.P.V.1
R.P.V.2
R.P.V.3
R.P.D.2
V.D.1
Yellow Cells
V.D.2
V.D.3
E gp.
F gp.
G. gp.
Neurones of the visceral ganglion
H cells
I cells
J cells
Kcells
L gp.
M gp.
V.V.1
V.V.2
Excitation
Inhibition
Biphasic response
Electrical junction

matophora, is less well developed and less decentralised compared with that of other molluscs (Bullock and Horridge, 1965). A number of central neurones involved in the behaviour have been described.

A single, probably unpaired, high-order putative withdrawal interneurone on the ventral side of the pedal G cluster (R.Pe.V.G neurone), is capable of inducing withdrawal behaviour (Fig. 4.2; Syed and Winlow, 1989c). It is active when this behaviour is induced by either mechanical stimulation of the head–foot complex or by photic stimuli, and depolarisation of the R.Pe.V.G neurone in a semi-intact preparation causes whole animal withdrawal. The L./R.Ce.A cluster neurones have recently been described as withdrawal motoneurones (Ferguson, 1984; Benjamin *et al.*, 1985), but Winlow and Haydon (1986) showed that L./R.Ce.A cluster neurones are left and right body wall motoneurones and that they are electrically coupled to each other within each ipsilateral ganglion. These motoneurones subserve locomotor, withdrawal and egg-laying behaviour depending on the central neural or hormonal mechanisms activating them. The details of this central circuitry and its interactions with neurohormones remain to be elucidated.

4.4. **Locomotion**

In neutrally buoyant, aquatic pulmonates such as *Lymnaea*, ciliary beating produces the prime locomotor force (Jones, 1975), and directional

Fig. 4.1. (*facing page*) The multiganglionic neuronal network of *Lymnaea stagnalis* has known connections in all the central ganglia, and a number of pedal neurones share common synaptic inputs with buccal neurones (not shown) as outlined by McCrohan and Kyriakides in Chapter 3. Many of the neurones in the network participate in several behaviours, but the role of other cells is yet to be determined. However, the network is clearly incomplete and we expect that additional interneurones and groups of motoneurones will be added to it in the future. The wide-acting synaptic inputs were originally described by Winlow and Benjamin (1976) and extended by Benjamin and Winlow (1981) and Syed (1988).

The source of input 3 has recently been identified as a neurone of the ventrolateral surface of the right parietal ganglion by Syed *et al.* (1990). We propose that its formal nomenclature should be R.P.V.4 (right parietal ventral 4). A further identified interneurone, the ring neurone (see Chapter 2 for details) also has effects on the multiganglionic network: it inhibits the putative ciliomotoneurones (L./R.Pe.A-clusters) of the pedal ganglia and has excitatory and inhibitory effects on motoneurones of the columellar muscle (L./R.Pe.D and G-clusters). It appears to be activated by the cerebral caudodorsal cells which produce egg-laying hormone.

The neurones L./R.Pe.D.11 lie within the pedal G-clusters as does an unpaired high-order putative withdrawal interneurone known as the R.Pe.V.G neurone. L./R.Pe.D.11 are active during withdrawal and co-ordinate the activities of many neurones within the network.

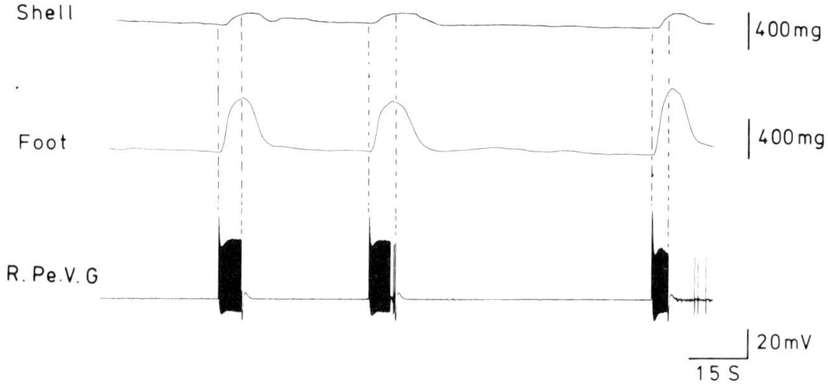

Fig. 4.2. Effects of electrical stimulation of the R.Pe.V.G. cell whose activity is sufficient to cause whole animal withdrawal. Contractions of the foot precede shell movements, which are equivalent to the withdrawal response in semi-intact animals (from Syed and Winlow, 1989c).

When the R.Pe.V.G. neurone is hyperpolarised (not shown) to prevent its activity, the withdrawal response can still be induced, but is weaker and delayed relative to the normal response. Thus, although the R.Pe.V.G. neurone is not necessary to the completion of the withdrawal response, it is clearly of considerable importance in the release of the behaviour.

changes are achieved by muscular contractions of the foot and body wall. Rhythmic shell movements generated by the columellar muscle are usually associated with locomotion (Haydon and Winlow, 1986; Winlow and Haydon, 1986), but their role in this behaviour is unclear. Central neurones implicated in controlling locomotion are of two types: putative ciliomotoneurones (Syed *et al.*, 1988; Syed and Winlow, 1989a; Deliagina and Orlovsky, 1990) and motoneurones known to control the musculature of the foot (Winlow and Haydon, 1986). In *Lymnaea*, the putative ciliomotoneurones form two paired clusters, identified as the pedal A-cluster neurones, each containing 30 cells and lying on the medial faces of the pedal ganglia. They contain serotonin and project into the tissues of the foot where they eventually innervate both the ciliated and mucus-containing cells of the foot sole (for review see Winlow and Syed, 1991). Deliagina and Orlovsky (1990) have recently described another set of ciliomotoneurones in *Planorbis corneus*, which is closely related to *Lymnaea*. Here four pairs of efferent locomotor neurones control the activity of the ciliated epithelium. These neurones are responsible for differential control of locomotor activity in various zones of the ciliated epithelium during different behavioural acts. Whether or not similar neurones exist in *Lymnaea* remains to be determined.

The muscle motoneurones are relatively small and lie in the left and right pedal D, G and F clusters. Neurones of the D and G clusters innervate the body wall and column, whilst those of the F clusters innervate the body wall and foot. All these motoneurones elicit reproducible contractions of the foot musculature and participate in locomotory behaviour in semi-intact preparations. Thus many locomotor motoneurones have been identified, but the mechanisms through which their activity is selected by interneurones are unclear. However, there is good evidence that stereotypic activity of locomotor motoneurones, recorded from semi-intact preparations, is driven by a central pattern generator lying entirely within the central ganglia (Syed and Winlow, 1988c).

4.5. **There are common synaptic inputs to the feeding and locomotor systems**

Elegant studies on the feeding system of *Lymnaea* have been carried out over many years in Benjamin's laboratory (e.g. Benjamin, 1983) and more recently by McCrohan. Experiments by Kyriakides and McCrohan (1988 and see Chapter 3 of this book) indicate that cyclical synaptic inputs originating from neurones of the buccal central pattern generator are present in neurones of the pedal E-, F-, and M-clusters and also in nine pedal giant neurones (L./R.Pe.D.1, L./R.Pe.D.2, L./R.Pe.D.3, L.Pe.D.5, L./R.Pe.D.10) as well as a putative pedal interneurone. These inputs were mainly inhibitory and occurred predominantly during the radula retraction phase of the feeding cycle. The function of these inputs may be to suppress locomotion at precisely the time when the radula is in contact with the substrate.

4.6. **Egg-laying is a composite behaviour**

Egg-laying behaviour is a highly complex composite of several behaviours and involves aspects of feeding and locomotory behaviour, movements of the shell and extrusion of egg-masses from the female reproductive tract (see Chapter 2 for details). These activities occur both serially and synchronously at various times during the behavioural act. It is clear that egg-laying behaviour will only be fully understood when the neural and neurohormonal mechanisms of the various component behaviours are understood. Furthermore, egg-laying behaviour may be interrupted when the animal moves to the surface to ventilate, only to recommence on returning to the initial site. How, one wonders, is one complex behaviour terminated, replaced by another complex behavioural act, involving many of the same neurones, and then restarted once again?

4.7. **Respiratory behaviour**

4.7.1. *Normal respiratory behaviour in* Lymnaea

In their natural environment, freshwater pond snails use both lung and skin respiration (Berg and Ocklemman, 1959; Jones, 1961). Lung ventila-

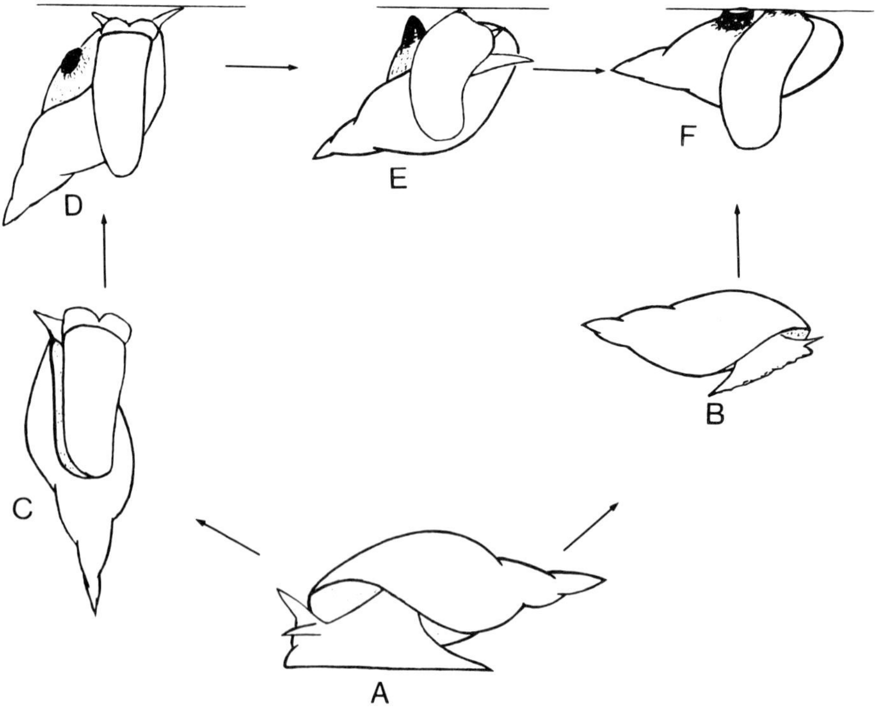

Fig. 4.3. Stages of normal respiratory behaviour in *Lymnaea stagnalis*. A, Normal locomotion; B, C, negative geotaxis: active locomotion (B) and rapid surfacing behaviour (C); D, E, F, respiratory act consisting of a series of stereotyped movements (see text for details) (from Moroz, 1991).

tion is activated when pO_2 is lowered either in the lung cavity and/or in the surrounding water. Changes of pO_2 are detected by oxygen-sensitive receptors, which are mainly localised in the region of the pneumostome (lung valve) and in the lung cavity (Janse, 1981, 1982).

Respiratory behaviour is complex, involving two main components (Fig. 4.3): first, movement to the water surface and, secondly, ventilation of the lung. The animal reaches the water surface either by active locomotion or by rapid surfacing behaviour. Active locomotion (negative geotaxis) is most often used as a way of reaching the surface (Janse, 1981, 1982) and usually occurs when the animals are browsing on the stems and leaves of plants close to the water surface. In rapid surfacing behaviour the animal ceases to adhere to the substrate, and, because it is buoyant due to air contained in the lung, it floats to the surface where lung ventilation can commence. During ventilation, locomotion ceases entirely and this may, in part, be due to the activities of the ring neurone, an identified, multiganglionic interneurone which projects to putative ciliomotoneurones (Jansen and Bos, 1984; Jansen and Ter Maat, 1985 and see Fig. 2.3B).

According to Moroz (1990) the respiratory act itself consists of a series of stereotyped complex reactions: (i) *shell movements*, i.e. anticlockwise turning of the shell and upward movements of the head. These movements are started when the edges of the lip touch the surface film and are often accompanied by inhibition of locomotion (see also Syed, 1988); (ii) *pneumostome erection*, which occurs in parallel with the shell movements; (iii) *opening of the pneumostome* by radial muscles, accompanied by ejection of a small air bubble which breaks the surface film. This sequence of movements takes 10–15 s. The duration of lung ventilation with fresh air varies from several seconds to 2–5 min (usually 45–90 s). Within 5–6 s after inspiration has ceased, the pneumostome is closed and the animal usually leaves the surface. The frequency of the respiratory act varies between 1–2/min to 1–2/h depending on environmental conditions (temperature, pO_2, etc; see Jaeger *et al.*, 1979; Janse, 1981, 1982) and localisation of the snails with respect to the surface, since weak tactile stimulation of the tentacles and the pneumostome induce respiratory movements.

4.7.2. *A network of interneurones underlies ventilatory behaviour*

Three interacting identified synaptic inputs underlie ventilatory behaviour in *Lymnaea*: R.Pe.D.1, Input 3 and VD4. Details of these synaptic inputs are reviewed by Winlow and Syed (1991) and the circuitry has now been reconstructed by Syed *et al.* (1990). Briefly, the giant dopamine-containing neurone, R.Pe.D.1, is an interneurone with multiple postsynaptic actions which lies in the right pedal ganglion and is the source of input 1 (Winlow and Benjamin, 1976; Benjamin and Winlow, 1981; Winlow *et al.*, 1981). It has follower cells in the right parietal and visceral ganglia, and input 3 has excitatory effects upon it (Benjamin and Winlow, 1981). Strong electrical stimulation of R.Pe.D.1 can initiate input 3 by postinhibitory rebound excitation (Winlow *et al.*, 1981; Syed *et al.*, 1990).

Input 3 is a sporadically occurring, two-component, compound p.s.p. lasting for several seconds. Its source has recently been localised to the right parietal ganglion (Syed *et al.*, 1990) and in addition a single neurone located in the visceral ganglion (within the boundary of G-group) appears to reproduce some of the effects of input 3 (Moroz, 1991), including pneumostome opening and reciprocal excitation of interconnections with R.Pe.D.1. The well characterised effects of input 3 occur on follower cells in the right parietal and visceral ganglia (Benjamin and Winlow, 1981) and more recently effects of input 3 have been found in many neurones of the pedal ganglia (Syed, 1988). In summary, input 3 excites neurones involved in pneumostome opening and expiration whilst inhibiting locomotor motoneurones. It also inhibits interneurone VD4, which is known to drive pneumostome closure.

VD4 (neurone visceral dorsal 4), like R.Pe.D.1, is an interneurone with widespread effects and was first described by Benjamin (1984) as the visceral white interneurone (VWI) and later termed VD4 by Janse *et al.* (1985). It contains FMRFamide (Benjamin *et al.*, 1988) and has reciprocal inhibitory connections with R.Pe.D.1; these connections are known to be monosynaptic (Syed, 1988; Syed *et al.*, 1990). Apart from driving pneumostome closure, VD4 inhibits locomotor motoneurones and withdrawal motoneurones, and has connections to neurones in the cerebral ganglia and to almost all the pedal neurones with the exceptions of the giant cells L./R.Pe.D.5, 6 and 7.

Experiments on freshly isolated brains indicate that the respiratory rhythm arises from a central pattern generator. Furthermore the rhythm can be initiated by intracellular stimulation of R.Pe.D.1 which causes postinhibitory rebound excitation of input 3 (Fig. 4.4; Winlow *et al.*, 1981; Syed *et al.*, 1990).

During lung ventilation input 3 causes opening of the pneumostome, inhibits VD4 and causes delayed inhibition of the putative ciliomotoneurones of the foot sole. When input 3 terminates, VD4 fires a burst of action potentials causing pneumostome closure and activation of the putative ciliomotoneurones. Detailed evidence for the role of VD4 in respiration is presented elsewhere (Syed *et al.*, 1990).

4.7.3. *Respiratory motoneurones and neuromodulatory cells*

Using a newly developed semi-intact preparation, Syed *et al.* (1990) have defined muscles which cause either opening or closure of the pneumostome. When the pneumostome is fully open, another muscle group, the mantle cavity muscle, contracts to exert pressure on the entire mantle cavity including the lung. During this phase of the respiratory cycle, air is expelled from the lung cavity. The identified visceral J and K neurones were found to be motoneurones to the pneumostome opener and closer muscles respectively, whereas the mantle cavity muscle was found to be innervated by the right parietal A-group neurones. These motoneurones are active during spontaneously occurring respiratory behaviour but their precise role in ventilation remains to be established since, during input 3, the synaptic connections to these cells are temporarily inactivated (Fig. 4.5). In addition Moroz (1990) has demonstrated that several visceral neurones induce opening (VPo — possibly small G-group neurones) or closing (VPc — possibly M-group cells) of the pneumostome at short latency. There are no synaptic connections between these cells. It was not determined whether these cells were genuine motoneurones, and they are termed effector neurones.

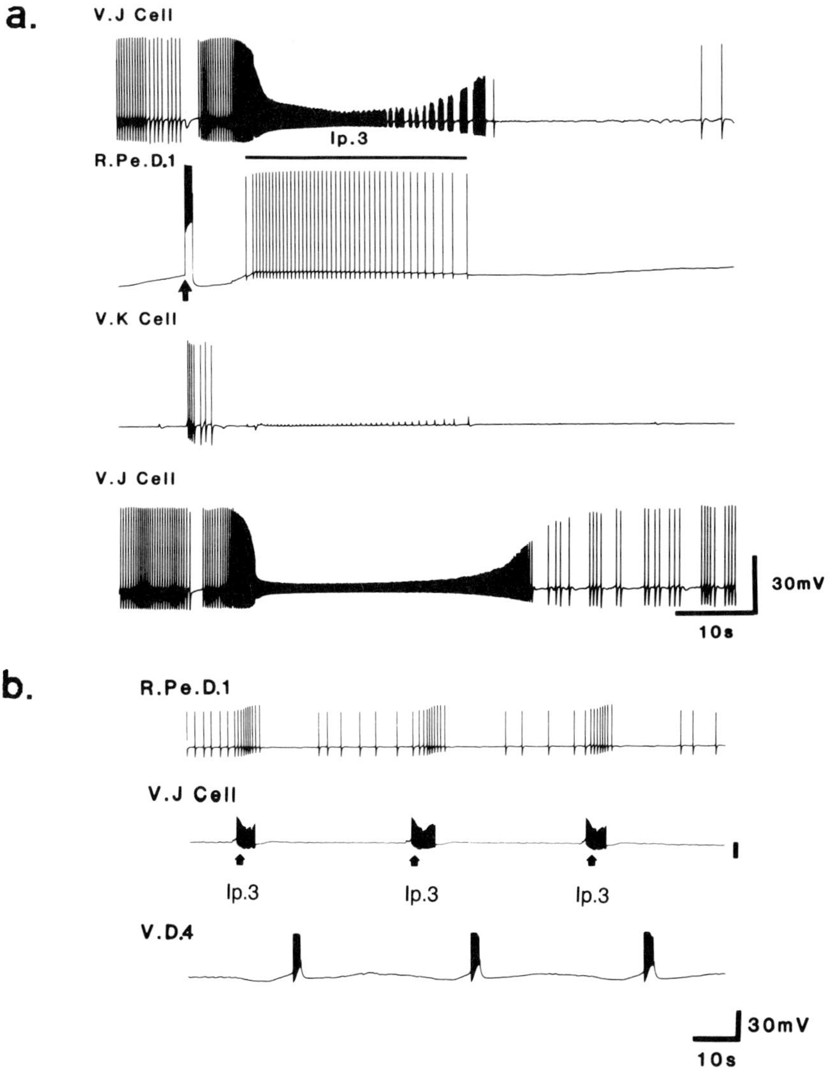

Fig. 4.4. (a) Interactions between R.Pe.D.1 and Input 3. The visceral J and K cells (V.J and V.K cells, respectively) are used to monitor the activity of input 3 whose location has only recently been discovered (Syed *et al.*, 1990). Electrical stimulation of R.Pe.D.1 (at arrow) inhibited its follower V.J cell whilst exciting a V.K cell. Stimulation of R.Pe.D.1 also initiated the activities of Input 3 neurones due to postinhibitory rebound excitation of Input 3. Once activated the Input 3 discharge (bar) excited both V.J cells and R.Pe.D.1. Although the effect of Input 3 on V.K cells is known to be inhibitory, it is not obvious in this preparation because this cell was hyperpolarised to prevent its spontaneous activity. (b) Respiratory rhythm recorded from an isolated brain preparation. Spontaneously occurring Input 3 discharges caused the excitation of R.Pe.D.1 and a V.J cell (arrows) while inhibiting V.D.4. Soon after the inhibitory effects of Input 3, V.D.4 recovered, fired a burst of action potentials and the cycle repeated several times (Syed and Winlow, unpublished).

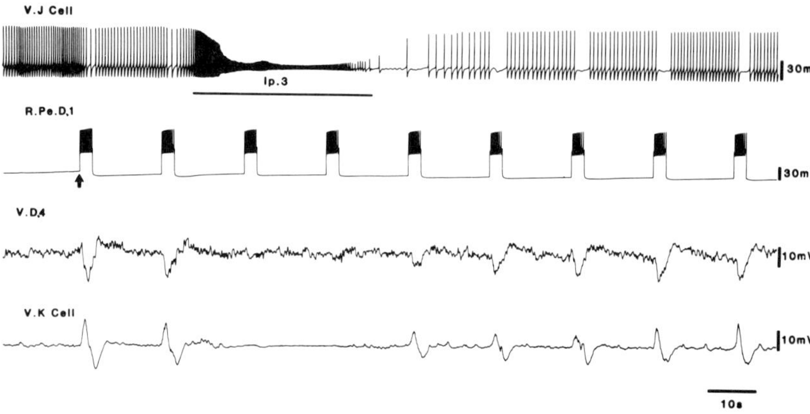

Fig. 4.5. Synaptic plasticity of R.Pe.D.1 connections with its follower cells during the Input 3 discharge. A depolarising square pulse of 4 s duration at 15 s intervals was injected into R.Pe.D.1 (at arrow). Stimulation of R.Pe.D.1 not only inhibited the V.J cell and V.D.4, and had biphasic effects on a V.K cell, but also switched on input 3 discharge (Ip.3, bar). Both V.D.4 and the V.K cell were hyperpolarised to prevent spiking. Although current injection was maintained throughout the Input 3 discharge, R.Pe.D.1 failed to induce its usual effects on its follower cells during Input 3. Once the powerful discharge of Input 3 ceased, the physiological contact between R.Pe.D.1 and its follower cells was re-established. It is possible that Input 3 causes presynaptic inhibition of connections between R.Pe.D.1 and follower cells (Syed and Winlow, unpublished).

A second class of cells, acting as modulatory neurones, include the electrically coupled pair of cells, V.D.1 and R.P.D.2 (van der Wilt *et al.*, 1988, and Moroz, unpublished), containing ACTH-like peptide (Boer *et al.*, 1979) and the 5-HT-containing neurone L.Pe.D.1. There were no detectable pneumostome movements when these cells were stimulated intracellularly. However, with hyperpolarisation and subsequent abolition of the action potentials either the frequency of occurrence of pneumostome opening cycles increased (see also van der Wilt *et al.*, 1988) or low-amplitude apparently chaotic movements of the pneumostome were depressed in the resting state.

4.8. **Effects of transmitters on behavioural selection**

Exogenously applied transmitter substances can induce patterned motor outputs analogous to those in the normal behavioural repertoire in a variety of preparations (Weiland and Gelperin, 1983; Trimble and Barker, 1984; Flamm and Harris-Warrick 1986a,b; Tuersley and McCrohan, 1986, 1987, 1988; Moroz, 1991). In many cases these transmitter-induced reactions correspond closely to the normal behaviour of the animal (Livingstone *et al.*, 1980; Sakharov, 1983; Lent and Dickinson, 1984; Sombati and Hoyle, 1984; Kabotyanski and Sakharov, 1988; Moroz, 1988, 1989; Marder and Weimann, Chapter 1, this volume).

Any of the three common catecholamines dopamine, noradrenaline or adrenaline (in decreasing order of effectiveness) when injected into the haemocoel of *Lymnaea* result, perhaps by mimicking low pO_2, in partial expression of respiratory behaviours conforming closely to those in the normal respiratory behavioural repertoire of the animal. The typical behavioural reactions induced by catecholamines are: strong constant or repeated erection of the pneumostome even in the absence of pneumostome contact with the surface film or when the animals are completely submerged; extrusion of bubbles of air; decrease in the amplitude of the tactile responses; inhibition and even arrest of locomotion; wrinkling of the foot; and decrease of adhesion to substrates. Following restoration of normal locomotion, normal shell movements as described by Haydon and Winlow (1986) are accompanied by clockwise turning of the shell and its movement to a position directly over the head, and there is also an increased sensitivity of the snail to tactile stimulation of the skin. A qualitative description of these responses is presented elsewhere (Moroz, 1991). Taking into consideration that dopamine is the most abundant catecholamine in *Lymnaea* (Werkman, 1989; Hetherington, 1990), the present data indicate that the respiratory reactions are primarily dopamine dependent.

Further support for the hypothesis that catecholamines play a leading role in the initiation and integration of the respiratory programme has been provided by the finding that all the transmitters tested on minimally dissected semi-intact preparations, consisting of the CNS with all effector nerves and organs left intact, only catecholamines and serotonin (5-HT) were effective in eliciting pneumostome movements. Adding them to the bathing solution activated repeating cycles of pneumostome erection and opening. Each of the monoamines induces a distinct pattern of activity of the pneumostome. For example, 5-HT (10^{-5} M or greater) mainly increased the frequency of the respiratory acts, while dopamine (10^{-6} M or greater) primarily increased the force of contraction and the duration of the open state of the pneumostome. Thus dopamine begins to act at lower concentrations than 5-HT and its effects are 1.5–2 times longer than those induced by 5-HT. However, on application to the denervated pneumostome, only 5-HT caused low-amplitude apparently chaotic movements suggesting that the action of monoamines is primarily on the CNS. The experiments on reduced semi-intact preparations have shown that only the visceral and right parietal ganglia and their peripheral connections via the intact right internal and external parietal and anal nerves are required to trigger and express the monoamine-dependent rhythmic reactions of the pneumostome.

Other transmitters either have no effect on respiratory behaviour or disturb some of its individual components. These effects are mainly observed as inhibition of locomotion (L-glutamate, GABA, ACh, octopamine) and changes in muscle tone (glycine, L-glutamate, octopamine). Nevertheless, it is significant that different transmitters induce specific behavioural effects which cannot be mimicked by other transmitters. For example, ACh (in doses of 10 μg/g or greater) induces the whole body withdrawal response (Sakharov, 1991) and inhibits the feeding rhythm (McCrohan *et al.*, 1989) as do 5-HT and FMRFamide. GABA (2, 10, 50 μg/g) induces specific radular and buccal movements, whilst dopamine reliably activates the feeding rhythm generator (Kyriakides and McCrohan, 1989; McCrohan *et al.*, 1989). Thus, a particular combination of characteristic reactions can be determined for each of the transmitters.

4.9. **Electrically coupled coordinating interneurones within the multiganglionic network**

A pair of interneurones in the pedal ganglia, L./R.Pe.D.11, have been identified, which co-ordinate the activities of many of the cells within the multiganglionic network, including the left and right body wall motoneurones and foot and column motoneurones (Syed and Winlow, 1991). They also make connections with putative heart motoneurones (Benjamin *et al.*, 1988; Syed and Winlow, 1988a) and either excite or inhibit many cells involved in locomotion or ventilation.

L./R.Pe.D.11 lie within the paired G-clusters of the pedal ganglia (Slade *et al.*, 1981) and their morphology was first reported by Ferguson (1984). Iontophoretic injection of Lucifer Yellow reveals that L./R.Pe.D.11 have similar morphologies to one another and encircle the lower ganglionic ring (paired pedal, pleural, parietal and single median visceral ganglia), as well as entering the cerebral ganglia, where they are strongly electrically coupled to one another and to the cerebral A-cluster neurones (Haydon, 1982; Ferguson, 1984), which are activated during withdrawal behaviour (see above). The 20–25 cerebral A-cluster neurones within each ganglion are electrically coupled to the pedal G-cluster neurones and also to the multibranched pleural D-cluster neurones (Haydon and Winlow, 1982). L./R.Pe.D.11 are electrically coupled to other foot motoneurones in the L./R.Pe.G clusters, which innervate the left and right side of the foot respectively. Even though the main axons of L./R.Pe.D.11 do not exit via the pedal nerves (which innervate the foot, body wall and column musculature), electrical stimulation of these neurones causes contraction of the foot and body wall.

L./R.Pe.D.11 make apparent chemical synaptic contacts with a wide variety of neurones in the visceral ganglion, including all the E-group neurones, two of which are known to be involved in the control of the

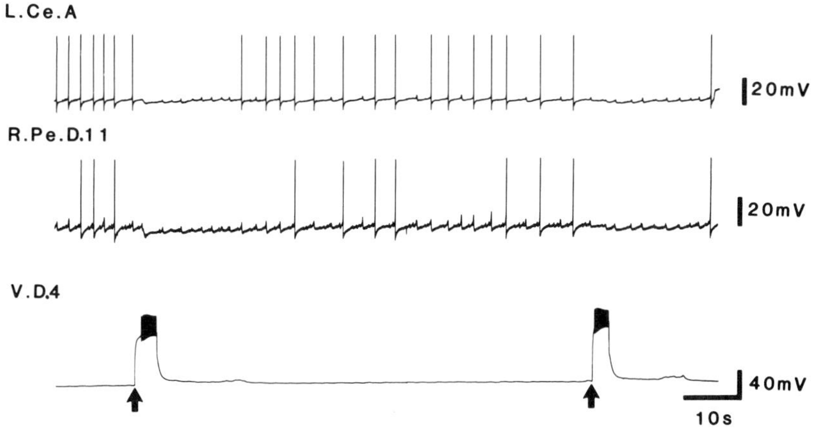

Fig. 4.6. Inhibition of R.Pe.D.11 and L.Ce.A-cluster neurone by respiratory interneurone V.D.4. The interneurone V.D.4 is involved in the expiratory phase of respiratory behaviour and induces the closure of the pneumostome. Electrical stimulation of V.D.4 (at arrows) inhibited both L.Ce.A-cluster neurones and R.Pe.D.11. The e.p.s.p.s seen in both neurones are produced by other electrically coupled neurones (Syed and Winlow, 1991).

heart (Benjamin *et al.*, 1988) and the visceral F-group neurones. It is not yet known whether these connections are monosynaptic or polysynaptic. Stimulation of the respiratory interneurone V.D.4 inhibits interneurones L./R.Pe.D.11 (Fig. 4.6) and the motoneurones coupled to them, i.e. L./R.Pe.G., L./R.Pl.D and L./R.Ce.A cluster neurones, thereby preventing some body-wall and foot movements during respiration (Syed and Winlow, 1991). They are active during whole animal withdrawal and may serve to coordinate symmetrical contractions of the foot and body wall during this behaviour and during the rhythmic shell movements correlated with locomotion (see above). Both L./R.Pe.D.11 have strong excitatory effects on another pair of electrically coupled neurones, V.D.1 and R.P.D.2 of the visceral and parietal ganglia; these contain ACTH-like substances and are sensitive to changes in external pO_2 (Boer *et al.*, 1979; van der Wilt *et al.*, 1988; Moroz, unpublished; see above for their role in respiratory behaviour).

4.10. **The behavioural hierarchy in *Lymnaea***

In this chapter we have described various types of behaviour in *Lymnaea*. These are classified as defensive reactions, arousal reactions and behaviours exhibited by aroused animals. Defensive reactions consist of localised 'passive' withdrawal responses and an active whole animal withdrawal response whereby the animal withdraws into its shell. Aroused animals

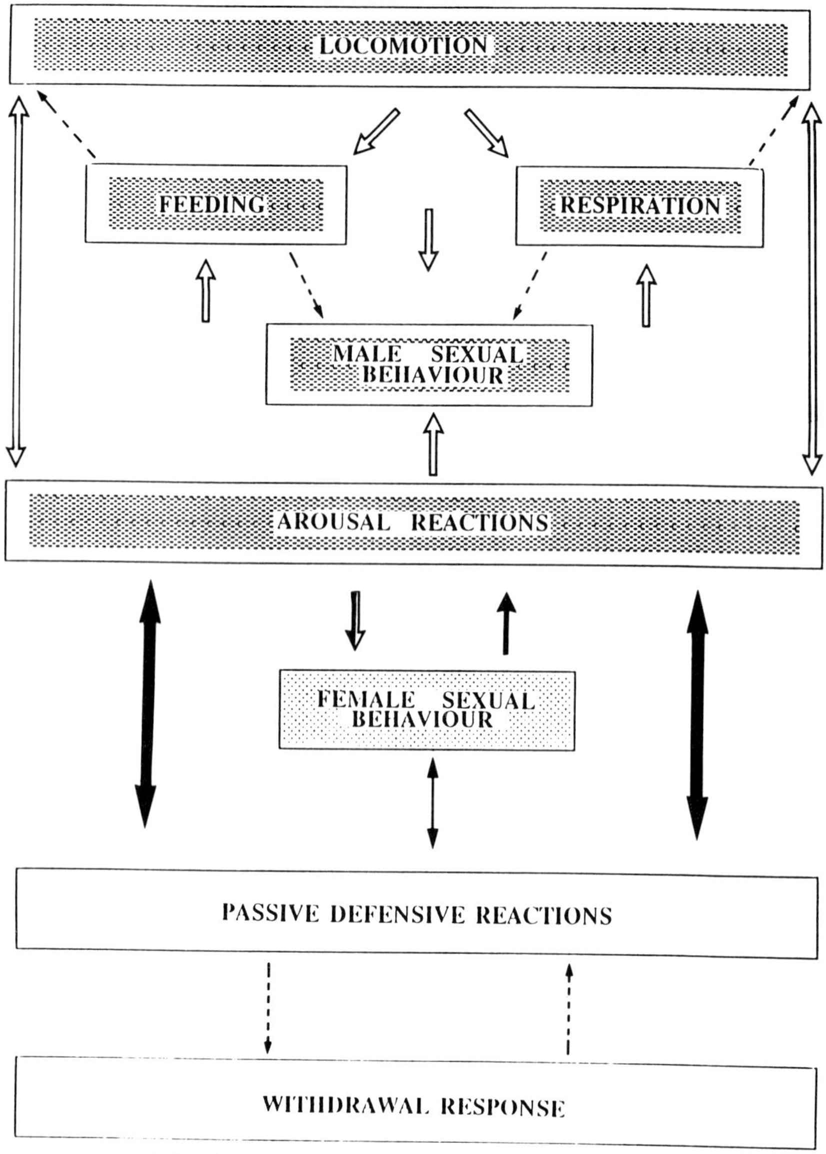

Fig. 4.7. The behavioural hierarchy in *Lymnaea*. We have classified behaviour in *Lymnaea* as defensive reactions, arousal reactions and behaviours exhibited by aroused animals. Defensive reactions are incompatible with those of aroused animals and vice versa. Female sexual behaviour presents something of a problem in the hierarchy in that it is incompatible with both arousal reactions and defensive reactions. It is therefore classified separately.

Aroused animals perform the normal range of behaviours, and it should be clear from the diagram that whilst locomotion and feeding are inextricably tied together, locomotion ceases momentarily during particular phases of the feeding are clearly discernible from those in the various defence postures since the foot is fully extended and the shell is held upright. Aroused animals are capable of respiratory behaviour, locomotion, feeding, etc. All these behaviours fit into a behavioural hierarchy in which some of the behaviours are mutually exclusive and some are not (Fig. 4.7). For example, the whole animal withdrawal response is incompatible with locomotion and vice versa whereas feeding and locomotion can occur simultaneously. Female sexual behaviour is entirely incompatible with other arousal reactions and is classified separately.

Passive defensive behaviours may be induced by exogenous application of catecholamines (Sakharov, 1990), whilst the withdrawal response may be triggered by application of acetylcholine (Moroz, 1989, 1991). Rapid surfacing behaviour (see above) often occurs naturally and may appear when the animal is endangered. In this case it may be considered to be both an active defensive behaviour and an arousal reaction. However, under all other circumstances defensive reactions and arousal reactions are wholly incompatible.

Dopamine appears to be involved in the release of both defensive and respiratory behaviours, and induces respiratory responses similar to those normally induced by lowering peripheral pO_2 (see above). A similar role for the dopaminergic system in passive defensive behaviour has been suggested (Sakharov and Salanki, 1980; Salanki-Rozsa *et al.*, 1983; Kabotyanski and Sakharov, 1988; Sakharov, 1990) implying that common releasing mechanisms underlie these two behavioural programmes. It is important that the circuitry underlying these behaviours should be understood, since it may provide a clue to the mechanisms of switching between passive defensive behaviour and arousal reactions.

---

cycle (hence the dashed arrow). Similar logic applies to respiration with respect to locomotion. Feeding and respiration are compatible only when the animal is browsing on vegetation close to the water surface. Male sexual behaviour occurs in aroused animals following locomotion. It is incompatible with respiration, feeding and locomotion.

'Passive' defensive reactions are localised withdrawals of the foot or tentacles due to low-intensity stimuli. A more intense or noxious stimulus leads to whole animal withdrawal, but passive withdrawal cannot occur during whole animal withdrawal and only recurs as the animal slowly re-emerges from the shell. Following re-emergence the animal goes into an aroused state where the foot is fully extended and the shell is held in its normal upright position.

Symbols: solid arrows are shown between incompatible behaviours; broken arrows are shown between behaviours which sometimes or always occur sequentially; open arrows are shown between compatible behaviours; the part-solid part-open arrow to female sexual behaviour from arousal reactions indicates that aroused animals may act as females only when appropriate.

4.11. **Conclusions**

It is clear from the above data that behaviour in *Lymnaea* is multi-ganglionic in origin and that the behaviours themselves fit into a complex hierarchy of compatible and incompatible behaviours. Furthermore, many of the behaviours consist of a number of components; e.g. respiratory behaviour clearly consists of several different locomotory reactions, shell movements, erection of the pneumostome and ventilation of the lung. Respiration is regulated by the oxygen content of the water, food, sex arousal, osmotic state, etc., and the network controlling respiration must logically include the neurone ensembles underlying its component behaviours. Thus, 600–700 functionally identified neurones will be involved in this one behavioural programme, making it one of the largest neuronal networks with identified connections so far described. It is quite clear that egg-laying behaviour involves an even larger number of cells, since it is even more complex than respiratory behaviour and utilises many of the same network components.

A number of very specific neurones appear to be involved in the phenomenon of behavioural choice, but this must be seen against a background of the actions of neuromodulatory substances which continuously mould and modify the actions of both individual cells and large portions of the network. Given the number of cells and connections so far identified in *Lymnaea*, it ought to be possible to begin to understand the interplay of neural and humoral actions influencing this complex, distributed, neural network. Identifying the circuitry is merely the first step and has already taken a great deal of time, and many more interconnections remain to be discovered. Coming to terms with the emergent properties of that circuitry will take considerably longer!

Acknowledgements

We gratefully acknowledge the support of a Science and Engineering Research Council grant to WW. Thanks are also due to Dr V. T. Shalatonin for support of this work, to Mr D. Johanson for his help with the photography, to D. Yu Budko for his skilful technical help and construction of amplifiers, as well as for drawing some of illustrations and to Mr D. Harrison for his expert technical assistance.

References

Altman, J. S. and Kien, J. (1989) New models for motor control. *Neural Computation* **1**: 173–183.

Benjamin, P. R. (1983) Gastropod feeding: behavioural analysis of a complex, multicomponent system. In: *Neural Control of Rhythmic Movements*,

A. Roberts and B. L. Roberts, eds. Cambridge University Press, Cambridge, pp. 159–153.

Benjamin, P. R. (1984) Interneuronal network acting on snail neurosecretory neurones (yellow cells and yellow green cells of *Lymnaea*). *J. exp. Biol.* **113**: 165–185.

Benjamin, P. R., Buckett, K. J. and Peters, M. (1988) Neurones containing FMRFamide-like peptides in the model invertebrate system, *Lymnaea. Symp. Biol. Hung.* **36**: 247–259.

Benjamin, P. R., Elliott, C. J. H. and Ferguson, G. P. (1985) Neural network analysis in the snail brain. In: *Model Neural Networks and Behaviour*, A. Selverston, ed. Plenum Press, New York, pp. 87–108.

Benjamin, P. R. and Winlow, W. (1981) The distribution of three wide-acting synaptic inputs to identified neurones in the isolated brain of *Lymnaea stagnalis* (L). *Comp. Biochem. Physiol.* **70A**: 293–307.

Berg, K. and Ocklemman, K. W. (1959) The respiration of freshwater snails. *J. exp. Biol.* **36**: 690–708.

Boer, H. H., Geraerts, W. P. M. and Joosse, J. (eds) (1987) *Neurobiology: Molluscan Models*. North Holland, Amsterdam.

Boer, H. H., Schot, L. P. C., Roubos, E. V., Maat, A., Lodde, J. C. and Reichelt, D. (1979) ACTH-like immunoreactivity in two electrotonically coupled giant neurons in the pond snail *Lymnaea stagnalis*. *Cell Tissue Res.* **202**: 231–240.

Bullock, T. H. and Horridge, G. A. (1965) *Structure and Function of the Nervous Systems of Invertebrates*, Vol. II, Freeman, San Francisco.

Deliagina, T. G. and Orlovsky, G. N. (1990) Control of locomotion in the freshwater snail *Planorbis corneus*: II. Differential control of various zones of the ciliated epithelium. *J. exp. Biol.* **152**: 405–423.

Ferguson, G. P. (1984) Neurophysiological analysis of whole body withdrawal in *Lymnaea stagnalis*. PhD Thesis, University of Sussex.

Flamm, R. E. and Harris-Warrick, R. M. (1986a). Aminergic modulation in lobster stomatogastric ganglion. 1. Effects on motor pattern and activity of neurons within the pyloric circuit. *J. Neurophysiol.* **55**: 847–865.

Flamm, R. E. and Harris-Warrick, R. M. (1986b) Aminergic modulation in lobster stomatogastric ganglion. 2. Target neurons of dopamine, octopamine, and serotonin within pyloric circuit. *J. Neurophysiol.* **55**: 866–881.

Getting, P. A. and Dekin, M. S. (1985) *Tritonia* swimming. A model system for integration within rhythmic motor systems. In: *Model Neural Networks and Behaviour*, A. I. Selverston, ed. Plenum Press, New York, pp. 3–20.

Haydon, P. G. (1982) An electrophysiological study of the nervous control of locomotion in the pond snail *Lymnaea stagnalis* (L.). PhD Thesis, University of Leeds.

Haydon, P. G. and Winlow, W. (1982) Multipolar neurones of *Lymnaea stagnalis*. I. Multiple spike initiation sites and propagation failure allow neuronal compartmentalisation. *J. Comp. Physiol.* **14A**: 503–510.

Haydon P. G. and Winlow, W. (1986) Shell movements associated with locomotion of *Lymnaea* are driven by a central pattern generator. *Comp. Biochem. Physiol.* **83A**: 23–25.

Hetherington, M. S. (1990) A quantitative analysis of the biogenic amines in the central ganglia of the pond snail, *Lymnaea stagnalis* (L.). PhD Thesis, University of Leeds.

Jaeger, J. C., Middelburg-Frielin, K. M., Mooij-Vogelaar, J. W. and Steen, W. J. (1979) Effect of oxygen and food location on behaviour in the freshwater snail *Lymnaea stagnalis*. *Proc. Kon. Ned. Acad. Wet.* **82**: 177–180.

Janse, C. (1981) The effect of oxygen on gravity orientation in the pulmonate snail *Lymnaea stagnalis*. *J. Comp. Physiol.* **142**: 51–59.

Janse, C. (1982) Sensory systems involved in gravity orientation in the pulmonate snail *Lymnaea stagnalis*. *J. Comp. Physiol.* **A145**: 311–319.

Janse, C., van der Wilt, C. J., van der Plas, J. and van der Roest, M. (1985) Central and peripheral neurones involved in oxygen perception in the pulmonate snail *Lymnaea stagnalis* (Mollusca, Gastropoda). *Comp. Biochem. Physiol.* **82A**: 459–469.

Jansen, R. F. and Bos, N. P. A. (1984) An identified neuron modulating the activity of the ovulation hormone producing caudo-dorsal cells of the pond snail, *Lymnaea stagnalis*. *J. Neurobiol.* **15**: 161–167.

Jansen, R. F. and Ter Maat, A. (1985) Ring neuron control of columellar motor neurones during egg-laying behaviour in the pond snail. *J. Neurobiol.* **16**: 1–14.

Jones, H. D. (1975) Locomotion. In: *Pulmonates*, Vol. I, V. Fretter and J. Peake, eds. Academic Press, London, pp. 1–32.

Jones, J. D. (1961) Aspects of respiration in *Planorbis corneus* (L.) and *Lymnaea stagnalis* (L.) (Gastropoda, Pulmonata). *Comp. Biochem. Physiol.* **4A**: 1–29.

Kabotyanski, E. A. and Sakharov D. A. (1988) Monoamine-dependent behavioural states in the pteropod mollusc *Clione limacina*. *Symp. Biol. Hung.* **36**: 463–477.

Kien, J. and Altman, J. S. (1991) Decision-making in the insect nervous system: a model for selection and maintenance of motor programmes. In: *Neurobiology of Motor Programme Selection: New Approaches to the Study of Behavioural Choice*, J. Kien, C. R. McCrohan and W. Winlow, eds. Manchester University Press, Manchester, pp. 147–169.

Kyriakides, M. A. and McCrohan, C. R. (1988) Central coordination of buccal and pedal neuronal activity in the pond snail *Lymnaea stagnalis*. *J. exp. Biol.* **136**: 103–123.

Kyriakides, M. A. and McCrohan, C. R. (1989) Effect of putative neuromodulators on rhythmic buccal motor output in *Lymnaea stagnalis*. *J. Neurobiol.* **20**: 635–650.

Kyriakides, M., McCrohan, C. R., Slade, C. T., Syed, N. I. and Winlow, W. (1989) The morphology and electrophysiology of the neurones of the paired pedal ganglia of *Lymnaea stagnalis* (L.). *Comp. Biochem. Physiol.* **93A**: 861–876.

Lent, C. M. and Dickinson M. H. (1984) Serotonin integrates feeding behaviour of the medicinal leech. *J. Comp. Physiol.* **154A**: 457–471.

Lever, J. and Boer, H. H. (1983) *Molluscan Neuroendocrinology*. North Holland, Amsterdam.

Livingstone, M. S., Harris-Warrick, R. M. and Kravitz, E. A. (1980) Serotonin and octopamine produce opposite postures in lobsters. *Science* **208**: 76–79.

Marder, E. and Weimann, J. M. (1991) Modulatory control of multiple task processing in the stomatogastric nervous system. In: *Neurobiology of Motor Programme Selection: New Approaches to the Study of Behavioural Choice*, J. Kien, C. R. McCrohan and W. Winlow, eds. Manchester University Press, Manchester, pp. 3–19.

McCrohan, C. R. and Kyriakides, M. A. (1991) Motor programme selection and the control of feeding in the snail. In: *Neurobiology of Motor Programme Selection: New Approaches to the Study of Behavioural Choice*, J. Kien, C. R. McCrohan and W. Winlow, eds. Manchester University Press, Manchester, pp. 37–51.

McCrohan, C. R., Kyriakides, M. A. and Tuersley, M. D. (1989) Initiation and

modification of rhythmic motor output in the isolated central nervous system of *Lymnaea stagnalis*. *J. Moll. Stud.* **55**: 183–192.
McCrohan, C. R. and Winlow, W. (1985) Interganglionic co-ordination and bilateral symmetry in the nervous systems of gastropod molluscs. In: *Bilateral and Intersegmental Co-ordination*, B. M. H. Bush and F. Clarac, eds. Cambridge University Press, Cambridge, pp. 33–62.
Moroz L. L. (1988) Monoaminergic mechanisms of behaviour integration of freshwater pulmonate snails: pharmacological and cellular analysis. In: *Integrative Action of Neuron: Molecular Basis*, K. V. Sudacov, ed. Nauka, Moscow, pp. 52–53 (in Russian).
Moroz, L. L. (1989) Monoaminergic mechanisms of freshwater pulmonate snail's behaviour: pharmacological and cellular analysis. PhD Thesis, Institute of Developmental Biology of the USSR Academy of Science, Moscow (in Russian).
Moroz, L. L. (1990) Dopaminergic mechanisms of integration of respiratory behaviour in the pond snail, *Lymnaea stagnalis* (L.). *J. Physiol. (Lond.)* **430**: 57P.
Moroz, L. L. (1991) Monoaminergic control of respiratory behaviour in the freshwater pulmonate snail, *Lymnaea stagnalis* (L.). In: *Signal Molecules and Behaviour*, W. Winlow, O. V. Vinogradova and D. A. Sakharov, eds. Manchester University Press, Manchester, pp. 101–123.
Roubos, E. W., Schmidt, E. D., van Heumen, W. R. A. and van Ven, A. M. H. (1988) Structural and functional aspects of the egg laying controlling caudodorsal cells of *Lymnaea stagnalis*. *Symp. Biol. Hung.* **36**: 233–245.
Sakharov, D. A. (1983) Organization of heterogeneous neuronal systems. In: *XIVth Congr I.P.Pavlov Physiol. Soc.* Vol. 1. (Abstracts), Nauka, Leningrad, (in Russian), pp. 175–177.
Sakharov, D. A. (1990) Diversity of neurotransmitters: functional significance. *Zh. Evol. Biokhim. Fiziol.* **26** (in Russian).
Sakharov, D. A. (1991) Integration of high threshold whole-body withdrawal in the pond snail. In: *Signal Molecules and Behaviour*, eds. W. Winlow, O. V. Vinogradova and D. A. Sakharov, Manchester University Press, Manchester, pp. 124–130.
Sakharov, D. A. and Salanki, J. (1980) Effect of dopamine antagonist on snail locomotion. *Experientia* **38**: 1090–1091.
Salanki-Rozsa, K., Salanki, J. and Sakharov, D. A. (1983) Long-term effect of 6-hydroxy-dopamine on identified central neurons involved in control of visceral function in *Helix pomatia*. *Comp. Biochem. Physiol.* **76C**: 327–333.
Selverston, A. I. (1985) *Model Neural Networks and Behaviour*, Plenum Press, New York.
Slade, C. T., Mills, J. and Winlow, W. (1981) The neuronal organisation of the paired pedal ganglia of *Lymnaea stagnalis* (L). *Comp. Biochem. Physiol.* **69A**: 789–803.
Sombati, S. and Hoyle, G. (1984) Generation of specific behaviours in locust by local release into neuropil of the natural neuromodulator octopamine. *J. Neurobiol.* **15**: 481–506.
Syed, N. I. (1988) The neural control of locomotion in *Lymnaea*. PhD Thesis, University of Leeds.
Syed, N. I., Bulloch, A. G. M. and Lukowiak, K. (1990) *In vitro* reconstruction of the respiratory central pattern generator of the mollusk, *Lymnaea*. *Science* **250**: 282–285.
Syed, N. I., Harrison, D. and Winlow, W. (1988) Locomotion in *Lymnaea* — role

of serotonergic motoneurones controlling the pedal cilia. *Symp. Biol. Hung.* **36**: 387–402.

Syed, N. I. and Winlow, W. (1988a) A pair of electrically coupled interneurones coordinating locomotor, respiratory and cardiac neuronal networks in *Lymnaea. J. Physiol.* **400**: 35P.

Syed, N. I. and Winlow, W. (1988b) The role of central neurones in respiratory behaviour in *Lymnaea. J. Physiol.* **403**: 62P.

Syed, N. I. and Winlow, W. (1988c) A central pattern generator for locomotion in *Lymnaea. J. Physiol.* **406**: 78P.

Syed, N. I. and Winlow, W. (1989a) Morphology of and electrophysiology of neurons innervating the ciliated locomotor epithelium in *Lymnaea stagnalis* (L). *Comp. Biochem. Physiol.* **93A**: 633–644.

Syed, N. I. and Winlow, W. (1989b) Evidence for a multiganglionic neuronal substrate underlying several behaviours in *Lymnaea stagnalis* (L). *J. Physiol.* **412**: 61P.

Syed, N. I. and Winlow, W. (1989c) An identified, high-order neurone driving the whole animal withdrawal response in *Lymnaea. J. Physiol.* **418**: 72P.

Syed, N. I. and Winlow, W. (1991) Coordination of locomotor and cardiorespiratory networks of *Lymnaea stagnalis* by a pair of identified interneurones. *J. exp. Biol.* **158**: 37–62.

Ter Maat, A., van Duivenboden, Y. A. and Jansen, R. F. (1987) Copulation and egg-laying behaviour of *Lymnaea stagnalis.* In: *Neurobiology, Molluscan Models*, H. H. Boer, W. P. M. Geraerts and J. Joose, eds., Mon. Kon. Ned. Akad. Wetensch. North Holland, Amsterdam, pp. 255–260.

Trimble, D. L. and Barker, D. L. (1984) Activation by dopamine of patterned motor output from the buccal ganglia of *Helisoma trivolvis. J. Neurobiol.* **15**: 37–48.

Tuersley, M. D. and McCrohan, C. R. (1986) Modulation by dopamine and 5-hydroxytryptamine of rhythmic activity in buccal motoneurones of the pond snail *Lymnaea stagnalis.* In: *Neurobiology of Dopamine Systems*, W. Winlow and R. I. Markstein, eds. Manchester University Press, Manchester, pp. 116–119.

Tuersley, M. D. and McCrohan, C. R. (1987) Food arousal in the pond snail *Lymnaea stagnalis. Behav. Neur. Biol.* **48**: 222–236.

Tuersley, M. and McCrohan, C. R. (1988) Serotoninergic modulation of patterned motor output in *Lymnaea stagnalis. J. exp. Biol.* **135**: 473–486.

van der Wilt, G., van der Roest, M. and Janse, C. (1988) The role of two peptidergic giant neurons in modulation of respiratory behaviour in the pond snail, *Lymnaea stagnalis. Symp. Biol. Hung.* **36**: 377–386

Weiland, S. J. and Gelperin A. (1983) Dopamine elicits the feeding motor programme in *Limax maximus. J. Neurosci.* **3**: 1735–1745.

Werkman, T. R. (1989) Dopaminergic neurotransmission in the central nervous system of *Lymnaea stagnalis*. PhD Thesis, Free University, Amsterdam.

Winlow, W. and Benjamin, P. R. (1976) Neuronal mapping of the brain of the pond-snail, *Lymnaea stagnalis* (L.). In: *Neurobiology of Invertebrates, Gastropoda Brain*, J. Salanki, ed. Akademiai Kiado, Budapest, pp. 41–59.

Winlow, W. and Haydon, P. G. (1986) Locomotion of *Lymnaea stagnalis*: a behavioural and neuronal analysis. *Comp. Biochem. Physiol.* **83A**: 13–21.

Winlow, W., Haydon, P. G. and Benjamin, P. R. (1981) Multiple postsynaptic actions of the giant dopamine-containing neurone R.Pe.D.1 of *Lymnaea stagnalis* (L.). *J. exp. Biol.* **4**: 137–148.

Winlow, W. and Syed, N. I. (1991) Modulation of behaviour by a multiganglionic neuronal ensemble. In: *Signal Molecules and Behaviour*, W. Winlow, O. V. Vinogradova and D. A. Sakharov, eds. Manchester University Press, Manchester, pp. 85–100.

5 *Stephen R. Soffe*

To flex, swim or struggle? Behavioural selection in *Xenopus* embryos

5.1. Introduction

Most and perhaps all animals are capable of a range of behavioural responses with which they can meet the demands of their environment. Simpler animals, whether in terms of their phylogeny or their current stage of development, generally show narrower repertoires of response. It is not surprising, therefore, that among the vertebrates the repertoires of embryos and larvae of 'lower' groups are more limited even than those of adults of the same group. Indeed it has been this very limitation, in combination with the relative simplicity of the nervous systems of such animals, that has allowed considerable progress to be made in recent years towards understanding how behaviours can be generated and maintained.

5.2. Patterns of behavioural selection

To understand the neural basis for behaviour, there is much to be gained from concentrating on single identifiable motor acts, and particularly on those that are repetitive or stereotyped and therefore predictable. This approach, of course, allows only a certain degree of progress since it ignores the fact that behavioural responses, even very simple ones, often involve more than one component motor act and therefore must involve changes from one act to another: i.e. behavioural selection.

When considering the question of behavioural selection, a broad division can be drawn between motor acts that involve different neuromuscular systems and those that involve the same one. In the first case, different motor acts can in principle be turned on or off, or modulated, quite independently. In practice, there may be a greater or lesser degree of 'coupling' or interaction between them. In the second case, there can be no such independence; different motor acts must be mutually exclusive and therefore only one may be selected at a time (though intermediate forms could occur). It is this second case that I shall discuss with respect to amphibian embryos.

5.3. Behaviours of amphibian embryos

Amphibian embryos possess only a single major neuromuscular system: the segmental myotomes on the two sides of the trunk and tail and the motoneurones that drive them. Their repertoire of behavioural responses is therefore constrained by what they can do using this system and, in fact, probably all the most obvious possibilities for movement have been investigated over the small range of amphibian species that have been studied. In the past, some of these movements have been used to designate particular 'physiological' stages of development (Coghill, 1928). Broadly, though, they can be split into two groups: non-rhythmic behaviours and rhythmic behaviours.

The simplest non-rhythmic movements seen in embryos of the urodele *Triturus* (Soffe *et al.*, 1983) and the anurans *Rana* and *Xenopus* (Soffe, 1991a) are twitches, usually termed flexions. The smallest of these are probably driven by weak contractions of the myotomes on one side (Fig. 5.1A,Bi). Stronger flexion responses can involve a tight bending to one side (often termed a C-flexion), as in *Triturus*, or even coiling as in *Rana*. Again the responses are driven by contractions on one side only and may or may not produce a change of body direction. *Triturus* embryos can also show S-flexions. Here contraction of myotomes rostrally on one side is associated with contraction caudally on the opposite side.

The most obvious rhythmic movement is swimming (Fig. 5.1Bii). In all embryos studied so far, this involves alternate contractions of myotomes on the two sides (Kahn *et al.*, 1982; Soffe *et al.*, 1983; Soffe, 1991a). The pattern of contraction is such that a propulsive wave of bending passes along the animal from head to tail. A new wave starts at the head on one side just as the previous wave reaches the end of the tail on the opposite side. The precise relationship between the timing of muscle contractions and bending is not yet clear in these small animals, but muscle activity also propagates from head to tail. Lastly, embryos can show stronger rhythmic movements: *Xenopus* shows 'struggling' (Fig. 5.1Biii) (Kahn and Roberts, 1982b), *Rana*, 'lashing' (Soffe, 1991a), and *Triturus*, 'slow alternation' (Soffe *et al.*, 1983). The common feature of all these behaviours is that they involve high amplitude movements but generally little if any forward motion. For this reason they are probably responses designed to free embryos from capture rather than to assist locomotion. In both *Rana* and *Triturus* the strong movements are essentially more exaggerated forms of swimming, and again involve a wave of bending that propagates from head to tail. Struggling in *Xenopus* is rather different in that it involves a reversed wave of bending that passes from tail to head (Fig. 5.1Biii). This makes it qualitatively as well as quantitatively distinct from swimming.

More complex temporal sequences of behaviour result from a combination of these basic responses so that, for example, touching the head in *Triturus* or *Rana* can evoke a flexion that leads to turning of the body, followed by swimming (Soffe *et al.*, 1983; Soffe, 1991a).

5.4. **Central programming of amphibian embryo behaviour**

An important feature of all the behaviours described above is that they are largely centrally programmed. This means that, for each response, the neural machinery necessary to drive appropriate patterns of motoneurone

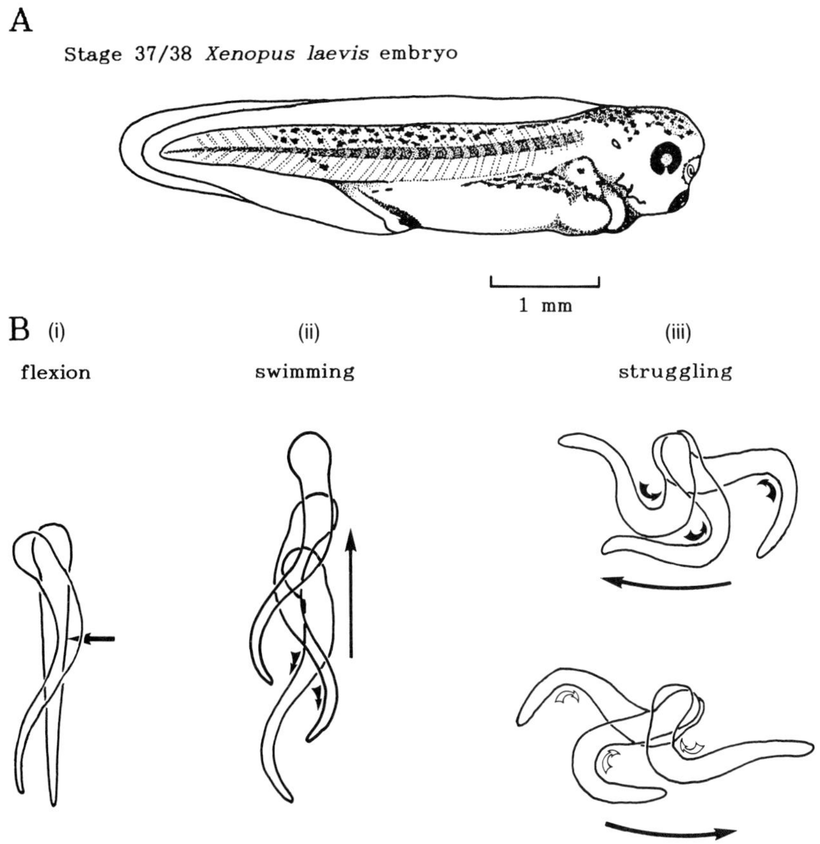

Fig. 5.1. Behavioural responses of *Xenopus* embryos. (A) Stage 37/38 (Nieuwkoop and Faber, 1956) *Xenopus laevis* embryo. (B) Diagrams of three main responses in stage 37/38 embryos. (i) In rare cases touching the trunk on one side (arrow) evokes a contraction of myotomes on the opposite side, producing a *flexion*. (ii) During *swimming*, waves of bending (small arrows) pass from head to tail as the embryo moves forward (large arrow). (iii) During *struggling* in restrained embryos, waves of bending (small arrows) pass from tail to head (large arrows indicate movement). (i) is hypothetical; (ii) and (iii) are based on high speed cine records (from Kahn and Roberts, 1982b and Kahn *et al.*, 1982).

discharge and consequently muscle contraction exists within the central nervous system and can function in animals immobilised with neuromuscular blockers (Kahn and Roberts, 1982a; Soffe *et al.*, 1983; Soffe, 1991a; Soffe and Sillar, 1991). Thus it is feasible to answer behavioural questions by studying the responses of such immobilised animals.

What, then, determines the response that an embryo gives to a stimulus? To answer this we must ask at least two more specific questions. First, what stimulus produces a particular response? We can ask this because we are dealing here with animals in which movements occur mostly as a result of an obvious stimulus. Secondly, what is actually happening within the central nervous system during each response to ensure that motoneurone activity is appropriate to drive the required pattern of muscle activity? We are currently close to answering both of these questions in studies using *Xenopus* embryos. As a result we can start to ask what are the *changes* that take place centrally in order that a particular response be selected?

5.5. **Sensory stimulation of different responses in *Xenopus* embryos**

As outlined above, the main behavioural responses of *Xenopus* embryos are flexions, swimming and struggling (Fig. 5.1). How are these evoked? Basically the weakest stimulus produces the weakest response and the strongest stimulus the strongest response. But the likelihood of a particular response being evoked changes during development. The response evoked most readily in embryos around the time of hatching is swimming; either touching the body or dimming the illumination can evoke a sustained episode of swimming lasting many seconds (in practice this is usually until the embryo bumps into something, at which point swimming is actively terminated; Boothby and Roberts, 1988). In contrast the simple flexion response which was described for younger embryos (Muntz, 1964) is the most difficult to evoke at this time. A stimulus must be sufficiently strong to evoke the flexion but not strong enough to evoke swimming, and this is difficult to achieve with older embryos since most suprathreshold mechanical stimuli (usually hand-delivered strokes with a fine mounted hair) evoke full swimming. Struggling is easier to evoke than flexion; it is reliably produced by grasping the embryo firmly by the head or trunk and applying sustained pressure. If the embryo is then released, it usually swims off.

In embryos that have been immobilised with neuromuscular blockers, these same stimuli can be used to evoke suitable patterns of motor root discharge for each of the main behavioural responses, but whilst mechanical stimulation can be used experimentally, it is generally more convenient to simulate the skin electrically. If we consider only the trunk, stimulation is effective because the skin is innervated by the free nerve

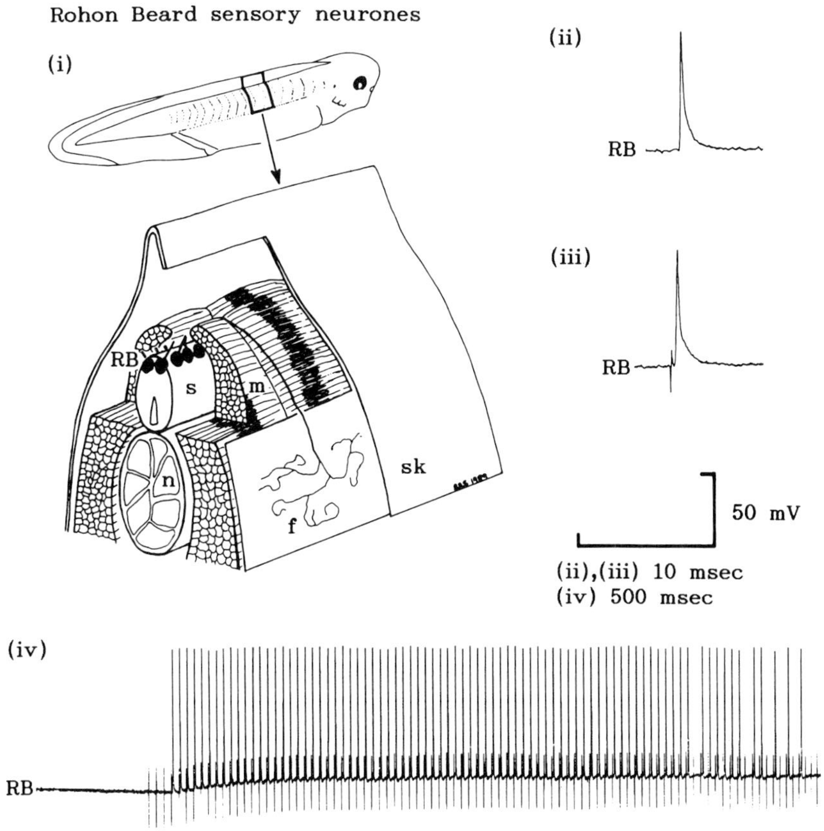

Fig. 5.2. Responses of Rohon–Beard neurones which are primary sensory neurones of the trunk skin. (i) Diagram of a portion of the trunk showing the cell body positions of Rohon–Beard neurones (RB) in the dorsal part of the spinal cord (s), which lies above the notochord (n). Each neurone has a neurite that passes out of the spinal cord, over the myotomes (m) to innervate the skin (sk) with free nerve endings (f). Rohon–Beard cells can respond with one (or a few) impulses to light touch (ii) or brief electrical stimulation (iii) of the trunk skin within the receptive field. Impulses recorded with an intracellular microelectrode from the cell body. (iv) Rohon–Beard neurones can follow repetitive electrical skin stimulation. (ii)–(iv) kindly supplied by Dr J. D. W. Clarke.

endings of sensory Rohon–Beard neurones (Roberts and Hayes, 1977; Clarke *et al.*, 1984), whose cell bodies lie within the spinal cord (Fig. 5.2) (Hughes, 1957; Roberts and Clarke, 1982). Brief mechanical or electrical stimulation of the skin characteristically evokes just one or a small number of impulses in Rohon–Beard neurones. However, repetitive electrical stimulation can evoke repetitive firing up to about 50 Hz (Fig. 5.2). Centrally, impulses may be relayed along the spinal cord by long ascending and descending axons. Rohon–Beard neurones appear to provide the only access for trunk skin sensory information directly into the spinal cord. If the CNS is transected caudal to the level of the Xth cranial root, other sensory inflow to the spinal cord, in particular from the light-sensitive pineal eye and from skin impulses, can be excluded. As we shall see, this operation does not prevent embryos making the same responses to trunk skin stimulation, which thus all appears to be evoked through the Rohon–Beard sensory pathway.

In the *Xenopus* embryo, then, different behavioural responses not only can be produced by the same neuromuscular system but also can be evoked by the same sensory pathway. How can this occur?

5.6. **To flex or swim?**

The simplest behavioural distinction in *Xenopus* is between flexing and swimming. As has been mentioned, flexions are more characteristic of younger embryos. In older embryos they are difficult to evoke. By the time of hatching, therefore, flexion responses are probably not a major option in terms of behavioural selection. In immobilised embryos, short latency flexion responses are also evoked most readily in younger embryos (Fig. 5.3A). At older stages, unsustained short latency responses are best evoked by transecting immobilised embryos at the level of the caudal brainstem to make them generally less excitable and therefore less likely to swim (Fig. 5.3B).

The pathway from Rohon–Beard sensory neurones to motoneurones that mediates short latency responses to trunk skin stimulation may involve only a single interposed cell type: sensory interneurones called dorsolateral commissural interneurones (Roberts and Clarke, 1982) which are excited by Rohon–Beard neurones (Clarke and Roberts, 1984; Sillar and Roberts, 1988) and in turn probably excite motoneurones on the opposite side of the spinal cord via crossing axons. This crossed excitation means that flexion responses or the short-latency responses of immobilised embryos are characteristically on the opposite side to the stimulus (Figs 5.1, 5.3).

In order for swimming to be evoked, excitatory and inhibitory premotor interneurones must be activated in sufficiently large numbers to start to generate the necessary self-sustaining circuit rhythmicity. This activation also probably involves an initial crossed excitation via dorsolateral commissural interneurones, but must also involve excitation of neurones on the stimulated side, perhaps via a second group of sensory interneurones with uncrossed axons. Each Rohon–Beard neurone probably contacts many sensory interneurones, more than one of which may then contact each motoneurone and each swimming premotor interneurone. With this potential amplification of sensory stimuli, it is easy to

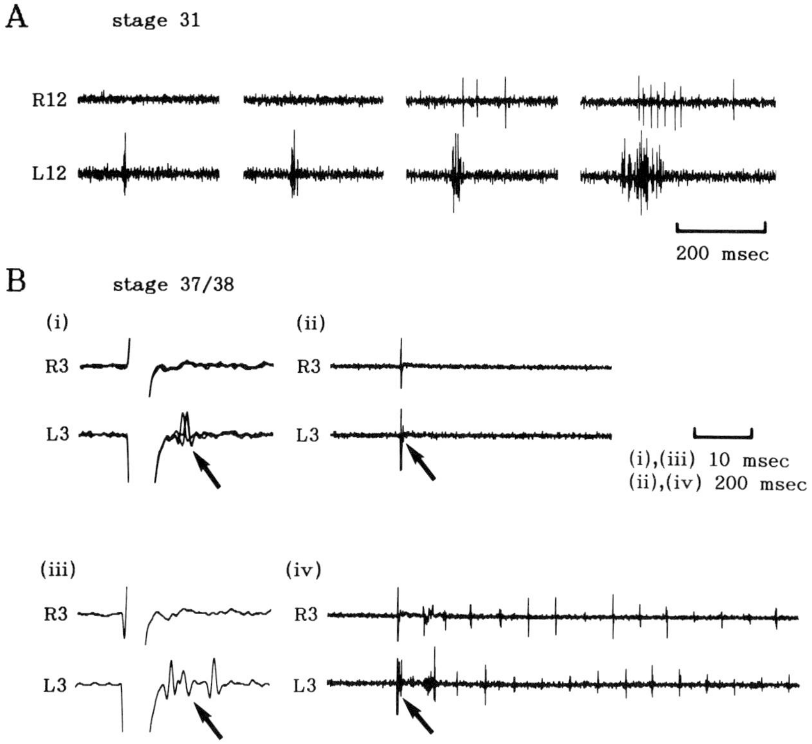

Fig. 5.3. Short-latency responses in immobilised embryos. (A) Gently touching the trunk skin on the right side of a young (stage 31) embryo evokes a motor root response on the opposite side (L12 = motor discharge recorded with a suction electrode from motor axons running in the 12th post-otic intermyotome cleft on the left side). Notice that stronger responses are associated with some activity, indicative of excitation, on the stimulated side (R12). (B) Responses in an older (stage 37/38) embryo whose CNS has been transected just caudal to the Xth root. (i) Short latency motor root spikes (arrowed, three overlapped traces) on the opposite side to a skin stimulus (at artefact). (ii) One of the traces in (i) (arrowed) compressed to show that there is no sustained response. (iii) Short-latency motor root response (arrowed) to a stronger stimulus on the opposite side. (iv) Trace in (iii) compressed to show that after the initial response (arrowed), there is sustained rhythmic swimming.

see how a stimulus of sufficient intensity to evoke spiking in neurones up to and including the motoneurones, as required for flexion, will usually be sufficient to evoke the full swimming response (Fig. 5.3B). There is no evidence to suggest that, once embryos have become capable of swimming, anything other than the strength of the stimulus and the degree of its subsequent amplification determines whether flexion or swimming results.

Once initiated, the swimming motor pattern appears to be driven by the cellular properties and circuit interactions of a single group of excitatory premotor interneurones and a single group of inhibitory interneurones together with the motoneurones themselves. Swimming is a self-sustaining pattern in that it requires only a brief stimulus to initiate a long episode of rhythmic motor discharge. Current thinking on the nature of a possible rhythm-generating system that drives swimming has been reviewed elsewhere (Roberts *et al.*, 1986; Roberts and Tunstall, 1990). The most recent findings point very much towards a number of overlapping mechanisms existing within the spinal cord and caudal brainstem (Soffe, 1989, 1991b). One involves interactions between the two sides of the CNS, mediated at least in part by reciprocal inhibition. Other mechanisms act within each side and utilise 'on-cycle' inhibition and N-methyl-D-aspartate (NMDA) receptor-mediated excitatory amino acid excitation.

5.7. **The synchronous pattern**

A somewhat anomalous pattern that occurs in the immobilised *Xenopus* embryo, though it has not been documented behaviourally (or indeed been seen in other amphibian embryos), is 'synchrony' (Roberts and Kahn, 1982). During synchrony, motor discharge is swimming-like but coincides on the two sides at approximately half the normal cycle period. Even during swimming, motor discharge on one side can occasionally appear in mid-cycle and therefore coincident with discharge on the opposite side. However, synchrony differs from this in lasting often for many cycles. Synchrony often occurs near the start of episodes of swimming when the level of excitation is high, and, like swimming, is self-sustaining. It is hard to imagine what behavioural significance such an apparently self-defeating pattern might have. It is also not at all clear what, other than perhaps a change in the level of self-generated excitation, triggers the switch between swimming and synchrony (though one possible sequence of events has been proposed, Roberts and Kahn, 1982). Synchrony appears to represent a second stable state for the swimming circuitry. Its main interest here, therefore, is that it demonstrates how switching between two different motor patterns (if not in this case realistic behaviours) *could* occur in a simple 'flip-flop' manner between two stable output states of the same rhythm-generating circuitry.

5.8. **To swim or struggle?**

The means for selecting between struggling and swimming is less clear than that between flexion and swimming, though it is probably more behaviourally significant to the hatched embryo. In immobilised embryos, a struggling motor pattern can be evoked, like the behaviour, by sus-

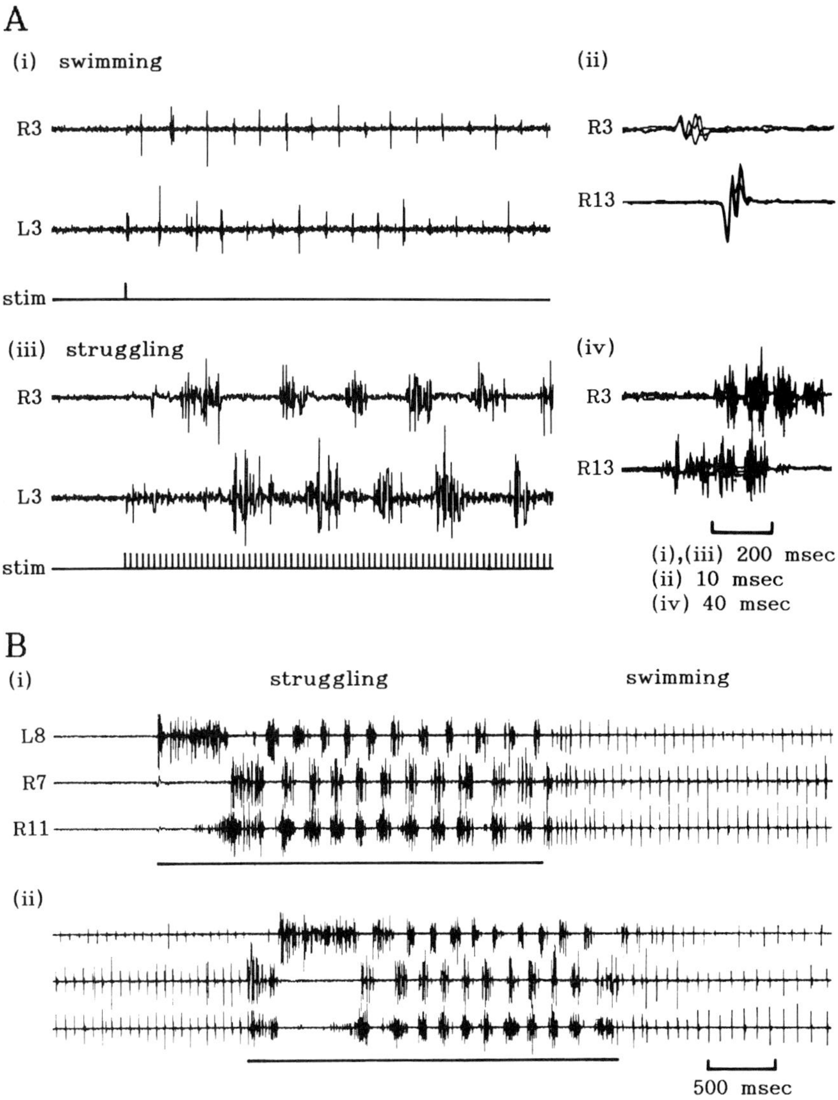

Fig. 5.4. Motor patterns for swimming and struggling in immobilised embryos, with the spinal cord transected caudal to the Xth cranial root. (A) (i) A single brief trunk skin stimulus (stim) evokes self-sustained swimming. Motor discharges alternate on the two sides (R3 and L3). (ii) Motor discharge in each cycle passes from front (R3) to back (R13) (three overlapped traces). (iii) Repetitive stimulation in the same location evokes struggling. Activity consists of bursts of motor root discharges that alternate on the two sides. (iv) Unlike in swimming, motor discharge in each cycle passes from back (R13) to front (R3) (three traces overlapped). Apparent burst fractionation is due to electronic blanking of repetitive stimulus artefacts. (B) Struggling only occurs during periods of repetitive skin stimulation (33 Hz, indicated by line). (i) Activity switches from struggling to swimming at the end of a period of repetitive stimulation. (ii) Repetitive stimulation switches in a period of struggling during an episode of swimming.

tained pressure to the head or rostral trunk. This stimulus is of course far from ideal experimentally but fortunately can also be mimicked electrically. In this case the effective stimulus is repetitive stimulation of the trunk skin at intervals of 20 to 50 ms (Fig. 5.4), a stimulus capable of evoking repetitive firing in Rohon–Beard neurones (Fig. 5.2). A major difference between swimming and struggling is that, whereas the former is self-sustaining in response to a brief stimulus, struggling apparently requires repetitive sensory stimulation and rarely outlasts this by more than a cycle. It is therefore better described as 'centrally patterned' than as 'centrally generated'.

The pattern of stimulation not only selects appropriate *initial* responses to stimulation. Both behaviourally, and particularly in immobilised preparations, switches between swimming and struggling can be made to occur *within* episodes of activity (Fig. 5.4B). Broadly, embryos struggle during periods of repetitive trunk skin stimulation, but otherwise they swim. At least in the paralysed preparation, the actual transitions between swimming and struggling can be rather 'messy' and frequently do not involve an instant switch. For example, synchrony often occurs between the end of struggling and the start of swimming after a period of repetitive stimulation (Fig. 5.5A). However, the different patterns are quite distinct and do not appear to grade from one to another.

What sensory pathway is involved? As mentioned above, Rohon–Beard neurones seem to represent the only direct route for trunk skin sensory information into the spinal cord. After transecting the CNS caudal to the Xth cranial root (thereby removing the influence of other forms of sensory inflow), repetitive electrical stimulation of the trunk skin is still effective in evoking struggling. We must, therefore, conclude that Rohon–Beard neurones can not only evoke flexing and swimming, but, if repetitively stimulated, can also evoke struggling.

5.9. A role for Rohon–Beard neurones in behavioural selection?

The main components of the behavioural repertoire of the *Xenopus* embryo can apparently be elicited by Rohon–Beard sensory neurones. Behavioural selection can occur at least in part, therefore, according to the pattern of stimulation of a single sensory pathway. Weak, brief stimulation may produce flexion; stronger brief stimulation can produce self-sustained swimming; and repetitive stimulation can produce struggling. It is not yet clear how a strong natural mechanical stimulus to the skin could evoke repetitive firing in Rohon–Beard neurones, since these normally respond to light strokes and are rapidly adapting. It may be that mechanical stimulation needs to damage the peripheral neurites partially in order to produce a sufficiently sustained response.

We have already seen how the *Xenopus* embryo possesses a single neuromuscular system for driving the whole range of body movements. We have now seen, in addition, that a single sensory system is capable of eliciting a major part of this behavioural repertoire. These, then, represent the output that has to be selected and one means by which selection can be made to occur. It does not, however, tell us *how* switching occurs. The patterns of firing in sensory Rohon–Beard neurones that result from different patterns of skin stimulation can hardly be expected to be *directly* responsible for the different patterns of drive to motoneurones that underlie each behaviour. Two other broad possibilities exist. First, different forms of stimulation could elicit activity in different premotor 'pattern generating' networks for swimming and struggling. Secondly, they might bring about changes in (or modulate) the output performance of a single such network. Evidence so far suggests that the latter is the case.

5.10. **Does the switch between swimming and struggling involve different premotor elements?**

Intracellular recordings made from presumed premotor interneurones in the spinal cord during swimming and struggling suggest that the same neurones tend to be active in both behaviours. This is supported by anatomical evidence suggesting that, other than those neurones driving swimming and responsible for sensory processing, the neuroanatomically simple spinal cord contains no major classes of neurones that are 'unaccounted for' and might therefore be available to drive struggling. However, it is quite possible that there is at least some diversity within classes of spinal cord neurones. It may well be, for example, that some members of the populations of excitatory or inhibitory premotor interneurones are more likely to be active during struggling than swimming because excitation (which they also receive) is stronger. Struggling could then involve not *different* populations of neurones but simply *larger* ones.

The one major addition during struggling is likely to be the second order sensory dorsolateral interneurones. These neurones, which are excited by Rohon–Beard neurones, are inhibited during swimming (Clarke and Roberts, 1984). For repetitive Rohon–Beard stimulation to be effective in evoking struggling, however, dorsolateral interneurones must be active during at least some of each struggling cycle. The pattern of this activity has yet to be determined.

5.11. **What cellular changes occur during behavioural switching?**

It is really too soon to consider the detailed changes that might occur at the level of individual neurones between swimming and struggling. However, some evidence exists for one important feature. The most striking difference between the motor patterns for swimming and strug-

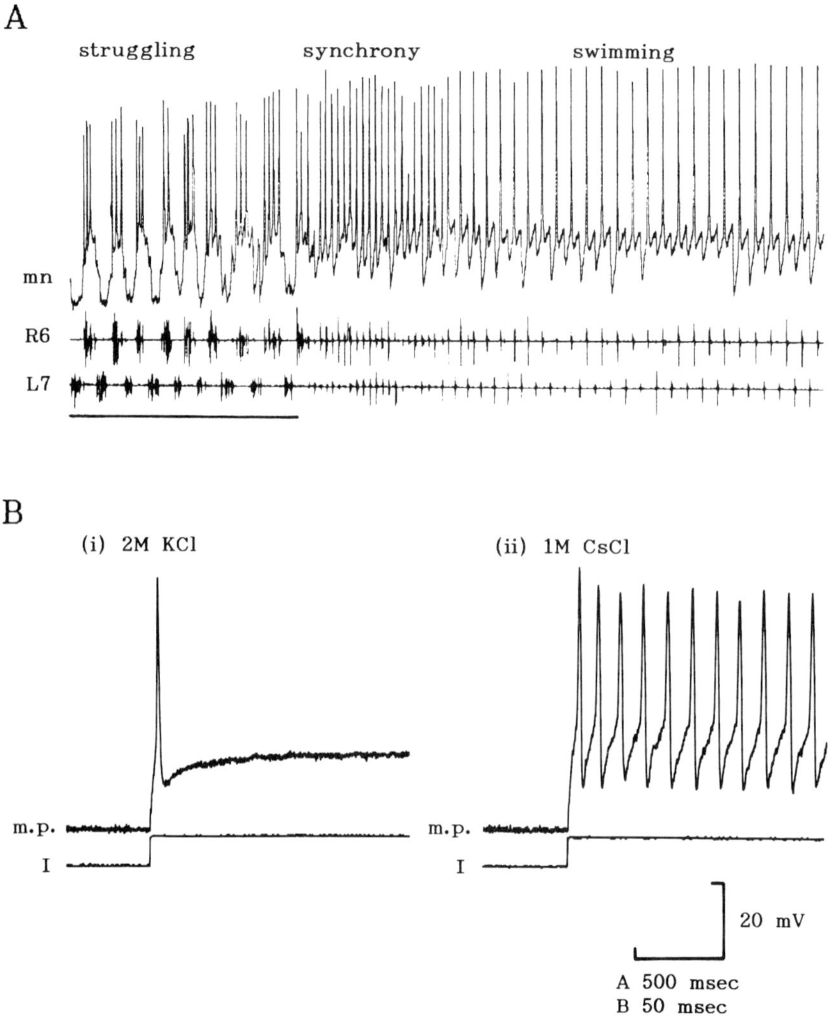

Fig. 5.5. Repetitive and single impulses in spinal neurones. (A) During struggling, spinal neurones fire several impulses in each cycle whereas during swimming they fire only once. Intracellular recording from a spinal cord motoneurone on the right side (mn) and right and left motor roots (R6 and L7). Notice the period of synchrony after struggling has ceased and before swimming resumes. Period of repetitive stimulation indicated by line. (B) Impulses evoked in spinal neurones (m.p.) by intracellular current injection (I). (i) Single impulse evoked by a depolarising step of injected current (0.42 nA: approximately twice threshold current for firing), recorded with a KCl-filled microelectrode. (ii) Repetitive firing in response to the same current (again approximately twice threshold) in a different neurone, this time using a CsCl filled microelectrode to weaken K^+ conductances.

gling is that, during swimming, motoneurones and premotor interneurones fire only one impulse per cycle whereas in struggling they fire repetitively (Fig. 5.5A). This difference is more significant than might at first appear. A recently proposed mechanism by which swimming could be generated within the spinal cord relies heavily on neurones firing only a single spike when depolarised (Roberts and Tunstall, 1990). This follows the observation that most motoneurones, when injected with a suprathreshold depolarising step of current, only generate a single action potential (Fig. 5.5B). Therefore the repetitive firing pattern of spinal neurones during struggling is not what would be predicted from the behaviour of individual neurones. It seems that motoneurones are prevented from firing repetitively on depolarisation by a relatively slowly activating, voltage-dependent, potassium conductance. This can be blocked by injecting caesium ions into a neurone through the recording microelectrode. If this is done, neurones are able to fire repetitively (Fig. 5.5B) (Soffe, 1990).

It seems reasonable to suppose that, during struggling, the firing behaviour of neurones may be altered by suppression of the potassium conductance that normally restricts them to a single spike. At present it is only possible to speculate on how this might occur. Where struggling is switched on by trunk skin stimulation, repetitive firing of Rohon–Beard neurones is presumably directly or indirectly responsible for modulation of neurone firing. An attractive possibility is that repetitive firing in Rohon–Beard neurones causes them to release substance P or a similar peptide (for which they stain immunocytochemically: Clarke *et al.*, 1984) in addition to their normal excitatory amino acid primary sensory neurotransmitter. This might then suppress the potassium conductance and allow neurones to fire bursts of impulses on each cycle. Substance P has been shown to have an appropriate effect on firing in frog sympathetic ganglion neurones by suppression of a somewhat similar potassium conductance, the M-current (Adams *et al.*, 1983).

The above suggestion is clearly only speculative, but does serve to illustrate an important point. In a situation where essentially the same sets of neurones are involved in driving different behaviours, and particularly where, as in *Xenopus*, these behaviours are activated by the same sensory pathway, modulation of the cellular properties of individual neurones is likely to play a very significant part in selecting particular motor patterns to bring about behavioural switching.

Acknowledgements

The author thanks the MRC for financial support, Dr Alan Roberts for early discussion, Dr Nick Dale for helpful comments on the manuscript and Dr Jon Clarke for the use of some previously unpublished records.

References

Adams, P. R., Brown, D. A. and Jones, S. W. (1983) Substance P inhibits the M-current in bullfrog sympathetic neurones. *Br. J. Pharmacol.* **79**: 330–333.

Boothby, K. M. and Roberts, A. (1988) Descending GABA neurones inhibit swimming in the amphibian embryo, *Xenopus laevis*. *J. Physiol.* **401**: 38P.

Clarke, J. D. W., Hayes, B. P., Hunt, S. P. and Roberts, A. (1984) Sensory physiology, anatomy and immunohistochemistry of Rohon–Beard neurones in embryos of *Xenopus laevis*. *J. Physiol.* **348**: 511–525.

Clarke, J. D. W. and Roberts, A. (1984) Interneurones in the *Xenopus* embryo spinal cord: sensory excitation and activity during swimming. *J. Physiol.* **354**: 345–362.

Coghill, G. E. (1928) *Anatomy and the Problem of Behaviour*. Cambridge University Press, Cambridge.

Hughes, A. F. W. (1957) The development of the primary sensory system in *Xenopus laevis*. *J. Anat.* **91**: 323–338.

Kahn, J.A. and Roberts, A. (1982a) The central nervous origin of the swimming motor pattern in embryos of *Xenopus laevis*. *J. exp. Biol.* **99**: 185–196.

Kahn, J. A. and Roberts, A. (1982b) The neuromuscular basis of rhythmic struggling movements in embryos of *Xenopus laevis*. *J. exp. Biol.* **99**: 197–203.

Kahn, J. A., Roberts, A. and Kashin, S. M. (1982) The neuromuscular basis of swimming movements in embryos of the amphibian *Xenopus laevis*. *J. exp. Biol.* **99**: 175–184.

Muntz, L. (1964) Neuro-muscular foundations of behaviour in embryonic and larval stages of the anuran *Xenopus laevis*. PhD Thesis, University of Bristol.

Nieuwkoop, P. D. and Faber, J. (1956) *Normal Tables of Xenopus laevis (Daudin)*. North-Holland, Amsterdam.

Roberts, A. and Clarke, J. D. W. (1982) The neuroanatomy of an amphibian embryo spinal cord. *Phil. Trans. Roy. Soc., Ser. B*, **296**: 195–212.

Roberts, A. and Hayes, B.P. (1977) The anatomy and function of free nerve endings in an amphibian skin sensory system. *Proc. Roy. Soc. B* **196**: 415–429.

Roberts, A. and Kahn, J. A. (1982) Intracellular recordings from spinal neurones during 'swimming' in paralysed amphibian embryos. *Phil. Trans. Roy. Soc. Ser. B*, **296**: 213–228.

Roberts, A., Soffe, S. R. and Dale, N. (1986) Spinal interneurones and swimming in frog embryos. In: *Neurobiology of Vertebrate Locomotion*, S. Grillner, P. S. G. Stein, D. Stuart, H. Forssberg and R. M. Herman, eds. Macmillan, London, pp. 279–306.

Roberts, A. and Tunstall, M. J. (1990) Mutual re-excitation with post-inhibitory rebound: a simulation study on the mechanisms for locomotor rhythm generation in the spinal cord of *Xenopus* embryos. *Eur. J. Neurosci.* **2**: 11–23.

Sillar, K. T. and Roberts, A. (1988) Unmyelinated cutaneous afferent neurones activate two types of excitatory amino acid receptor in the spinal cord of *Xenopus laevis* embryos. *J. Neurosci.* **8**: 1350–1360.

Soffe, S. R. (1989) Roles of glycinergic inhibition and N-methyl-D-aspartate receptor-mediated excitation in the locomotor rhythmicity of one half of the *Xenopus* embryo cns. *Eur. J. Neurosci.* **1**: 561–571.

Soffe, S. R. (1990) Active and passive membrane properties of spinal cord neurons that are rhythmically active during swimming in *Xenopus* embryos. *Eur. J. Neurosci.* **2**: 1–10.

Soffe, S. R. (1991a) Centrally generated rhythmic and non-rhythmic behavioural responses in *Rana temporaria* embryos. *J. exp. Biol.* **156**: 81–99.

Soffe, S. R. (1991b) Neuronal mechanisms for swimming in *Xenopus* embryos.

In: *Locomotor neural mechanisms in arthropods and vertebrates*, D. M. Armstrong and B. M. H. Bush, eds. Manchester University Press, pp. 61–72.

Soffe, S. R., Clarke, J. D. W. and Roberts, A. (1983) Swimming and other centrally generated motor patterns in newt embryos. *J. Comp. Physiol.* **152**: 535–544.

Soffe, S. R. and Sillar, K. T. (1991) Pattern of synaptic drive to ventrally located spinal neurones in *Rana temporaria* embryos during rhythmic and non-rhythmic motor responses. *J. exp. Biol.* **156**: 101–118.

6 *Chun Xiao Falk, Jian-Young Wu, Larry B. Cohen, Hans-Peter Hopp, Dejan Zecevic, Jill London and Vadim Roschin*

Many neurones in the *Aplysia* abdominal ganglion are active during the gill-withdrawal reflex

6.1. Introduction

This chapter describes and discusses the results of the application of optical methods for monitoring neurone activity to the abdominal ganglion of the sea hare, *Aplysia*. While the results thus far obtained do not allow any firm conclusions to be reached about the mechanisms of behavioural choice, they do provide a basis for further speculation.

The original motivation for developing optical methods for monitoring neuronal activity was the hope that such methods could be used to record the action potentials of all the neurones in some of the simpler invertebrate ganglia during behaviour (Davila *et al.*, 1973) and it was expected that the information obtained would be useful for working out the neurone interactions that underlie behaviour. However, measurements of the activity of many neurones simultaneously during the *Aplysia* gill-withdrawal reflex have raised new problems which, it seems to us, will force a more pessimistic view of the present understanding of the neuronal basis of gill withdrawal and other simple behaviours.

The neuronal basis of the *Aplysia* gill-withdrawal reflex has been studied extensively using microelectrode methods (Kandel *et al.*, 1967; Kupfermann and Kandel, 1969; Kupfermann *et al.*, 1971; Byrne *et al.*, 1978; Hawkins *et al.*, 1981; Byrne, 1983; Goldberg and Lukowiak, 1983; Frost *et al.*, 1988; Koester, 1989). Hawkins *et al.* (1981) described ten interneurones or interneurone groups in the gill-withdrawal circuitry and noted that the circuit was 'somewhat forbidding in its complexity'. Subsequently, many additional interneurones have been identified (Byrne *et al.*, 1978; Byrne, 1983; Frost *et al.*, 1988; Koester, 1989). From a literature survey Yadin Dudai (personal communication) determined that 80 neurones were already known to be involved in the gill-withdrawal reflex. While the final number is not yet known, it has been suggested that more interneurones are yet to be found (Frost, 1987). We have used optical recording methods in an attempt to provide an overview of the action potential activity in the abdominal ganglion that occurs during gill withdrawal.

6.2. **Optical recording methods**

Opisthobranch molluscs have been the preparation of choice for optical recordings because their central nervous systems have relatively few neurones, which are of relatively large size (Bullock and Horridge, 1965; Coggeshall, 1967; Boyle *et al.*, 1983; Cash and Carew, 1989) and it is thought that almost all cell bodies are fully invaded by the action potential (for exceptions see Arshavsky *et al.*, 1985; Susswein and Byrne, 1988). This last characteristic is important since the signal-to-noise ratio for an optical recording in a ganglion is not large and this ratio would be reduced if the cell bodies did not have full-sized action potentials.

In general, the optical monitoring technique is based on a system for monitoring changes in the absorption or fluorescence of a voltage-sensitive dye that has stained a piece of nervous tissue. In optical recordings from intact invertebrate ganglia, the signal-to-noise ratio for absorption has been larger than that for fluorescence for *Aplysia californica* (Zecevic *et*

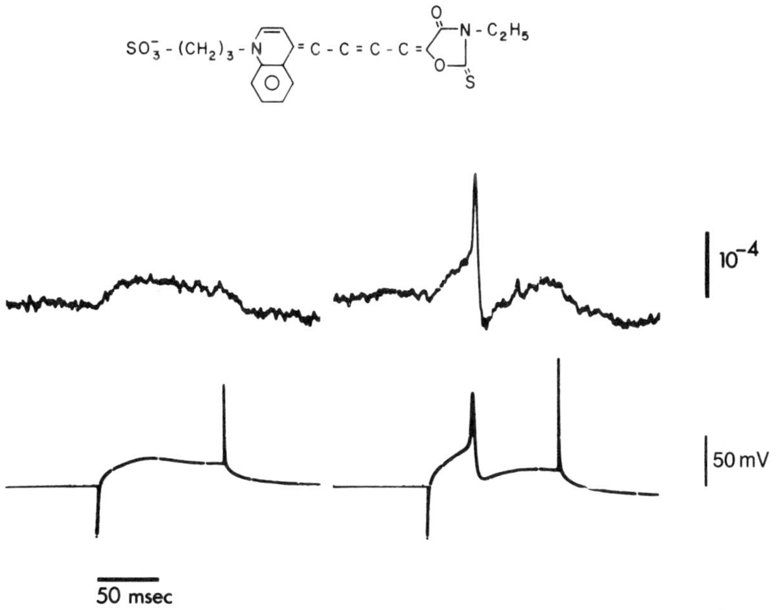

Fig. 6.1. Comparison of changes in light absorption with membrane potential changes in a stained neurone from a barnacle supra-oesophageal ganglion. These signals were recorded in single trials. For both subthreshold and action potential activity, the optical signals (top traces) have a shape that is essentially identical to the electrode signals (bottom traces). The electrode artefacts that appear at the beginning and end of the current step do not occur on the optical recording. The structure of the merocyanine dye used in this experiment is shown at the top. (From Salzberg *et al.*, 1977.)

al., 1985) and for the leech, *Hirudo medicinalis* (Elliott, Cohen, London, Xiao, Zecevic and Cohen, unpublished results; Grinvald *et al.*, 1982). Figure 6.1 compares absorption signals and microelectrode recordings from a single neurone in a barnacle ganglion. The changes in absorption mirrored both sub- and suprathreshold events recorded with the microelectrode. On the left, a small subthreshold membrane potential change as a result of injection of depolarising current is recorded electrically (bottom trace). A small change in the light absorption (top trace) of the cell is recorded simultaneously. When the cell is stimulated to give an action potential (on the right), the optical trace also reflects this activity.

For multiple site recording the ganglion is placed in the object plane of a microscope objective which then forms a magnified real image of the ganglion at its image plane. A 12×12 array of silicon photodiodes is positioned at this plane to record the light transmitted by the ganglion.

6.2.1. *Pharmacological and photodynamic effects of the dyes*

The intensity of these effects is species dependent as well as dye dependent. In experiments on a semi-intact *Navanax* preparation stained with 0.5 mg/ml of the oxonol dye, RH155 (NK3041), for 30 min, photodynamic damage and bleaching appeared to be insignificant (London *et al.*, 1987). Illumination of a preparation for 20 min at the light intensity used for the optical measurements resulted in little if any change in the pattern of neuronal activity or in signal size. The pharmacological effects of dyes were examined in behavioural experiments in *Navanax*. Response to food and subsequent feeding were compared with those in animals whose buccal ganglia either had or had not been stained. Little, if any, change in behaviour was observed as a result of staining.

In contrast, when the staining procedure for *Navanax* was used in experiments on the *Aplysia* gill-withdrawal reflex, there were apparent pharmacological effects in some preparations (Zecevic *et al.*, 1989). Also, the number of active neurones detected optically seemed to decrease by about 5–10% from trial to trial, suggesting that there was either photodynamic damage or dye bleaching. Our *Aplysia* experiments might have benefited from additional dye screening to look for dyes with reduced toxic effects. Whereas more than 1000 dyes have been tested on squid axons, only 50–60 dyes have been tested in *Aplysia*.

6.2.2. *Determining neurone activity from photodiode outputs*

When individual photodiodes were placed over the images of specific cell bodies, a large signal was recorded only in the detector positioned directly over the stimulated cell body indicating that the optical signal was confined to the expected area of the image (Salzberg *et al.*, 1977). How-

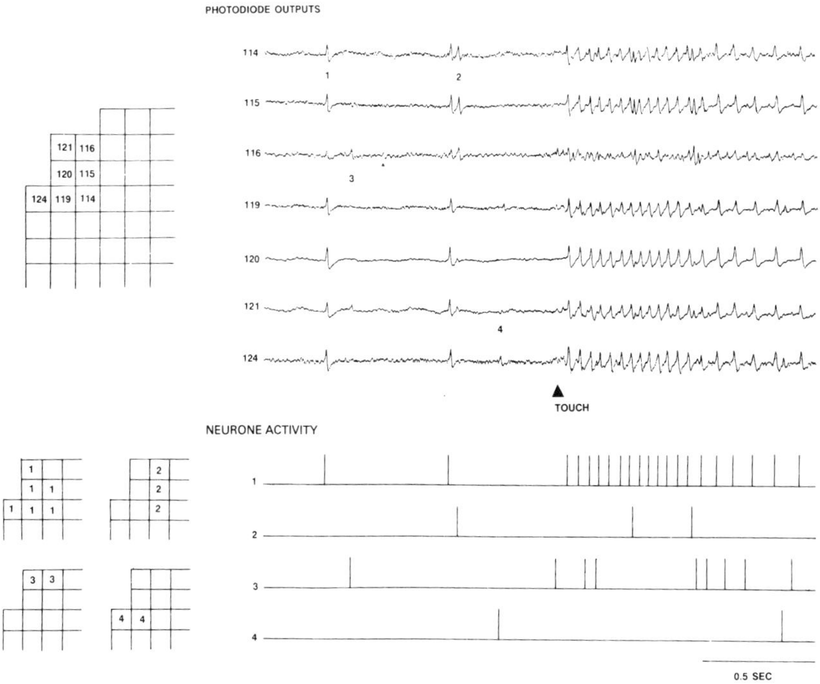

Fig. 6.2. Optical recordings from a portion of a 12 × 12 array from an *Aplysia* abdominal ganglion. The drawing to the left represents the relative position of the detectors whose activity is displayed. In the top section, the original data from seven detectors is illustrated. The number to the left of each trace identifies the detector from which the trace was obtained. In the bottom section the results of our analysis of these data show the spike activity of the four neurones (From Zecevic *et al.*, 1989.)

ever, when the image of a ganglion is formed on a 12 × 12 diode array, the correspondence of cells and detectors is lost. The light from larger cells will fall on several detectors and the activity of such a cell will be recorded as simultaneous events on neighbouring detectors. In addition, one detector may receive light from, and record the activity of, several cells. Finally, the images of some neurones will cover the entire area of a diode and the signal recorded on that detector will be relatively large while other detectors will receive only a small part of the cell's image and those signals will be small.

These possibilities are illustrated in Fig. 6.2 which shows a small portion of an array measurement from an *Aplysia* abdominal ganglion. In the top section, original recordings from seven of the photodiodes from the array are shown. At the number 1 there are synchronously occurring action potentials on all seven detectors. Similar synchronous events occur many times (>20), and we presume that each represents an action potential in a relatively large neurone. The activity of this cell is represented by the vertical lines on trace 1 of the analysed results shown at the bottom. The activity of a second cell (number 4 in Fig. 6.2) is indicated by a spike on 119 and a small signal on its neighbour, 120. The activity of two additional neurones (numbers 2 and 3 in Fig. 6.2) was similarly identified. Several of the problems associated with multicellular recordings are illustrated by this figure. One problem arises from the small signal-to-noise ratio. There may be an additional spike on detector 116 just before the stimulus (at the arrow), but the signal-to-noise ratio was not large enough to be certain. A second problem results from the fact that, following the stimulus, there is a great deal of activity, and it would be easy for this activity to obscure small signals and lead to an underestimation in the number of active neurones. A third problem is that the analysis is tedious and error prone. The kind of analysis illustrated in this figure has produced the data used to generate the raster shown in Fig. 6.3. More detailed discussions of optical recording methods have been published elsewhere (Cohen and Salzberg, 1978; Salzberg, 1983; Cohen and Lesher, 1986; Grinvald *et al.*, 1988; Cohen *et al.*, 1989).

6.3. **Recording from the *Aplysia* abdominal ganglion during the gill-withdrawal reflex**

Figure 6.3 illustrates the results of analysis of optical recordings from 124 detectors during two gill withdrawals in one preparation (for methods see Zecevic *et al.*, 1989). Each horizontal line represents the activity of one neurone, and each vertical tick represents one action potential in that neurone. The stimulus was delivered by a computer-controlled motor at the beginning of the third second of the recording. The recording on the left illustrates the results obtained during the 12th habituation trial and that on the right the results obtained 10 min after sensitising stimulation to the pleural–abdominal connective. For this experiment we also made extracellular electrode recordings from three nerves and two connectives. The activity recorded with the electrodes is indicated by the vertical black lines on the left of the traces. Activity was detected in 92 neurones (57 optically) in the habituated state and in 150 neurones (103 optically) in the sensitised state.

Optical recordings during the *Aplysia* gill-withdrawal reflex have been obtained by Satoru Shiono and collaborators (Nakashima *et al.*, 1989). They improved on our measurements in two ways. First, they used a diode array with 512 pixels, a number that would seem more appropriate for ganglia of 600–1100 neurones. With this array they detected activity in 50% more neurones. Secondly, they stained with a lower concentration of dye. The dye concentration we used suppressed the spontaneous activity

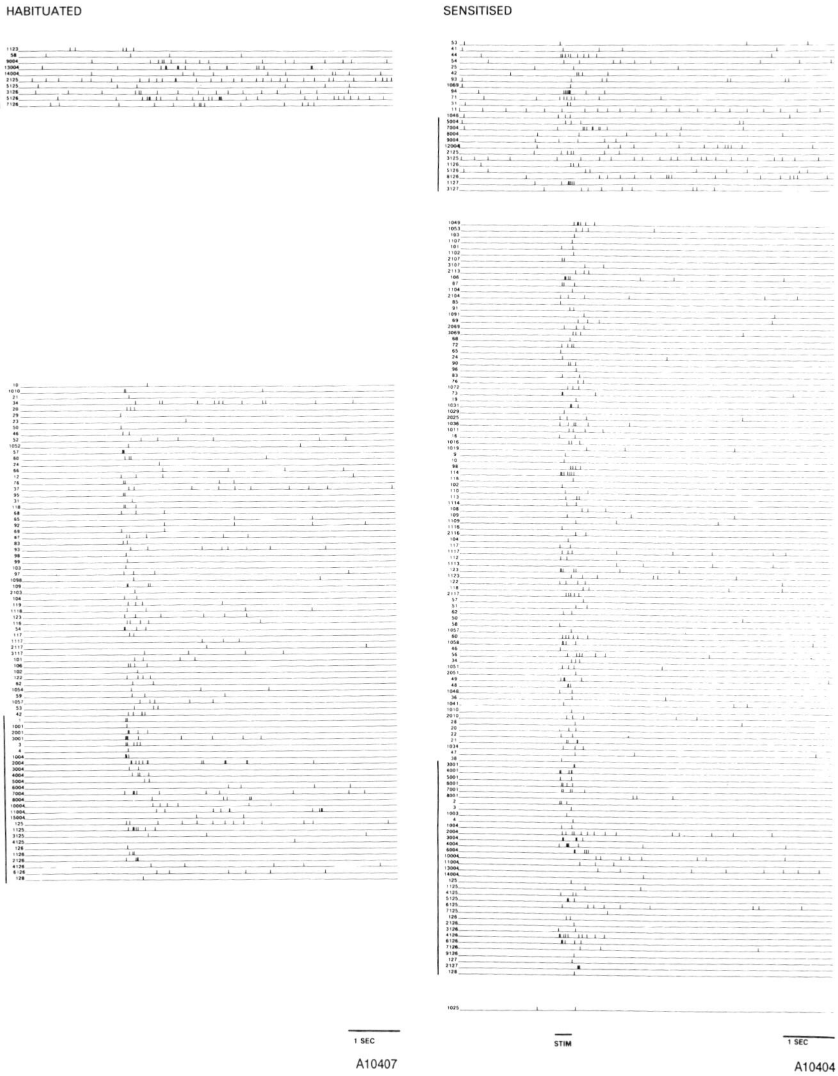

Fig. 6.3. The action potential activity recorded optically and electrically during two gill-withdrawals in one preparation. The recording on the left was taken during the 12th habituation trial. To habituate the response the stimulus (light touch to the siphon skin) was presented once every minute. On the right is the response obtained 10 min after a sensitising electrical stimulus of the pleuro-abdominal connective. The number of active neurones was 150 in the sensitised state and 92 in the habituated state. The action potential activity in the siphon, brachial and genital nerves and the pleuro-abdominal connectives was recorded with extracellular electrodes (indicated by the vertical black lines on the left). When the activity of a neurone was detected on both optical and electrode recordings, only the result from the optical recording is shown. The mechanical stimulus was delivered by a computer-controlled motor. The timing of the mechanical stimulus is indicated by the bar at the bottom. (From Zecevic *et al.*, 1989.)

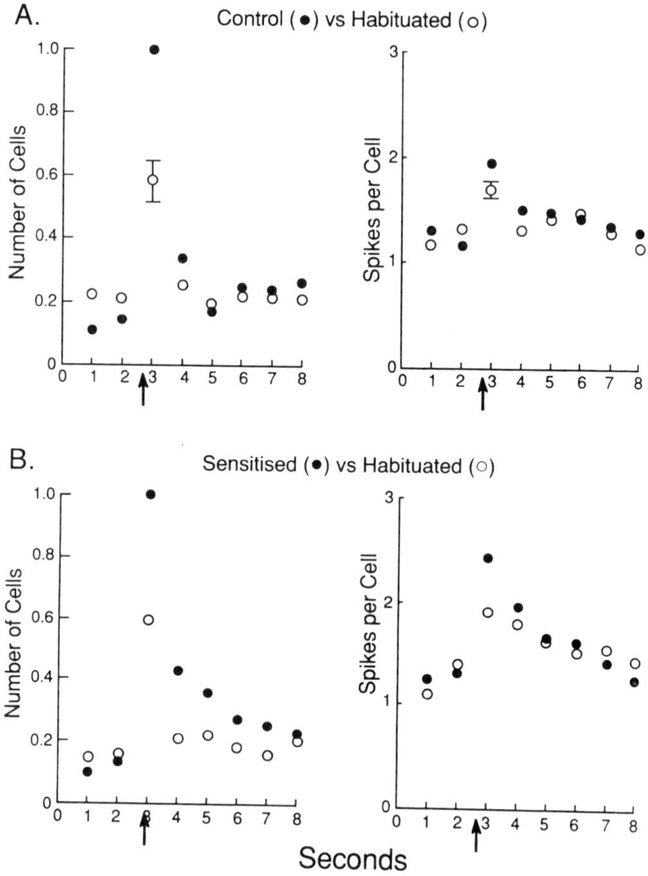

Fig. 6.4. Comparison of the number of active neurones (on the left) and the number of spikes per second (on the right) in control, habituated and sensitised preparations. The main effect of habituation and sensitisation is on the number of active neurones although there also appears to be a small effect on the number of spikes per cell. The touch to the siphon was delivered at the beginning of the third second (indicated by the arrows). For the plots of the number of active neurones (on the left), the results for the control and sensitised states in the third second were normalised to 1.0 for each preparation and this normalisation factor was used for the remaining values from that preparation. The spikes per cell per second are shown on the right. (From Zecevic *et al.*, 1989.)

in the ganglion. Since the ability to detect inhibition in these recordings depends on detecting a decrease in the activity of spontaneously active cells (Aersten and Gerstein, 1985), their protocol allowed them to detect more neurones whose activity was inhibited by the touch to the siphon.

Figure 6.3 illustrates results from a preparation where we detected a relatively large number of active neurones. In more than 10 additional preparations the number of neurones detected optically during the gill-withdrawal response was between 50 and 105 in the control state. However, in other preparations, the number of optically detected neurones was smaller. Because of factors which could cause the number of detected neurones to be artefactually reduced (incomplete staining, excess optical noise, preparation in poor condition), our prejudice is that the larger numbers are more representative of the normal neurone activity.

From the results presented in Fig. 6.3 it appears that habituation and sensitisation can effect large changes in the number of neurones whose activity is elicited by the touch to the siphon skin. To determine if there was also a change in the firing frequency of the neurones, we determined the number of action potentials per active neurone in each second of the optical recording. In Fig. 6.4A (top) there is a comparison of average values from control and habituated states and in B (bottom) a comparison of average values from sensitised and habituated states. The number of neurones that were active during each second is shown on the left and the number of spikes per second per neurone is shown on the right. The main effects of habituation and sensitisation are on the number of activated neurones although there also seems to be a small effect on the discharge rates.

Because of the abdominal ganglion's relatively high opacity and large number of neurones, the recordings in Fig. 6.3 were likely to be incomplete. The following method was used to estimate how many neurones were actually activated by the touch to the siphon. The number of neurones detected optically during a touch to the siphon in a sensitised preparation was determined. We then attempted to activate all of the neurones in the ganglion by sequentially giving electrical shocks to three nerves and two connectives, and again determined the number of neurones whose action potential activity could be detected optically. Dividing the first number by the second gives the fraction of detectable neurones that are active during gill withdrawal. In five preparations, this fraction was 35 ± 6% (mean ± standard error of the mean). Since the number of cell bodies present in ganglia of the size we used is either 1100 (Coggeshall, 1967) or 600 (Cash and Carew, 1989), the finding that 35% of the detectable neurones are active suggests that, assuming that large and small neurones are activated in the same proportions, the actual number of active neurones during the gill-withdrawal reflex is between 250 and 400, substantially more neurones than illustrated in Fig. 6.3.

6.4. **Discussion**

A large fraction of the neurones which are activated by the touch to the siphon discharge within 1 s of the application of the stimulus. Because of

Table 6.1. % of all *C elegans* neurones reached by one or two synapses by three sensory neurones; % of neurones that reach two motor neurones in one or two synapses

Sensory neurone	*% Monosynaptic*	*% Disynaptic*	*% monosynaptic*	*Motorneurone*
ADE-L	9	61		
OLL-L	5	36		
PHC-L	2	31		
	8	37	8	RMDD-L
	2	30	2	RID

Compiled from White *et al.*, 1986.

this extensive overlap in firing times, it is not easy to make inferences about function from the temporal sequence of spike activity in the *Aplysia* recordings.

The deduction that a light touch to a small section of the siphon skin activates 250–400 neurones in the abdominal ganglion demonstrates that the information about this mild sensory stimulus is widely distributed in the *Aplysia* abdominal ganglion. We calculate that as many as 30–40% of the cells in the ganglion are activated enough to make at least one action potential. This result might have been anticipated from anatomical studies in other systems. It is estimated that the P mechanosensory cell in the leech makes 2500 anatomical synapses on to the 390 postsynaptic neurones in a segmental ganglion (Muller and McMahan, 1976; Muller, personal communication). In the nematode *C elegans*, the ADE-L sensory neurone is monosynaptically connected to 9% of all other neurones in the animal and disynaptically connected to an additional 52%. Table 6.1 (compiled from White *et al.*, 1986) includes this result along with similar data for two other sensory neurones, OLL-L and PHC-L, where only slightly lower percentages were obtained. In addition, Table 6.1 presents information about the percentage of cells in the ganglion that make monosynaptic or disynaptic connections to two motoneurones, RMDD-L and RID. The percentage of *C elegans* neurones that make monosynaptic contacts to the motoneurones as well the percentage that make contact via one interposed neurone are similar to the percentages contacted by sensory neurones.

The conclusion that several hundreds of neurones may be active during the gill-withdrawal reflex suggests that it could be very difficult to obtain an understanding of the cellular basis of this behaviour. For example, the 250–400 neurones possibly activated by touch to the siphon are not necessarily all involved in gill withdrawal *per se*. Indeed they may be involved in many other behaviours or responses which can be elicited by touch to the siphon. Figures 6.5A and B illustrate two possible neuronal organisations, both of which imply that a large number of neurones would

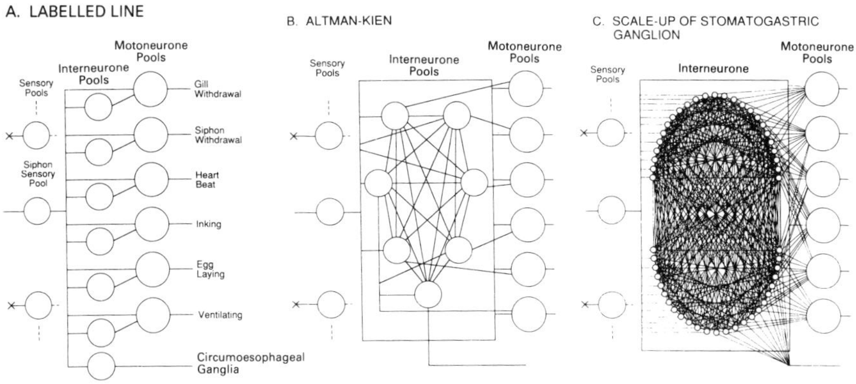

Fig. 6.5. (A, B) Schematic diagrams of two possible organisations of interneurones in the *Aplysia* abdominal ganglion. Both are consistent with the finding that several hundred neurones in the ganglion are activated by a light touch to the siphon skin. In both panels, the only sensory inputs indicated are from mechanosensory cells with endings in the siphon skin; the monosynaptic connections between these sensory neurones and motoneurones are also shown. (A) A 'labelled line' organisation is illustrated. Only a few of the interneurones in the ganglion are involved in the gill-withdrawal reflex. (B) A single distributed network drives gill withdrawal and other behaviours. (C) Schematic diagram of synaptic connections in a series of 50 interneurones, each of which makes synaptic contact with one-quarter of the other interneurones. About 300 synaptic interactions are illustrated. (From Zecevic *et al.*, 1989; (B) derived from Altman and Kien, 1987; (C) based on Miller and Selverston, 1982, Mulloney and Selverston, 1974; White *et al.*, 1986.)

be activated by touching the siphon. In A, on the left, a 'labelled-line' neuronal organisation is illustrated. Here the majority of the neurones activated by the touch to the siphon are directly involved not with the gill-withdrawal reflex but with other aspects of defensive withdrawal or in entirely different behaviours. However, because the gill movements are coordinated with a variety of other behaviours, this labelled-line circuit cannot be totally realistic (Altman and Kien, 1987); a schematic drawing showing interaction among interneurone pools to provide this coordination is shown in panel B. This view implies that a large fraction of the interneurones in the abdominal ganglion would be active at the same time, possibly to organise related responses and to coordinate them with the gill-withdrawal response. A more specified version of panel B is illustrated on the right in panel C. This version uses the results from the lobster stomatogastric ganglion where all possible pairs of interneurones have been examined by pairwise intracellular recordings (Mulloney and Selverston, 1974; Miller and Selverston, 1982). In the stomatogastric ganglion there are extensive synaptic interactions between neurones. In panel C, 50 interneurones are shown and 25% of the pairs have synaptic interactions. If this were a realistic representation of the amount of synaptic interaction among *Aplysia* interneurones, then a relatively large fraction of the interneurones in the ganglion would be active during generation of the gill-withdrawal reflex.

These anatomical and physiological findings certainly allow the speculation that the organisation of any particular motor output by the nervous system may activate a large fraction of the animal's interneurones. Understanding circuits that involve more than a few neurones is a challenge to neurobiologists.

6.5. **Difficulties and future directions**

6.5.1. *Signal-to-noise ratio*

The present signal-to-noise ratios are adequate for recording action potentials in cell bodies in simpler invertebrate ganglia and, with the presently available dyes, pharmacological effects and photodynamic damage are not overwhelming problems. However, it is clear that the signal-to-noise ratios for smaller potential changes or cells with smaller areas of membrane would not be adequate for recording synaptic activity in single trials or where only small electrotonically spread action potentials reached the cell body (Salzberg *et al.*, 1977; Grinvald *et al.*, 1981). Following Shiono and collaborators (Nakashima *et al.*, 1989), sensitivity may be increased by improving the spatial resolution of the apparatus through the use of an array with more elements. Using preparations with less scattering (Boyle *et al.*, 1983) would also improve the signal-to-noise ratio.

6.5.2. *Problems caused by time-consuming analysis*

The analysis of the data required to generate the raster displays, as in Fig. 6.3, is difficult, and time consuming, taking about 60 h to complete. This means that many experiments have not been attempted since the amount of data generated will be so large. There is therefore considerable motivation for developing automatic analysis programs.

6.5.3. *Too many cells active at once*

When many neurones are active together, there may be many action potential signals on every detector resulting in a noisy baseline in the optical recording. This excessive noise causes two kinds of problem. First, small signals will be obscured, resulting in an underestimate of the number of active cells. Secondly, this excessive noise, and for that matter all action potential signals, will obscure synaptic signals.

6.5.4. *Cell identification*

Clearly, there is a need to provide some sort of identification for the large number of cells that are active during the *Aplysia* gill-withdrawal reflex. One possibility is to monitor the direction of spike propagation in peripheral nerves and match their timing with soma recordings. Activity of cells with peripheral nerve spikes propagating outwards is possibly of motoneurone origin; spikes propagating towards the ganglion may be from sensory neurones. However, because only 20% of the activity recorded optically and electrically was matched (Zecevic *et al.*, 1989), this procedure may be useful for only a fraction of the active neurones.

6.5.5. *Answers to more general questions about nervous systems*

The ability to monitor the activity of a substantial fraction of the neurones involved during a behaviour allows one to ask more general questions about how nervous systems work. How reproducible is the neuronal response to the same stimulus (Wu *et al.*, 1989)? Is the nervous system very determined or is it chaotic? How similar is the organisation of the nervous system from animal to animal? How much does feedback from movements contribute to the neuronal activity first elicited by the touch to the siphon (Falk *et al.*, 1989)? What are the correlations between the spike times of individual cells (Schiminovich *et al.*, 1989)? It may now be possible to obtain answers to these and related questions.

Acknowledgements

We are grateful to Vincent Castellucci, William Frost and Eric Kandel for demonstrating the isolated siphon preparation. We thank Rina Hildesheim, Amiram Grinvald, Leslie Loew, Joe Wuskell and Alan Waggoner for kindly providing dyes. The research described was supported by Public Health Service Grant NS08437.

References

Aersten, A. M. H. J. and Gerstein, G. L. (1985) Evaluation of neuronal connectivity: sensitivity of cross correlation. *Brain Res.* **340**: 341–354.

Altman, J. S. and Kien, J. (1987) A model for decision making in the insect nervous system. In: *Nervous Systems in Invertebrates*, M. A. Ali, ed. Plenum Publishing Corp., New York, pp. 621–643.

Arshavsky Yu I, Beloozerova, I. N., Orlovsky, G. N., Panchin, Yu V and

Pavlova, G. A. (1985) Control of locomotion in marine mollusc *Clione limacina* II. Rhythmic neurons of pedal ganglia. *Exp. Brain Res.* **58**: 263–272.
Boyle, M. B., Cohen, L. B., Macagno, E. R. and Orbach, H. S. (1983) The number and size of neurons in the CNS of gastropod molluscs and their suitability for optical recording of activity. *Brain Res.* **266**: 305–317.
Bullock, T. H. and Horridge, G. A. (1965) *Structure and Function in the Nervous Systems of Invertebrates*, W.H. Freeman and Co, San Francisco.
Byrne, J. H. (1983) Identification and initial characterization of a cluster of command and pattern-generating neurons underlying respiratory pumping in *Aplysia californica. J. Neurophysiol.* **49**: 491–508.
Byrne, J. H., Castellucci, V. F. and Kandel, E. R. (1978) Contributions of individual mechanoreceptor sensory neurons to defensive gill-withdrawal reflex in *Aplysia. J. Neurophysiol.* **41**: 418–431.
Cash, D. and Carew, T. J. (1989). A quantitative analysis of the development of the central nervous system in juvenile *Aplysia californica. J. Neurobiol.* **20**: 25–47.
Coggeshall, R. E. (1967) A light and electron microscope study of the abdominal ganglion of *Aplysia californica. J. Neurophysiol.* **30**: 1263–1287.
Cohen, L. B., Hopp, H-P., Wu, J-Y., Xiao, C., London, J. and Zecevic, D. (1989) Optical measurement of neuron activity in invertebrate ganglia. *Ann. Rev. Physiol.* **51**: 527–541.
Cohen, L. B. and Lesher, S. (1986) Optical monitoring of membrane potential: methods of multisite optical measurement. *Soc Gen Physiol Ser.* **40**: 71–99.
Cohen, L. B. and Salzberg, B. M. (1978) Optical measurement of membrane potential. *Rev. Physiol. Biochem. Pharmacol.* **83**: 35–88.
Davila, H. V., Salzberg, B. M., Cohen, L. B. and Waggoner, A. S. (1973) A large change in axon fluorescence that provides a promising method for measuring membrane potential. *Nature, New Biol.* **241**: 159–160.
Falk, C. X., Hopp, H-P., Wu, J-Y. and Cohen, L. B. (1989) Blocking gill and siphon movements has little or no effect on the neuronal response to siphon touch in the *Aplysia* abdominal ganglion. *Soc. Neurosci. Abstr.* **15**: 1264.
Frost, W. N. (1987) Mechanisms contributing to short- and long-term sensitization in *Aplysia*. PhD Dissertation, Columbia University, New York.
Frost, W. N., Clark, G. A. and Kandel, E. R. (1988) Parallel processing of short-term memory for sensitization in *Aplysia. J. Neurobiol.*, **19**: 297–334.
Goldberg, J. I. and Lukowiak, K. (1983) Transfer of habituation between stimulation sites of the siphon withdrawal reflex in *Aplysia californica. Can. J. Physiol. Pharmacol.* **61**: 749–755.
Grinvald, A., Cohen, L. B., Lesher, S. and Boyle, M. B. (1981) Simultaneous optical monitoring of activity of many neurons in invertebrate ganglia using a 124-element photodiode array. *J. Neurophysiol.* **45**: 829–840.
Grinvald, A., Frostig, R. D., Lieke, E. and Hildesheim, R. (1988) Optical imaging of neuronal activity. *Physiol. Rev.* **68**: 1285–1366.
Grinvald, A., Hildesheim, R., Farber, I. C. and Anglister, L. (1982) Improved fluorescent probes for the measurement of rapid changes in membrane potential. *Biophys. J.* **39**: 301–308.
Hawkins, R. D., Castellucci, V. F. and Kandel, E. R. (1981) Interneurons involved in mediation and modulation of the gill-withdrawal reflex in *Aplysia*. I. Identification and characterization. *J. Neurophysiol.* **45**: 304–314.
Kandel, E. R., Frazier, W. T., Waziri, R. and Coggeshall, R. E. (1967) Direct and common connections among identified neurons in *Aplysia. J. Neurophysiol.* **30**: 1352–1376.

Koester, J. (1989) Chemically and electrically coupled interneurons mediate respiratory pumping in *Aplysia*. *J. Neurophysiol.* **62**: 1113–1126.

Kupfermann, I. and Kandel, E. R. (1969) Neuronal controls of a behavioral response mediated by the abdominal ganglion of *Aplysia*. *Science* **164**: 847–850.

Kupfermann, I., Pinsker, H., Castellucci, V. and Kandel, E. R. (1971) Central and peripheral control of gill movements in *Aplysia*. *Science* **174**: 1252–1256.

London, J. A., Zecevic, D. and Cohen, L. B. (1987) Simultaneous optical recording of activity from many neurons during feeding in *Navanax*. *J. Neurosci.* **7**: 649–661.

Miller, J. P. and Selverston, A. I. (1982) Mechanisms underlying pattern generation in lobster stomatogastric ganglion as determined by selective inactivation of identified neurons. IV. Network properties of the pyloric system. *J. Neurophysiol.* **48**: 1416–1432.

Muller, K. J. and McMahan, U. J. (1976) The shapes of sensory and motor neurons and the distribution of their synapses in ganglia of the leech: a study using intracellular injection of horseradish peroxidase. *Proc. Roy. Soc. Lond. B* **194**: 481–499.

Mulloney, B. and Selverston, A. I. (1974) Organization of the stomatogastric ganglion of the spiny lobster. I. II. III. *J. Comp. Physiol.* **91**: 1–78.

Nakashima, M., Yamada, S., Shiono, S. and Maeda, M. (1989) A 448 channel optical monitoring of neural signals from *Aplysia* ganglion. *Abstr. Soc. Neurosci.* **15**: 1046.

Salzberg, B. M. (1983) Optical recording of electrical activity in neurons using molecular probes. In: *Current Methods in Cellular Neurobiology*, J. L. Barker and J. F. McKelvy, eds. Wiley, New York, pp. 139–187.

Salzberg, B. M., Grinvald, A., Cohen, L. B., Davila, H. V. and Ross, W. N. (1977) Optical recording of neuronal activity in an invertebrate central nervous system: simultaneous monitoring of several neurons. *J. Neurophysiol.* **40**: 1281–1291.

Schiminovich, D., Cohen, L. B., Cohen, A. I., Hopp, H-P., Falk, C. X. and Wu, J-Y. (1989) A search for correlations in the spike activity of the *Aplysia* abdominal ganglion during the gill withdrawal reflex. *Biol. Bull.* **177**: 325.

Susswein, A. J. and Byrne, J. H. (1988) Identification and characterization of neurons initiating patterned neural activity in the buccal ganglion of *Aplysia*. *J. Neurosci.* **8**: 2049–2061.

White, J. G., Southgate, E., Thomson, J. N. and Brenner, S. (1986) The structure of the nervous system of the nematode *Caenorhabditis elegans*. *Phil. Trans. Royal Soc. London* **314**: 1–340.

Wu, J-Y., Falk, C. X., Hopp, H-P. and Cohen, L. B. (1989) Trial-to-trial variability in the neuronal response to siphon touch in the *Aplysia* abdominal ganglion. *Soc. Neurosci. Abstr.* **15**: 1264.

Zecevic, D., London, J. A. and Cohen, L. B. (1985) Simultaneous optical recording from many cells from *Aplysia* abdominal ganglia using fluorescence. *Neurosci Lett.* **22**: 370.

Zecevic, D., Wu, J-Y., Cohen, L. B., London, J. A., Hopp, H-P., and Falk, X. C. (1989) Hundreds of neurons in the *Aplysia* abdominal ganglion are active during the gill-withdrawal reflex. *J. Neurosci.* **9**: 3681–3689.

Part II

Distributed neural networks and motor programme selection

7 *R. Matthias Hennig*

Mechanisms of motor pattern switching in crickets: stridulation and flight

7.1. **Introduction**

A behavioural decision is the response of an animal to changes in the internal and external information it receives (McFarland, 1977). Such a decision may include terminating a current motor programme and selection and execution of a new or additional motor programme. The choice made by the animal may involve switching between (1) behaviours which involve only minor adjustments in motor control for a change of the motor output (such as different gaits in terrestrial locomotion: Grillner, 1975); (2) different behaviours which share the same final motor pathway (such as scratching and walking, Sherrington, 1947; Berkinblit *et al.*, 1978); or (3) entirely different behaviours which do not share any of the final motor pathways involved (such as feeding and locomotion (escape): Davis *et al.*, 1974; Bellman and Krasne, 1983). In particular, instances where behaviours compete for the same final motor pathway raise important questions concerning motor organisation. For example: to what extent is interneuronal control shared by competing behaviours? Are there central pattern generators that are shared by more than one behaviour? Can the same motoneurones and muscles (the common final path; Sherrington, 1947) be employed by two entirely separate interneuronal networks, and, if so, can these two networks operate simultaneously?

In order to address those questions I have studied two different behaviours of the male cricket, one of which involves switching between different aspects of the same behaviour and the other, different behaviour patterns utilising the same motoneurones and muscles. In both behaviours the final common motor pathway involves the forewings, and in the first case a switch occurs during the execution of one behaviour, stridulation, which is similar to a change in gait. In the second a switch occurs between two distinct behaviours, stridulation and flight, which compete for the same final motor pathway. Are there cellular events which might provide insights as to where and how in the nervous system the switching occurs?

7.2. The neuromuscular organisation of the forewing (mesothoracic) segment

An investigation of the neuronal organisation underlying switching between behaviours first requires a description of the respective movements and the different patterns of motor activities. This section outlines the neuromuscular organisation of the mesothoracic segment of the cricket where the two forewings insert; these latter are activated during both stridulation and flight (Fig. 7.1).

The forewings insert at the mesothoracic segment of the cricket via a number of sclerites. The forewing itself is a sclerotised structure with a special morphology adapted for efficient sound production (Nocke, 1971). The wing hinge allows the forewing to rotate around two positions: (1) a raised position employed for the inward and outward movement during stridulation (Fig. 7.1A), and (2), an outward position employed for the up-and-down movement during flight (Fig. 7.1B; Bentley and Kutsch, 1966; Kutsch, 1969). The wing and the wing hinge are innervated by many sense organs, including the stretch receptor, the tegula, campaniform sensilla, and mechanoreceptive hairs, which monitor the wing position and movements, and which may play important roles in stridulation and flight (Huber, 1963; Kutsch and Huber, 1970; Möss, 1971; Elliott and Koch, 1983; Schäffner and Koch, 1987). The wing is moved by direct as well as indirect muscles. There is a distal layer of muscles (98, 99) which depress the wing during flight and open the wing during stridulation, and a proximal layer of muscles (89, 90) which elevate the wing during flight and close the wing during stridulation (Fig. 7.1A,B; Huber, 1960; Bentley 1969a,b; Kutsch, 1969). The dorsal longitudinal muscle acts as an indirect depressor during flight, but does not function as a single entity during stridulation (Kutsch, 1969). Each muscle is innervated by between two and five motoneurones (Kutsch, 1969; Wang and Robertson, 1989). The organisation of the metathoracic segment is similar to that of the mesothoracic segment, but the hindwings are active only during flight and not during stridulation (Kutsch, 1969).

During stridulation each closing movement of the forewings generates a sound pulse, and the fast repetition of opening and closing strokes results in a calling song with a sound pattern specific for *Teleogryllus commodus*. The calling song contains two alternating rhythms, the chirp and the trill (Fig. 7.1C; Bentley and Hoy, 1972). The 'slow' chirp rhythm is instantly followed by the 'fast' trill. The muscles which act as openers/closers during stridulation depress/elevate both pairs of wings during flight (Fig. 7.1A,B,E). Switching from stridulation to flight may be caused by a wind stimulus (Fig. 7.1E).

In order to record intracellularly from central neurones in the cricket during stridulatory and flight activity the nervous system has to be exposed by dissection to allow microelectrode access and stabilised on a platform.

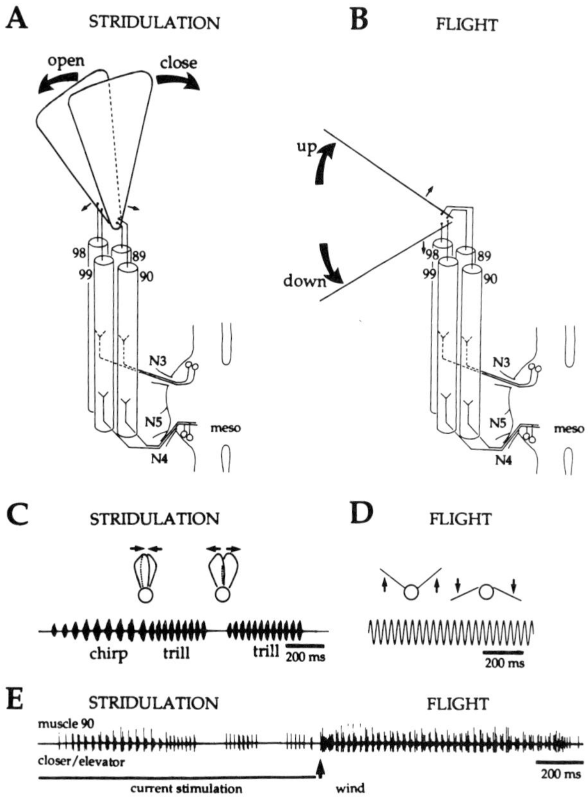

Fig. 7.1. Schematic presentation of the functional anatomy of the mesothoracic segment of the cricket and the forewing movements and movement patterns during stridulation and flight. (A) During stridulation the forewings are raised and moved rhythmically against one another. The main muscles accomplishing this movement and their action are indicated for the left side of the mesothoracic segment. Muscles 98 and 99 open the wings, whereas muscles 89 and 90 close them generating sound pulses as the scraper rubs against the file on the underside of the upper right wing (Huber, 1960; Kutsch, 1969; Koch *et al.*, 1988). The projections of the motoneurones innervating those muscles in the mesothoracic ganglion are indicated (Elliott, 1983; Hennig, 1989; Wang and Robertson, 1989). There are other muscles active during stridulation as well (for details see Kutsch, 1969). (B) The major muscles active during flight are the same as during stridulation although their action differs. Muscles 98 and 99 depress the wings, and muscles 89 and 90 elevate them. (C) The opening and closing movements during stridulation generate a species-specific sound pattern which in *Teleogryllus commodus* contains two different alternating rhythms (Hennig, 1990b). The sound recording of the calling song of *Teleogryllus commodus* shows a phrase; each phrase consists of one chirp and a variable number of trills. The 'slow' chirp rhythm is instantly followed by the 'fast' trill. (D) During flight the wings (including the hindwings which are not shown here) are unfolded and beat regularly up and down. (E) It is possible to switch stridulation to flight by a wind stimulus. The wings are unfolded and beat rhythmically up and down. The recording shown stems from muscle 90 which is a wing closer during stridulation and a wing elevator during flight.

A suction electrode is placed on the brain, and current is passed to elicit the stridulatory motor pattern (Hennig, 1989). A brief wind puff directed at the cerci is usually sufficient to elicit flight (Robertson, 1987; Hennig, 1990a). To monitor the two motor patterns, two extracellular electrodes are placed on nerve 3A3 (innervating muscle 89 in the mesothorax, Fig. 7.1A,B) and nerve 1 (innervating the dorsal longitudinal muscle in the metathorax). Recordings from these positions allow unambiguous recognition of the two motor patterns which involve the meso- and metathoracic musculature in the cricket (Hennig, 1990b).

7.3. **Stridulation: switching between chirp and trill**

The species-specific calling song of crickets attracts only conspecific females and thus serves as a pre-mating isolating mechanism (Hill *et al.*, 1972). Among the most complex song patterns in crickets is the stridulatory pattern of *Teleogryllus commodus* (Bentley and Hoy, 1972). It consists of a continuous repetition of phrases, each phrase possessing two distinct rhythms traditionally termed the chirp and the trill (Fig. 7.1C; Bentley and Hoy, 1972). The trill contains pulses produced at a higher rate (30 Hz) than those in the chirp (18 Hz; Bentley, 1971; Hennig, 1989). The unmistakable switching between these two rhythms occurs repeatedly and regularly during calling song production (Bentley and Hoy, 1972). The changes in frequency are fast but continuous (Fig. 7.2; Hennig, 1989) and it may take two or three cycles only to shift from one stable rhythm (chirp) to the next (trill; Fig. 7.2), suggesting that there is one pattern generator for both chirp and trill. The pattern of synaptic input to motoneurones also takes two cycles before a steady state is reached during the trill confirming a fast but not instantaneous transition between the two rhythms, and intermediate bursts, as judged from burst durations and the synaptic input, occur at the transition (arrows, Fig. 7.3A,B; Hennig, 1989). All interneurones recorded to date participate in both the chirp and the trill (Hennig, unpublished results). They show the transition observed in motoneuronal recordings (Fig. 7.3C), although the changes are not always as pronounced. However, the pattern of synaptic input to motoneurones suggests that some interneurones are active during the trill only, some are active during the chirp only, and some are active during both the trill and the chirp (Hennig, 1989) (Fig. 7.3).

What are the possible neuronal mechanisms controlling the switching from chirp to trill? Proprioceptors play an important role in sound production by ensuring the proper adjustment and pressure of the wings against each other (Kutsch and Huber, 1970; Elliott and Koch, 1983; Schäffner and Koch, 1987) but they do not appear to influence switching. Switching between chirp and trill occurs regularly in deafferented preparations with the same time course as in intact preparations (Fig. 7.2;

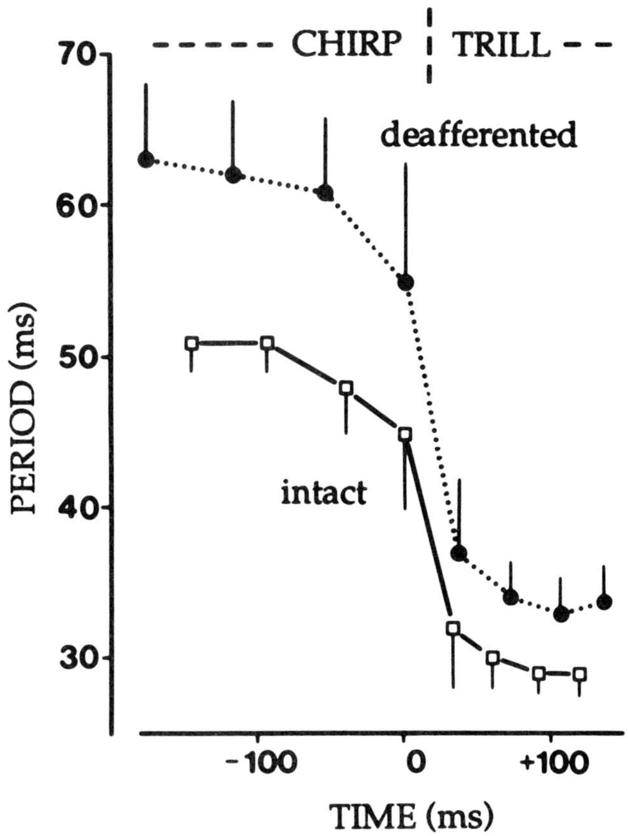

Fig. 7.2. Changes in cycle-period duration measured at the chirp/trill transition from sound recordings in intact animals and from nerve recordings in deafferented animals. Periods measured just prior to and after the transition from chirp to trill in intact and deafferented animals show that in both cases the same fast but continuous transition occurs. In deafferented preparations the rhythm is slower due to the lack of proprioceptive feedback (Hennig, 1989).

Hennig, 1989). Motoneurones also do not seem to influence the switching mechanism, which remains unperturbed even when the amplitude of bursts and the numbers of spikes generated in them vary (Hennig, 1989). This suggests that purely interneuronal mechanisms control the switching.

Bentley (1977) was able to elicit stridulation by electrical stimulation of single plurisegmental interneurones in the ventral nerve cord of the cricket *Teleogryllus oceanicus*, a species closely related to *T. commodus*, which also displays chirps and trills. Bentley (1977) observed normal switching from chirp to trill in his experiments, although the rate of stimulation was kept constant. The same effect is observed under constant DC stimulation

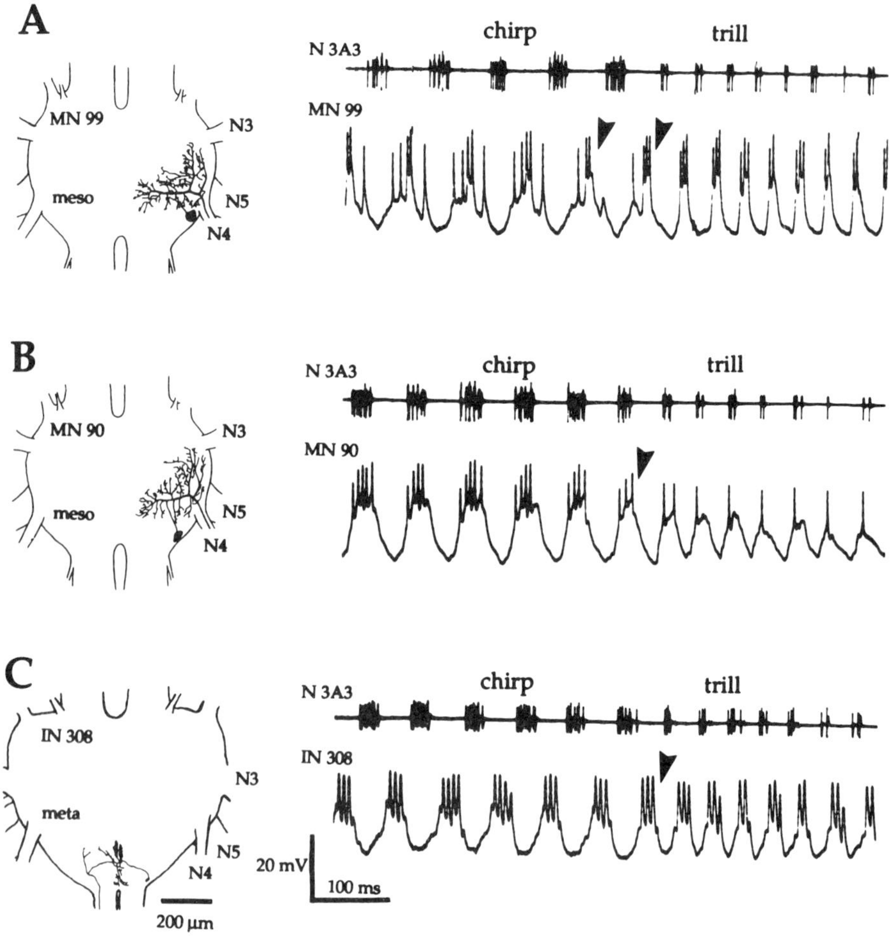

Fig. 7.3 The switch from chirp to trill as observed in intracellular recordings from motoneurones and interneurones. A plan drawing of the respective neurones in whole mount is given to the left of each recording trace. (A) Opener motoneurone 99 (lower trace) shows the switch from chirp to trill and the change in the burst shape (arrows). There appears to be different synaptic input during chirp and trill. The upper trace is an extracellular recording from nerve 3A3 of the closer activity of muscle 89. (B) The same change for the motoneurone 90 (closer) which is synergistically active with muscle 89 (see Fig. 7.1) and antagonistically active to opener motoneurone 99. The last chirp burst already shows a change in burst shape (see arrow; Hennig, 1989). (C) Interneurones also show different burst durations for chirp and trill and intermediate burst forms at the transition (see arrow), although the changes are not always as pronounced as seen in motoneuronal recordings (Hennig, unpublished).

of the brain with a suction electrode (Hennig, 1989). These observations suggest that switching is not a function of descending intersegmental interneurones, but may be a property of a stridulation pattern generating network.

The changes from chirp to trill largely reflect changes in cycle period (i.e. the reciprocal of the pulse rate; Fig. 7.2). A faster oscillation of the pattern generator during trill may automatically exclude some interneurones which are unable to follow a faster rhythm due to their slow integration or refractive times. Such a 'phasing out' of interneurones could be responsible for the change in synaptic input seen in motoneuronal recordings (Fig. 7.3A,B).

How circuit modification is achieved, possibly as a result of activity changes in just a few interneurones, remains to be established. Nevertheless, it appears likely that only one network generates the chirp and trill, and this network may also be responsible for the generation of the rhythms in aggressive and courtship song since in the cricket *Teleogryllus commodus* the chirp is a structure common to all three songs (Loher and Rence, 1978).

7.4. **Switching between stridulation and flight**

The switching between stridulation and flight is entirely different from that described for the chirp and trill of stridulation in the previous section (7.3). Whereas in stridulation virtually only one parameter of the motor pattern (i.e. the cycle period) changed during the continued performance of the *same* behaviour, in switching from stridulation to flight a switch *between* behaviours occurs involving not only changes in the rhythm and the phases of antagonistic muscle activity but also changes in wing position, wing movement, and wing coordination of both pairs of wings. Muscular activity switches from the mesothoracic segment during stridulation to both meso- and metathoracic segments during flight.

The switching between the two motor patterns may occur within 250 ms (Fig. 7.4C). After wind stimulus, the ongoing stridulatory motor pattern stops and the flight motor pattern starts. It is a common observation that flight overrides stridulation, whereas stridulation was never observed to switch off flight activity (Hennig, 1990b). Thus flight appears to be the dominant motor pattern, consistent with its important role in the escape behaviour of orthopteran insects. The main features distinguishing stridulation and flight are summarised schematically in Figs 7.1 and 7.5 (Hennig, 1990b). During stridulation the forewings are raised from their resting position and are rubbed rhythmically against one another in order to generate a specific pulse pattern (Figs 7.1C and 7.3); the hindwings remain stationary in their resting position. Thus recordings from hindwing nerves (meta; N1, Fig. 7.5, left-hand column) show no rhythmical activity

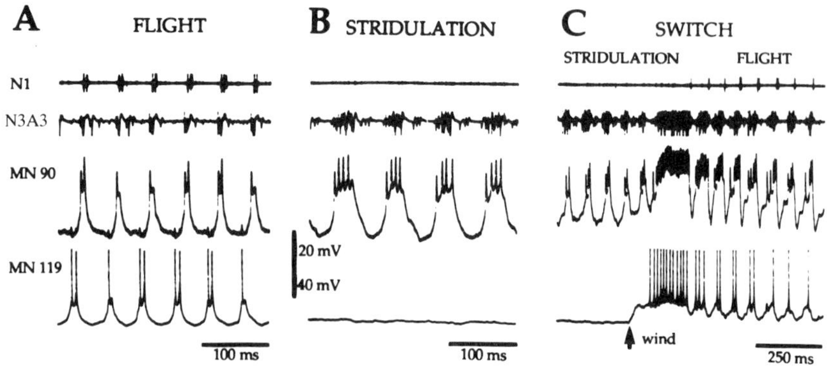

Fig. 7.4 Comparison of the activity patterns of serially homologous meso- and metathoracic motoneurones (MN 90 and MN 119) in stridulation and flight. (A) Both motoneurones function as elevators during flight, and their bursts are terminated when a depressor burst in the metathoracic dorsal longitudinal muscle (recorded on nerve 1) occurs. (B) During stridulation MN 90 shows strong oscillations in membrane potential while MN 119 displays no phasic synaptic pattern. (C) These differences in neuronal activity (A, B) become very obvious when flight is induced by a wind stimulus (see arrow) during ongoing stridulation: the motor pattern expressed by the animal switches from stridulation to flight and the synaptic input pattern to both motoneurones changes accordingly (Hennig, 1990b).

in the dorsal longitudinal hindwing muscle during stridulation. There is, however, bursting activity in nerve 3A3 (Figs 7.1C and 7.5) which is different from the pattern observed in this nerve during flight. During flight both pairs of wings are moved rhythmically up and down (Figs 7.1D and 7.5) and rhythmical activity may be recorded from selected nerves to the hind and forewing (meso: N3A3; meta: N1; Fig. 7.5 right-hand column).

7.4.1. *Recordings from motoneurones during stridulation and flight*

Some mesothoracic muscles are activated during stridulation *and* flight and are termed bifunctional muscles (Bentley, 1969a,b; Kutsch, 1969). Intracellular recordings from their motoneurones show that these are also bifunctional (Fig. 7.5A,B: MN 90 (remotor coxae: closer and elevator) and Fig. 7.5C,D: MN 99 (subalar: opener and depressor); Hennig, 1989, 1990a,b). However, these motoneurones receive different patterns of synaptic input during stridulation and flight as shown in Fig. 7.5. During stridulation MN 90 shows an i.p.s.p. followed by a strong excitation which then slowly decays (Fig. 7.5A) but during flight the depolarisation peaks in a shorter burst of spikes (Fig. 7.5B). During flight, MN 99 also shows a depolarisation supporting spikes which is followed by an i.p.s.p.

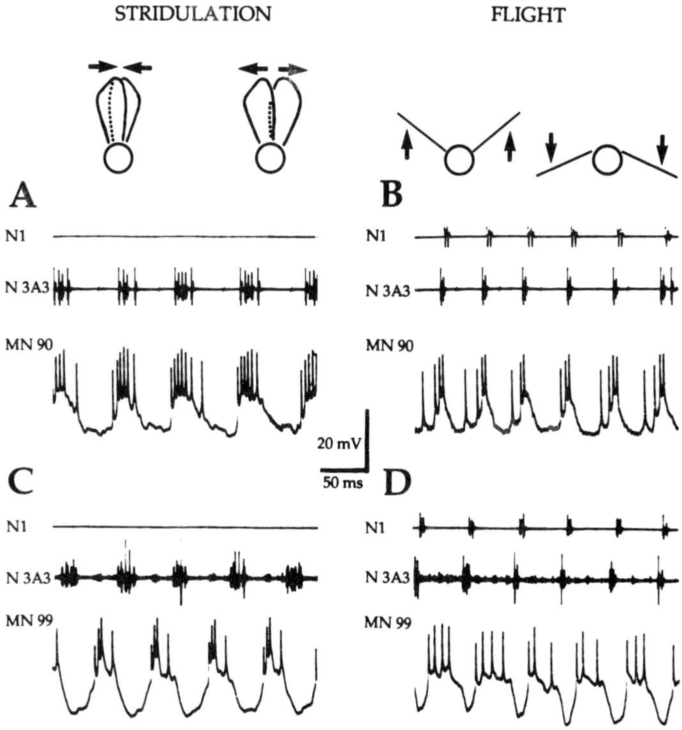

Fig. 7.5 Comparison of the discharge pattern of bifunctional mesothoracic motoneurones during stridulation (left column) and flight (right column); upper trace: N1 innervating the metathoracic dorsal longitudinal muscle (wing depressor during flight); middle trace: nerve 3A3 innervating elevator (closer) muscle 89 of the forewing; lower trace: intracellular recordings of mesothoracic motoneurones 90 and 99. Both MN 90 (A, B) and MN 99 (C, D) show different burst patterns during stridulation and flight (Hennig, 1989, 1990a,b).

approximately coincident with metathoracic DL activity (Fig. 7.5D). By contrast, during stridulation the depolarisation in MN 99 is faster and generates a spike doublet which is present in both the chirp and trill (Fig. 7.5C). These different patterns of synaptic input during stridulation and flight suggest that bifunctional motoneurones may be driven by largely different pools of premotor interneurones during the expression of the two behaviours (Hennig, 1990b).

7.4.2. *Recordings from interneurones during stridulation and flight*

Interneurones show depolarising oscillations in membrane potential which are in phase with either the stridulatory or the flight rhythm

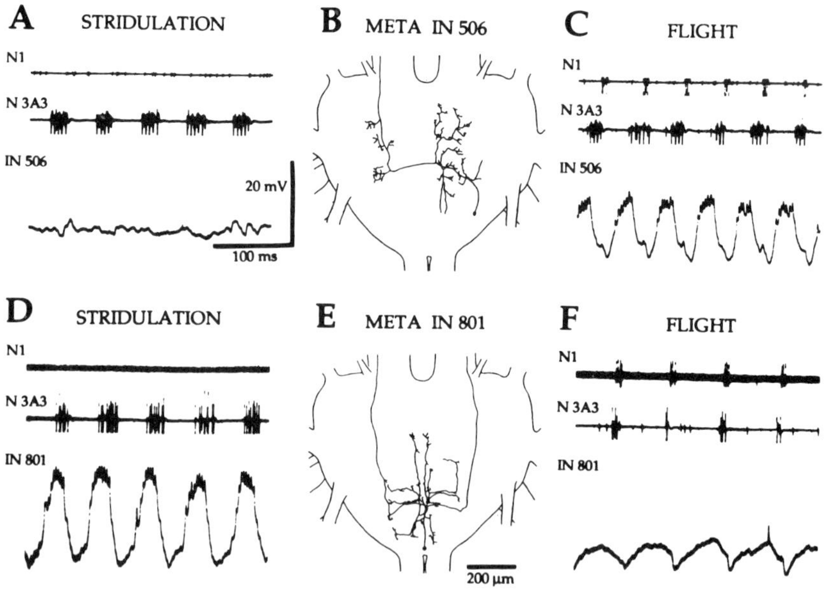

Fig. 7.6 Activity recorded in metathoracic interneurones (IN) and motor nerves (N1 and N3A3) during stridulation (left column) and flight (right column, Fig. 7.5) and the respective morphology of the interneurones as drawn from whole mounts. (A) During stridulation IN 506 receives no rhythmic excitatory input, but in some cases there occurs a weak phasic i.p.s.p. approximately in phase with the bursting indicating stridulation in nerve 3A3. (B) Morphology of IN 506 in the metathoracic ganglion. (C) IN 506 shows rhythmical bursting activity in phase with the flight rhythm. (D) During stridulation IN 801 shows strong oscillations in membrane potential approximately in phase with the closer burst recorded in nerve 3A3. (E) Morphology of IN 801 in the metathoracic ganglion. (F) IN 801 exhibits strong i.p.s.p.s approximately at the time of activation of the hindwing depressor muscle as indicated by activity in N1 (Hennig, 1990b).

(Hennig, 1990b). 'Flight' interneurones, including those capable of re-setting the flight rhythm, do not receive any rhythmical excitation while the stridulatory motor rhythm is being expressed. Instead a general inhibition and sometimes phase coupled i.p.s.p.s are seen (Fig. 7.6A, B,C). Similarly the 'stridulatory' interneurones are inhibited during flight activity (Fig. 7.6D,E,F). Phase coupled i.p.s.p.s (Fig. 7.6F) indicate that there is a mutual inhibition between the interneurones of the stridulatory and flight circuitry. It is important to note that both populations of interneurones occur in the meso- *and* the metathoracic ganglion. Thus, while switching from flight to stridulation involves switching segmental activity, the interneuronal control for both behaviours is distributed over both ganglia.

If there are indeed different populations of interneurones subserving each of the two patterns, then the motoneurones innervating muscles of the hindwing, active only during flight, should not receive excitatory input during stridulation. This is generally observed (Fig. 7.4; Hennig, 1990b), and sometimes an inhibition similar to that reported for interneurones may be seen. Therefore the metathoracic motoneurones do not seem to be connected to the circuitry organising the rhythm for stridulation, and this includes the motoneurones innervating muscle 112 (dorsal longitudinal, metathorax) which receive their synaptic input in the close vicinity of the mesothoracic motoneurones MN 90 and MN 99.

The likely existence of two populations of interneurones and the lack of excitatory input to metathoracic motoneurones during stridulation suggest that the central circuits for stridulation and flight are separate. If this is so, then simultaneous activation of both pathways should result in superposition of their synaptic input on to bifunctional motoneurones and the superposition of both rhythms. The sequence in Fig. 7.7A initially shows only flight activity which may be recognised from the rhythmical bursts in the extracellular nerve recording (metathoracic nerve 1D) and the concomitant i.p.s.p. in the intracellular record of MN 99. Stridulatory activity was evoked during flight activity, and this is recognisable by the high-frequency discharges, which are never seen during flight activity in this motoneurone. Then, as flight ends, the record shows only stridulatory activity. The depolarisations associated with stridulation are not precisely phase locked to the flight cycle indicated in Fig. 7.7C,D, although they do occur only during the excitatory bursts. The occurrence of superimposed rhythms in their outputs suggests that two functionally separate circuits for flight and stridulation are active at the same time (Hennig, 1990b).

There is also indirect supporting evidence for a separation of the circuitry generating the rhythm for stridulation and flight. Flight and stridulation cannot be evoked by stimulating the same sites in the brain; this suggests separate interneurones are involved in the initiation of each rhythm (Huber, 1960; Hennig, 1990b). Further, the same proprioceptors of the forewings are excited by wing movement during flight and stridulation (Möss, 1971), although their phases of activity and the magnitudes of their inputs must be different. However, although the information is carried centrally by the same afferents (only one in the case of the stretch receptor; Altman and Tyrer, 1977), the removal of proprioceptive feedback has entirely different effects on the two motor patterns: whereas stridulation is hardly perturbed (Bentley, 1969a; Kutsch and Huber, 1970; Hennig, 1989), flight may slow its rhythm by 50% (Robertson, 1987; Hennig, 1990a). Therefore peripheral information appears to be differently gated, supporting the idea of separate networks.

The evidence presented in this section suggests that there are separate neuronal networks which control stridulation and flight. Although it cannot be excluded with certainty that some interneurones may be shared

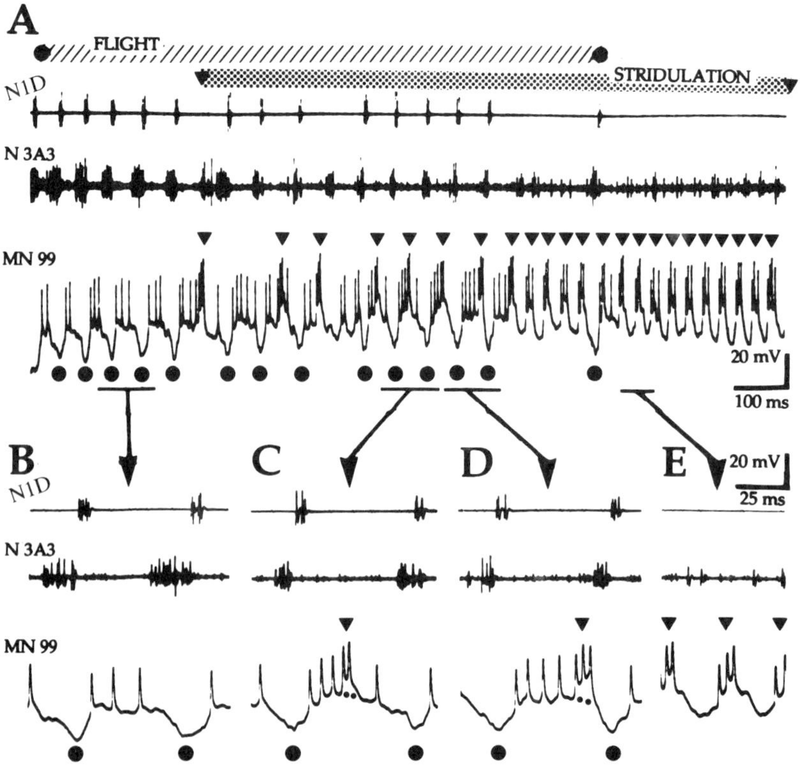

Fig. 7.7 The motor patterns for flight and stridulation may be expressed simultaneously. The recording shown is taken from motoneurone 99 (cf. Fig. 7.5 for a comparison of the synaptic input to this motoneurone during the expression of the two motor patterns; upper trace: metathoracic nerve 1D; middle trace: mesothoracic nerve 3A3; lower trace: intracellular recording of MN 99). (A) Simultaneous expression of flight and stridulatory activity in a continuous recording of motoneurone 99. Initially only the flight motor pattern is expressed as is apparent from the rhythmical activity in nerve 1D and the typical burst pattern in the intracellular recording of MN 99 (i.p.s.p.s coinciding with nerve 1D bursts are marked by a filled circle). Then the stridulatory motor pattern becomes superimposed on the flight motor pattern, as may be recognised from the typical phasic bursts (marked by filled triangles) which occur exclusively during stridulation in this motoneurone. The sequence ends with expression of the stridulatory rhythm alone after cessation of the flight rhythm (lack of bursting activity in the recording of metathoracic nerve 1D). (B, C, D, E) Enlarged bursts of selected sections of the recording in (A). (B) A flight cycle. (C) A section with flight and stridulatory inputs to MN 99. The stridulatory burst occurs halfway during the flight cycle. The normal flight input is indicated by the dotted line. (D) Here the stridulatory burst occurs at the end of the flight cycle. (E) A stridulatory cycle (Hennig, 1990b).

by both circuits, it is unlikely that such neurones have important roles in rhythm generation or initiation.

7.5. Conclusions — comparison of the switching patterns

In the two types of switching described here the same neuromuscular sets and the same appendages are involved but the types of switching and the likely neuronal mechanisms are quite different. The switching from chirp to trill is comparatively minor, since it occurs during the continued performance of the same behaviour and involves merely a change in cycle; i.e. the circuit simply runs faster. It is possible that additional interneurones also help speeding the rhythm (Hennig, 1989). In contrast, switching from stridulation to flight involves changes not only in period duration but also in wing position, coordination, segmental activity and motoneuronal and interneuronal activity patterns, and in the processing of peripheral feedback (Hennig, 1990b). Therefore the neuronal switching mechanisms for the two cases described here are likely to be quite different: a circuit modification might effect the change from chirp to trill; in contrast, switching between stridulation and flight may involve whole networks.

Different networks may compete for the final common motor pathway (scratch and flexion: Sherrington, 1947), or the same network may take on different configurations in order to fulfil different tasks (Getting and Dekin, 1985: polymorphic or multifunctional network). Two criteria can be used to distinguish between these two forms of organisation:

(a) *Sharing neurones*. The sharing of 'initiating' and/or 'resetting' interneurones by different motor patterns points to a multifunctional network organisation whereas separation of interneurones indicates distinct networks. Multifunctional networks have been conclusively demonstrated for the swimming and escape system of *Tritonia* (Getting and Dekin, 1985), for the stomatogastric rhythms of the lobster (Marder *et al.*, 1987; Selverston, 1988) and possibly for the shortening, crawling and swimming of the leech (Kristan *et al.*, 1988). In contrast, a separation of some interneurones was shown for the walking and flight system of the locust where some muscles are bifunctional (Ramirez and Pearson, 1988). In electric fish there are two different diencephalic systems forming distinct entities for different controls of pacemaker frequency (Heiligenberg, 1989). These systems are capable of producing the jamming avoidance response and chirping independently from each other while employing the same musculature (Kawasaki *et al.*, 1988; Heiligenberg, 1989). Separation of interneurones provides a strong indication of a separation of networks. The sharing of receptors and motoneurones is a less significant criterion, since two independent networks may well make different decisions given the same sensory receptors and motoneurones (see electric fish jamming avoidance response and chirping: Heiligenberg, 1987).

(b) *Simultaneous occurrence of two patterns*. This is a good indication for the presence of two distinct networks. It was observed for the bifunctional motoneurones of the locust flight and walking system (Ramirez and Pearson, 1988) as well as the jamming avoidance response and chirping (both activating the pacemaker nucleus which drives the electric organ discharge) of electric fish (Heiligenberg, 1989). On the other hand, it was not reported for *Tritonia* swimming (Getting, 1983; Getting and Dekin, 1985), the stomatogastric network of lobster (Marder *et al.*, 1987; Selverston, 1988), and the leech motor patterns (Kristan *et al.*, 1988). However, in those cases of simultaneous expression of both motor patterns it still has to be demonstrated that the same oscillator does not generate both rhythms, e.g. by demonstrating the absence of a strict phase coupling between the two rhythms.

These two criteria might help to assess the form of neuronal organisation without having to record from every neurone involved in the pattern generation of both rhythms. They are applicable to the two behavioural switches demonstrated in this study:

(i) The chirp and the trill may be viewed as changes in the output of a polymorphic network, since interneurones are apparently shared and the two rhythms do not occur simultaneously.
(ii) Stridulation and flight may be viewed as outputs of two separate networks, since interneurones are apparently not shared and the two rhythms may occur simultaneously.

What are the advantages of each of the two forms of organisation? Two major factors for the design of neuronal organisation are the *optimised function* (as indirectly selected for by behavioural performance), and *the evolutionary history* (Dumont and Robertson, 1986; Heiligenberg, 1987). Considering optimised function: the strength of a multifunctional network might be the flexibility of the different outputs it can generate. This organisation allows for adjustments, transitional and intermediate movements, and 'new' adaptive forms of motor patterns. Such flexibility appears to be present in a number of systems where multifunctional networks appear to account for a range of movements, e.g. escape and swimming in *Tritonia* (Getting and Dekin, 1985); scratching and walking in the cat (Berkinblit *et al.*, 1978); stomatogastric rhythms in the lobster (Marder *et al.*, 1987); and different forms of stridulation in the cricket (Loher and Rence, 1978). On the other hand flawless execution of movements which use the same limb (appendage) in different and mutually exclusive forms of coordination may warrant a largely separate neuronal control so as to avoid interference. This is likely for cricket stridulation and flight (Hennig, 1990b). A further major determinant of neuronal organisation is evolutionary history: the proposed organisation of separate circuits for flight and stridulation in the cricket presents an unexpected difficulty in terms of evolution. It is generally accepted that flight in insects evolved monophyletically (Kukalova-Peck, 1983) and stridulatory behaviour followed 'later', as suggested by the different mechanisms of sound production in orthopteran insects (Sharov, 1971). Any suggested preadaptive function for 'stridulation' in crickets would have to assume wing movement, and the only neural circuit known which was capable of operating the wings at that time in evolution is the flight circuit. Thus, it is thought that the evolution of stridulation is behaviourally and neurónally related to flight (Zeuner, 1939; Alexander, 1962; Huber, 1962; Bentley and Kutsch, 1966). It therefore comes as a surprise that the neural circuits controlling stridulation and flight should be functionally independent. However, flexibility and transitions between the two behaviours are not needed, if not actually undesirable. Furthermore, one has to bear in mind that stridulation and flight are performed during different internal and external states (e.g. mating versus migration or escape) of the animal, involving very different sets of stimuli for the generation of each motor pattern. This separation of the performance of the two motor patterns in time might favour a separation of the neural control mechanisms, and crickets have had 200 million years of evolution to perfect it.

7.6. **Summary**

The mechanism of switching between motor patterns involving the same effector organ is examined. Calling song stridulation and flight in the cricket both involve the forewings. Switching between two rhythms during the generation of one behaviour (stridulation) is considered first. During stridulation the calling song pattern switches between two rhythms, the slow chirp and the fast trill. Pattern generating interneurones are most likely to be responsible for this switch. Secondly, switching between two different behaviours, flight and stridulation, is described. Although the same set of muscles moves the forewings in both motor patterns, there appear to be two separate central neural networks which generate these distinct motor patterns, and switching between these behaviours involves switching between these two networks.

Acknowledgements

I thank Drs G. S. Boyan, E. E. Ball and E. J. Warrant for their valuable comments on the manuscript.

References

Alexander, D. R. (1962) Evolutionary change in cricket acoustical communication. *Evolution* **16**: 443–467.

Altman, J. A. and Tyrer, N. M. (1977) The locust wing hinge stretch receptors: I. Primary sensory neurons with enormous central arborizations. *J. Comp. Neurol.* **172**: 409–430.

Bellman, K. L. and Krasne, F. B. (1983) Adaptive complexity of interactions between feeding and escape in crayfish. *Science* **221**: 779–781.

Bentley, D. R. (1969a) Intracellular activity in cricket neurons during generation of song patterns. *Z. vergl. Physiol.* **62**: 267–283.

Bentley, D. R. (1969b) Intracellular activity in cricket neurons during the generation of behaviour patterns. *J. Insect Physiol.* **15**: 677–699.

Bentley, D. R. (1971) Genetic control of an insect neuronal network. *Science* **174**: 1139–1141.

Bentley, D. R. (1977) Control of cricket song patterns by descending interneurons. *J. Comp. Physiol.* **116**: 19–38.

Bentley, D. R. and Hoy, R. R. (1972) Genetic control of the neuronal network generating cricket (*Teleogryllus*, *Gryllus*) song patterns. *Anim. Behav.* **20**: 478–492.

Bentley, D. R. and Kutsch, W. (1966) The neuromuscular mechanism of stridulation in crickets (Orthoptera: Gryllidae). *J. exp. Biol.* **45**: 151–164.

Berkinblit, M. B., Deliagina, T. G., Feldman, A. G., Gelfand, I. M. and Orlovsky, G. N. (1978) Generation of scratching. II. Nonregular regimes of generation. *J. Neurophysiol.* **41**: 1058–1069.

Davis, W. J., Mpitsos, G. J. and Pinneo, J. M. (1974) The behavioral hierarchy of the mollusk *Pleurobranchea*. I. The dominant position of the feeding behaviour. *J. Comp. Physiol.* **90**: 207–224.

Dumont, J. P. C. and Robertson, R. M. (1986) Neuronal circuits: an evolutionary perspective. *Science* **233**: 849–853.

Elliott, C. J. H. (1983) Wing hair plates in crickets: physiological characteristics and connections with stridulatory motor neurons. *J. exp. Biol.* **107**: 21–47.

Elliott, C. J. H. and Koch, U. T. (1983) Sensory feedback stabilizing reliable stridulation in the field cricket *Gryllus campestris* L. *Anim. Behav.* **31**: 887–901.

Getting, P. A. (1983) Neural control of swimming in *Tritonia*. In: *Neural Origin of Rhythmic Movements*, A. Roberts and B. L. Roberts eds. Cambridge University Press, Cambridge, pp. 89–128.

Getting, P. A. and Dekin, M. S. (1985) *Tritonia* swimming: a model system for integration within rhythmic motor systems. In: *Model Neural Networks and Behavior*, A. I. Selverston, ed. Plenum Press, New York, pp. 3–20.

Grillner, S. (1975) Locomotion in vertebrates: central mechanisms and reflex interaction. *Physiol. Rev.* **55**: 247–304.

Heiligenberg, W. (1987) Central processing of sensory information in electric fish. *J. Comp. Physiol. A* **161**: 621–631.

Heiligenberg, W. (1989) Motor program selection in gymnotiform electric fish: separate representations and shared circuits. In: *Neural Mechanisms of Behavior*, J. Erber, R. Menzel, H.-J. Pflüger and D. Todt, eds. Paul Parey, Hamburg, pp. 180–181.

Hennig, R. M. (1989) Neuromuscular activity during stridulation in the cricket *Teleogryllus commodus*. *J. Comp. Physiol. A* **165**: 837–846.

Hennig, R. M. (1990a) Neuronal organisation of the flight motor pattern in the cricket, *Teleogryllus commodus*. *J. Comp. Physiol. A* **167**: 629–639.

Hennig, R. M. (1990b) Neuronal control of the forewings in two different behaviours: stridulation and flight in the cricket, *Teleogryllus commodus*. *J. Comp. Physiol. A* **167**: 617–627.
Hill, K. G., Loftus-Hills, J. J. and Gartside, D. F. (1972) Pre-mating isolation between the Australian field crickets *Teleogryllus commodus* and *T. oceanicus* (Orthoptera: Gryllidae). *Aust. J. Zool.* **20**: 153–163.
Huber, F. (1960) Untersuchungen über die Funktion des Zentralnervensystems und insbesondere des Gehirnes bei der Fortbewegung und der Lauterzeugung der Grillen. *Z. vergl. Physiol.* **44**: 60–132.
Huber, F. (1962) Central nervous control of sound production in crickets and some speculations on its evolution. *Evolution* **16**: 429–442.
Huber, F. (1963) The role of the central nervous system in Orthoptera during the co-ordination and control of stridulation. In: *Acoustic Behaviour of Animals*, R. G. Busnel, ed. Elsevier, Amsterdam, pp. 440–488.
Kawasaki, M., Maler. L., Rose. G. J. and Heiligenberg, W. (1988) Anatomical and functional organisation of the prepacemaker nucleus in gymnotiform electric fish: the accommodation of two behaviors in one nucleus. *J. Comp. Neurol.* **276**: 113–131.
Koch, U. T., Elliott, C. J. H., Schaeffner, K. H. and Kleindienst, H. U. (1988) The mechanics of stridulation of the cricket *Gryllus campestris*. *J. Comp. Physiol. A* **162**: 213–223.
Kristan, W. B., Wittenberg, G., Nusbaum, M. P. and Stern-Tomlinson, W. (1988) Multifunctional interneurons in behavioral cricuits of the medicinal leech. *Experientia* **44**: 383–389.
Kukalova-Peck, J. (1983) Origin of the insect wing and wing articulation from the arthropodan leg. *Can. J. Zool.* **61**: 1618–1669.
Kutsch, W. (1969) Neuromuskuläre Aktivität bei verschiedenen Verhaltensweisen von drei Grillenarten. *Z. vergl. Physiol.* **63**: 335–378.
Kutsch, W. and Huber, F. (1970) Zentrale versus periphere Kontrolle des Gesanges von Grillen (*Gryllus campestris*). *Z. vergl. Physiol.* **67**: 140–159.
Loher, W. and Rence, B. (1978) The mating behavior of *Teleogryllus commodus* (Walker) and its central and peripheral control. *Z. Tierpsychol.* **46**: 225–259.
McFarland, D. J. (1977) Decision making in animals. *Nature* **269**: 15–21.
Marder, E. E., Hooper, S. L. and Eisen, J. S. (1987) Multiple neurotransmitters provide a mechanism for the production of multiple outputs from a single neuronal circuit. In: *Synaptic Mechanisms*, G. M. Edelman, W. E. Gall and W. M. Cowan, eds. Wiley, New York, pp. 305–327.
Möss, D. (1971) Sinnesorgane im Bereich des Flügels der Feldgrille (*Gryllus campestris* L.) und ihre Bedeutung für die Kontrolle der Singbewegung und die Einstellung der Flügellage. *Z. vergl. Physiol.* **73**: 53–83.
Nocke, H. (1971) Biophysik der Schallerzeugung durch die Vorderflügel der Grillen. *Z. vergl. Physiol.* **74**: 272–314.
Ramirez, J. M. and Pearson, K. G. (1988) Generation of motor patterns for walking and flight in motoneurons supplying bifunctional muscles in the locust. *J. Neurobiol.* **19**: 257–282.
Robertson, R. M. (1987) Interneurons in the flight system of the cricket *Teleogryllus oceanicus*. *J. Comp. Physiol. A* **160**: 431–445.
Schäffner, K. H. and Koch, U. T. (1987) Effects of wing campaniform sensilla lesions on stridulation in crickets. *J. exp. Biol.* **129**: 25–40.
Selverston, A. I. (1988) Switching among functional states by means of neuromodulators in the lobster stomatogastric ganglion. *Experientia* **44**: 376–383.
Sharov, A. G. (1971) *Phylogeny of the Orthopteroidea*. Israel Program for Scientific Translations, Jerusalem.

Sherrington, C. S. (1947) *The Integrative Action of the Nervous System*, Yale University Press, New Haven (1st edn 1906).

Wang, S. and Robertson, R. M. (1989) Morphological study of flight motor neurons in the cricket. *J. Comp. Neurol.* **279**: 272–280.

Zeuner, F. E. (1939) *Fossil Orthoptera Ensifera*. Brit. Mus. (Nat. Hist.), London, pp. 1–321.

8 *Barry L. Roberts and Willem Mos*

Neural circuits for speed change in swimming fish

8.1. Introduction

Much animal behaviour is based on rhythmically repeated or episodic movements and postures, and so the first steps in understanding how behaviour results from the operation of the nervous system have involved the identification and characterisation of the sensory, neural and effector systems participating in movement. Attention can then be focused on the complex question of how the complete behavioural repertoire of an animal is represented within the nervous system. For this we need to understand how the neural networks that generate the different motor programmes are organised and controlled and how their outputs are placed in appropriate sequences.

Animals seldom move at constant speed for any length of time, but rather change direction and velocity frequently. These transitions take place smoothly even if they involve multiple programmes: terrestrial animals, for example, vary their speed by using 'gaits' (e.g. walk, gallop) which have distinct muscle activation patterns and so may involve different motor programmes. In contrast, the forward velocity of a fish during steady swimming is primarily determined by the frequency of the tail beat (Bainbridge, 1958) and, consequently, by varying just this single parameter, fishes can swim at speeds varying over a wide range. It is interesting, therefore, to consider how the locomotory programmes underlying swimming in fishes are organised and adapted to cope with speed variations. We examine in detail the two neuromuscular systems used for swimming at different speeds, and will discuss possible interrelations between them.

8.2. Swimming patterns and their muscular basis

The majority of fish use oscillatory movements of the body and tail for swimming, and the speeds at which they can do so continually without fatigue ('cruising') are usually lower than half the maximum speed they can reach. They exhaust rapidly when swimming faster than their maximum cruising speed, which is about three body lengths per second (L/s) for cod and 6 L/s for mackerel (Videler, 1985), Higher speeds — Wardle (1975) gives a maximum of 26 L/s for sprat — can be reached but only for

Fig. 8.1. Superimposed cinephotographic images of swimming fish, showing the two basic patterns of movement. (A) Mackerel, 30 cm long, swimming steadily from left to right at 5.1 L/s; frames 10 ms apart (from Videler, 1985). (B) Trout, 28 cm long, during a startle response; frames every 5 ms (from Bosch, unpublished).

short periods and so as an energy-saving strategy some fish use a 'kick and glide' ('burst and coast') movement when swimming at high velocities, by making brief accelerations interspersed with no active body contraction (Videler and Weihs, 1982). The fastest movements of all, made during rapid turns, attacks or escapes, are always of brief duration. The swimming movements of fishes can therefore be considered as being of two basic patterns: (1) steady, continuous movement in which the body is bent periodically into a propulsive wave which passes backwards at a velocity that is close to the swimming speed, and (2) rapid, transient movement, such as accelerations and fast turns, involving large-amplitude movements (Fig. 8.1). The transient movements may be preceded and followed by steady movement whereupon the two locomotory patterns are combined into a smooth continuous sequence by the swimming fish.

Electromyographic recordings from steadily swimming fish have shown that the basis for the propulsive wave is a regular activity of the myotomal musculature. Muscle fibres on each side of the body at the same segmental level are alternately active and myotomes on the same side of the body contract at different times in a rostrocaudal sequence, so establishing a contraction wave. In contrast, many transient movements involve an effectively synchronous activation of the musculature on one side of the body (Nissarov and Eaton, 1989; van Leeuwen *et al.*, 1990).

The two basic movement patterns involve the activity of separate divisions of the myotomal musculature that contain muscle fibres with

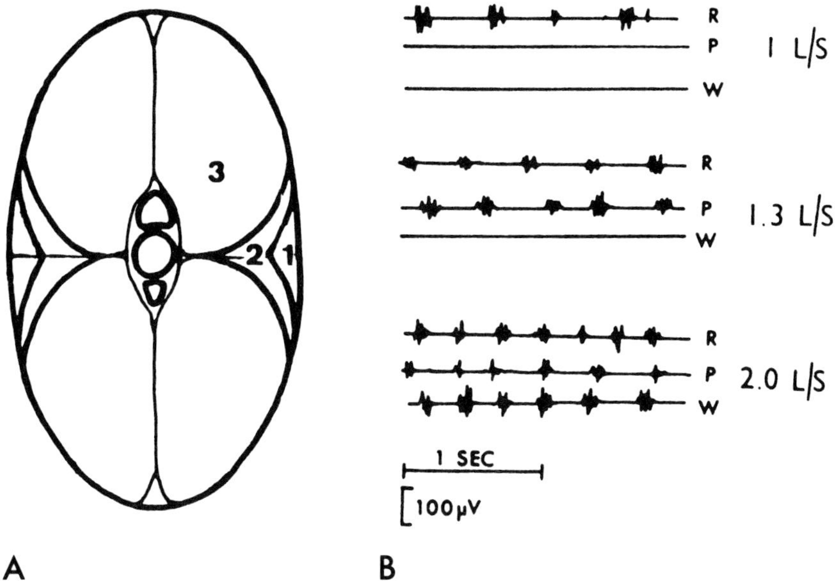

Fig. 8.2. Morphology and activity of fish myotomal muscle. (A) Diagram of a cross-section through the tail musculature of the zebrafish, showing the position of red (1), intermediate or pink (2), and white (3) muscle fibres (modified from van Raamsdonk *et al.*, 1982). (B) Activity of the red (R), pink (P), and white (W) muscle divisions of the myotomal muscle in the carp at increasing swimming speeds (lengths/s; from Johnston *et al.*, 1977).

distinctive properties (see Bone, 1978). Sustained swimming utilises the slowly contracting red muscle fibres that lie superficial to the mass of the white musculature which is used for transient movements and the fibres of which contract fast and fatigue rapidly. The two muscle divisions can often be subdivided and other fibre types can be recognised, e.g. intermediate fibres which have a position and metabolic properties that are intermediate between those of the red and the white fibres (van Raamsdonk *et al.*, 1982, Fig. 8.2A).

At low swimming speeds only the red muscle fibres are active, as was demonstrated in the electromyographic study of Bone (1966) for the spinal dogfish preparation (see below), and this fundamental observation has been confirmed for this preparation by subsequent workers (Roberts, 1969; Grillner, 1974; Williamson and Roberts, 1980). In the same preparation the white musculature is active only during rapid movements such as those evoked by cutaneous stimuli. A similar sharp division between the red muscle system used for cruising swimming over a range of speeds and the white muscle fibres that are active only for short-duration very rapid locomotion has also been observed in the herring

(Bone *et al.*, 1978). In the carp, slow swimming speeds are based on red muscle fibre activity but, as the swimming speed increases, the intermediate fibres also become active (Johnston *et al.*, 1977; Fig. 8.2B). The highest sustainable speeds in this species involve periodic accelerations, produced by intermittent activity of the fast-fatiguing white fibres (Fig. 8.2B) that are superimposed upon the continuing alternating activity of the red and intermediate fibres (Rome *et al.*, 1985). However, in other teleost fishes such as the trout, the white system becomes active at intermediate speeds of sustained swimming (Hudson, 1973) although this activity may involve specialised small muscle fibres with high levels of oxidative enzymes that are found within the white muscle system of this fish.

In summary we can say that the red fibres are designed for use in sustained swimming and that, although white muscle may be active for shorter or longer periods during sustained swimming, its primary role is in transient movement, namely in the execution of the first stroke in kick and glide movement (van Leeuwen *et al.*, 1990) and in rapid turns and escapes (Nissarov and Eaton, 1989).

8.3. **Muscle fibre innervation and motoneuronal organisation**

The major muscle divisions differ not only in the way they are employed during swimming but also, in many species, in the way they are innervated. The red fibres receive a polyneuronal and multiple innervation over the whole length of their membranes as do white fibres in most teleost fishes. However, in some teleost fishes such as the herring and eel and in all elasmobranchs, the white fibres are innervated focally at a single end plate (Bone, 1978). Focally innervated white muscle fibres could be expected to operate in the same manner as in other vertebrates with propagated action potentials, but the electrophysiological data on multiply innervated fish muscle fibres (reviewed by Mos *et al.*, 1988) suggest that their ability to generate action potentials is limited and that muscle action potentials are not essential for the mechanical performance. Because of this innervation pattern the red fibres have been assumed to operate by producing junctional potentials (JPs) at each motor synapse and to evoke graded local contractions, and multiply innervated white fibres were thought to produce JPs at low swimming speeds and action potentials during faster swimming (Bone, 1978). But the story is evidently not so simple and may need revision in the light of two recent reports. Altringham and Johnston (1988) have shown that both red and white myotomal fibres of *Myoxocephalus* produce action potentials even though their innervation is distributed, and Hidaka and Miyahara (1987) have observed hyperpolarising potentials in muscle fibres of the carp in response to spinal nerve stimulation, a result that implies that some of the input to the muscle fibres may be inhibitory. Consequently the significance of the multiple innervation is at present unclear but it lies perhaps in the fact that the innervation is provided by several motoneurones (i.e. it is polyneuronal). A single white muscle fibre in the sculpin (*Cottus*), for example, is innervated by four different spinal nerves (Hudson, 1969) and such a wide innervation would provide the opportunity for almost simultaneous multisegmental contraction which could be an important adaptation in those fishes that utilise high-speed escape movements.

All divisions of the myotomal musculature are innervated by spinal motoneurones, the somata of which are distributed in the ventral horn in a pattern that can be related to their targets. De Graaf *et al.* (1990) labelled motoneurones from small injections of horseradish peroxidase made into portions of the musculature of the zebrafish, and found that

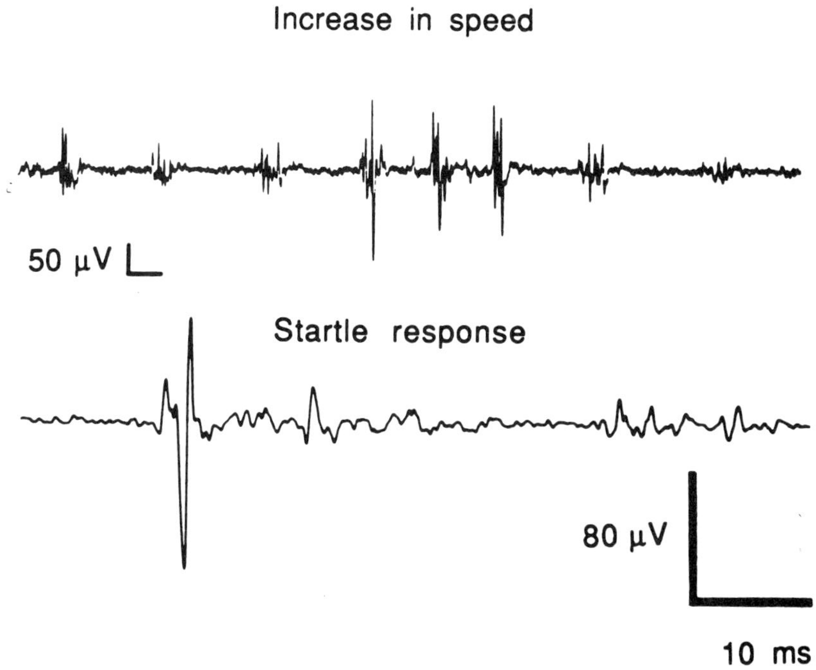

Fig. 8.3. Electromyographic recordings from the myotomal muscle of a restrained zebrafish during steady swimming as the tail-beat frequency increases and during a startle response. The large arrows indicate potentials assumed to arise from primary motoneurone activation of the musculature (from Liu and Westerfield, 1988).

the motoneurones innervating the different muscle divisions have somata that overlap in size and position. Nevertheless, on average, the motoneurones innervating the white muscle are the largest and lie the most dorsomedially in the motor column, whereas those innervating red muscle are the smallest and lie the most ventrolaterally. Comparable differences in size and distribution of the spinal motoneurones in relation to the musculature have been observed in dogfish (Mos and Williamson, 1986) and in goldfish (Fetcho, 1986), and this motoneurone topography is probably a standard feature of the spinal cord of fishes.

The largest, most dorsal motoneurones appear first in development ('primary' motoneurones) and are sufficiently distinctive to form an identifiable class that is recognisable from fish to fish. There are three such neurones in each spinal cord hemisegment in the zebrafish (Myers, 1985), eight to twelve in the goldfish (Fetcho, 1986) and seven to ten in the eel (Smit *et al.*, 1991). Each white muscle fibre in the zebrafish, as well as being innervated by several secondary motoneurones, is supplied by one of the three primary motoneurones that have been shown by Liu and Westerfield (1988) to become active at the faster swimming speeds (Fig. 8.3).

8.4. **The role of the spinal cord in movement**

It has been recognised for a long time that spinal dogfishes (*Scyliorhinus, Squalus*), in which the spinal cord has been completely severed close to the brain, perform continuous body movements that are similar in basic form, although slower and weaker, than those made by the free-swimming intact fish (e.g. Gray and Sand, 1936). Only a few other fish species, e.g. chronically spinal *Conger* (Gray and Sand, 1936) and *Anguilla* (von Holst, 1935), show such regular movements after the spinal cord has been transected, although it is possible to evoke some form of regular movement in spinal preparations of other species by pharmacological treatment (e.g. lamprey, Poon, 1980; stingray, Williams *et al.*, 1981). These observations indicate that the basic neuronal circuitry for producing locomotion in fishes is present within the spinal cord.

The movements of the 'swimming' spinal *Scyliorhinus* are very steady (on average 0.55 Hz, i.e. 1.8 s for a cycle period) and involve the regular activity of red muscle fibres (Bone, 1966; Williamson and Roberts, 1980) and certain motoneurones (Fig. 8.6A, Williamson and Roberts, 1986). The motoneurones discharge rhythmically (Fig. 8.6B) even after the musculature has been paralysed by an intravenous injection of curare which will also abolish any movement-related sensory input. Such persistent 'fictive' swimming has been observed in several other species (*Squalus*, Grillner *et al.*, 1976; *Dasyatis*, Droge and Leonard, 1983; lamprey, Cohen and Wallén, 1980; Poon, 1980).

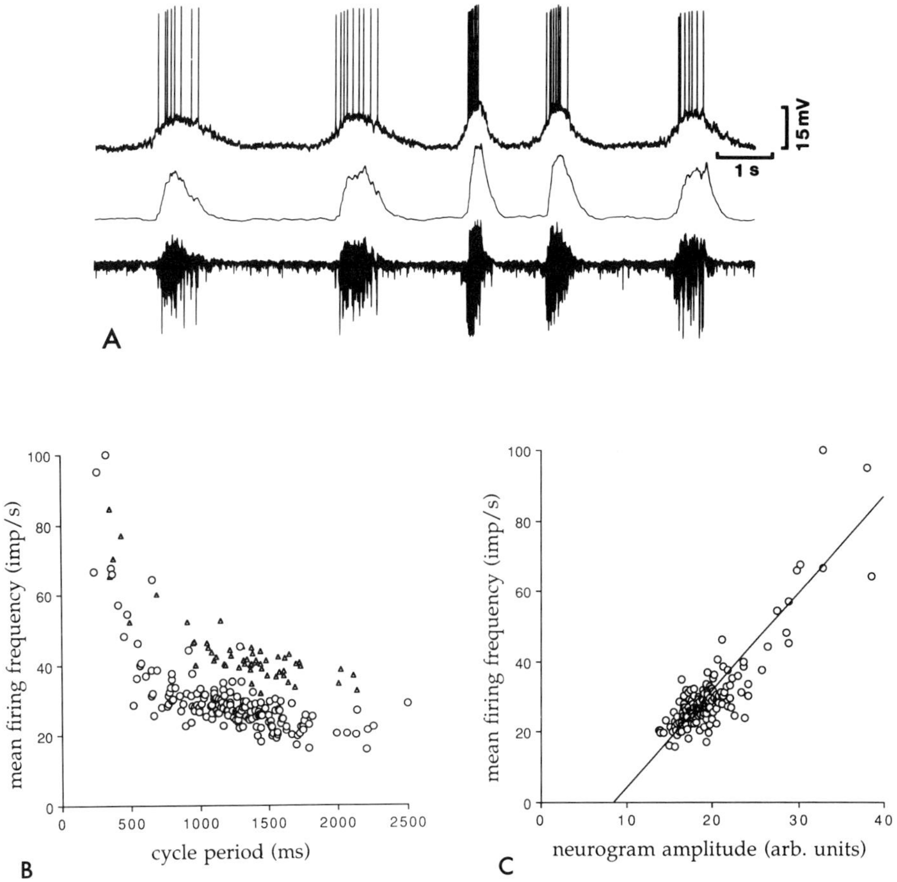

Fig. 8.4. Motoneuronal activity in the paralysed spinal dogfish (*Scyliorhinus*) during fictive swimming. (A) Simultaneous recordings from a motoneurone displaying bursts of action potentials superimposed upon e.p.s.p.s (upper trace) that are in phase with the activity recorded from the motor nerve of the same spinal cord segment (bottom trace). Middle trace: the rectified and filtered nerve signal, which shows larger amplitudes for the motoneuronal activity during shorter cycle periods of the motor rhythm, and resembles the shape of the e.p.s.p.s found in individual motoneurones indicating a common excitatory drive to these neurones. (B) Firing frequencies, averaged per motoneuronal burst, of two motoneurones which were recorded from the same spinal cord segment, plotted against the instantaneous cycle period, which is defined as the time between the onsets of two successive motor bursts. The two neurones show the same firing pattern, increasing their impulse frequency with decreasing cycle period. (C) Mean firing frequency of a motoneurone plotted against the amplitude of the rectified and filtered signal obtained from the motor nerve for 185 consecutive motor cycles ($R = 0.60$). Because these motoneurones are active also at long cycle periods during which in the unparalysed preparation no EMG activity from white muscle is recorded, we assume that they innervate the red musculature. (All data from Mos, Roberts and Williamson, 1990a.)

Microelectrode recordings from fictively swimming spinal *Scyliorhinus* indicate that only about 25% of the motoneurone population is regularly active although some of the inactive motoneurones will discharge in response to strong cutaneous stimulation. The active motoneurones, we believe, innervate the red muscle fibres because they discharge at the long cycle periods when only the red musculature is active in the swimming animal (Mos *et al.*, 1990a). They remain active as the cycle period of the rhythm changes, firing at higher frequency but for shorter periods as the cycle periods shorten (Fig. 8.4A). Although differing in their individual levels of firing (Fig. 8.4B), these neurones form a homogeneous population, changing their firing frequencies in ensemble. This can be seen when the firing frequencies of individual motoneurones recorded from the same spinal cord segment are plotted against the cycle period of the motor rhythm (Fig. 8.4B) and when the firing frequency of single motoneurones is plotted against the activity of the whole population as measured from the integrated whole nerve record (Fig. 8.4C).

The order in which the different muscle divisions are recruited — red for continuous and white for transient movement — correlates with the size ranges for the motoneurones that innervate the muscle divisions and suggests that the motoneurones could be activated in order of size. This would be in line with the 'size principle' formulated for mammalian motoneurones according to which the smallest motoneurones are the first to discharge when in receipt of a homogeneously distributed synaptic input of increasing strength. However, it is unlikely that this is the main way recruitment order is determined in fishes because the size distributions of the motoneurones overlap considerably. It is more likely that the different sets of motoneurones either receive a common excitatory interneuronal input that differs in strength across the motor pool or that they are driven by separate interneuronal circuits.

In any neural circuit the interneurones are usually the hardest to investigate, and we have very incomplete information about the interneuronal components of most locomotor circuits, particularly for vertebrate preparations. Only for the lamprey (Grillner *et al.*, 1986) and frog tadpole (Roberts *et al.*, 1986) have excitatory and inhibitory interneurones that are involved in motor output production been well characterised.

Many spinal interneurones in the paralysed spinal dogfish are active in association with fictive swimming and they can be categorised into three groups according to their discharge patterns (Mos, Roberts and Williamson, 1990b). Group I (about 30% of the active interneuronal population) comprises neurones that fire at a more or less constant frequency and are sometimes modulated in time with the motor output. However, their firing levels are unrelated to the frequency of the motor rhythm and, when this stops, these neurones continue to discharge. The interneurones belonging to group II (around 50%) discharge periodically in time with,

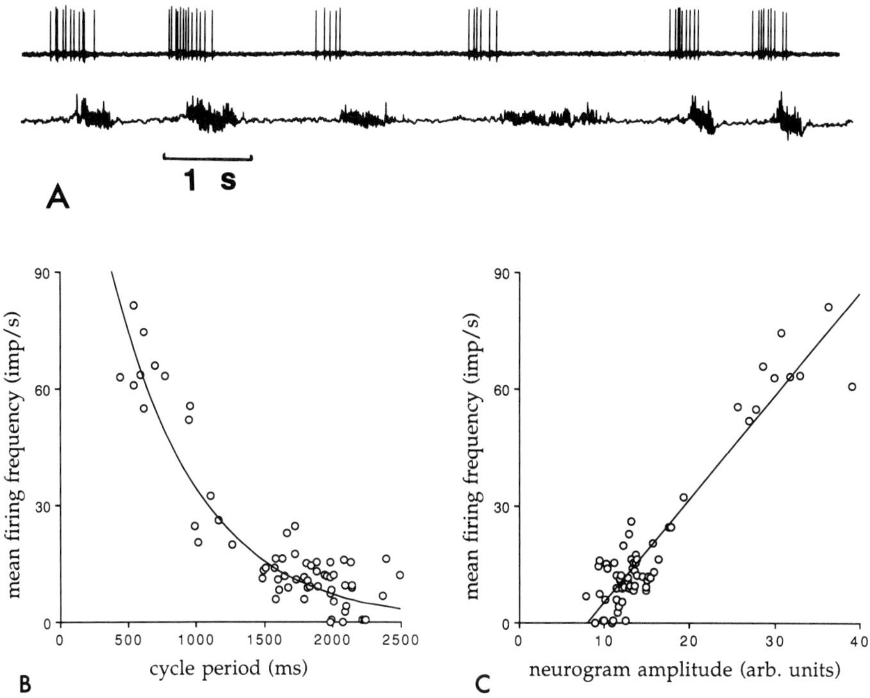

Fig. 8.5. Interneuronal activity in the paralysed spinal dogfish (*Scyliorhinus*) during fictive swimming. (A) Simultaneous recordings from an interneurone (upper trace) and the ipsilateral motor nerve (bottom trace). The interneurone fires at a fixed phase during the motor cycle, just before and during the first part of the motor burst. (B) Mean firing frequency during a burst of the interneurone shown in (A) plotted against the period of the cycle that starts during the firing of this neurone. The firing pattern is similar to that seen for motoneurones: firing frequency increases as cycles shorten. At slow motor rhythms (long cycles) this neurone failed to be recruited or fired just one spike (frequencies plotted as 0 or 0.5 respectively). (C) Mean firing frequency of the same interneurone plotted against motoneuronal activity, determined from the nerve recording as described in Fig. 8.4. ($R = 0.94$, frequency values 0 and 0.5 not included). All data are from Mos, Roberts and Williamson (1990b).

and at a fixed phase relationship to, the motor rhythm (Fig. 8.5A). Their behaviour with respect to changes in the motor rhythm is very similar to that of the red muscle-innervating motoneurones (Figs. 8.4B,C; 8.5B,C) and, when they stop discharging, the motor rhythm ceases. Group III neurones have intermediate firing patterns, discharging throughout the cycle of the motor rhythm just like group I neurones but increasing their discharge levels as the motor rhythm becomes faster, as do group II neurones. The major change seen in the behaviour of the interneurones as the fictive rhythm varies is an increase or decrease in the firing frequencies of the group II and III interneurones, suggesting that a network of these neurones could provide the appropriate drive to red muscle-innervating motoneurones at a variety of speeds of fictive swimming.

Further details of the interneuronal circuitry for *Scyliorhinus* are unknown, but much information is now available for the lamprey, the isolated spinal cord of which will generate organised rhythmical activity if it is placed in a bathing solution containing excitatory amino acids (Poon, 1980). The motoneurones supplying both the slow and fast muscle fibres receive phasic excitation and inhibition but only the motoneurones that are thought to innervate slow muscles discharge action potentials when the cycle period of the motor rhythm is long. The phasic excitation depends on the activation of kainate and N-methyl-D-aspartate (NMDA) receptors (Dale, 1986; Dale and Grillner, 1986), and slow rhythms (0.05–3 Hz) can be elicited by bath application of NMDA receptor agonists (Grillner *et al.*, 1981) while faster rhythms (1–8 Hz) are elicited by kainate agonists (Brodin *et al.*, 1985). In line with this, NMDA receptor antagonists block only the slow but not the fast rhythms (Brodin and Grillner, 1985).

There are several types of inhibitory interneurone in the lamprey and these appear to be glycinergic. Consequently, low concentrations of strychnine (a glycine blocker) in the bath result in an increase in the fictive rhythm but the application of GABA slows the rhythm (Grillner and Wallén, 1980). 5-Hydroxytryptamine (5-HT) is also present in the lamprey cord and prolongs the motor bursts and slows the rhythm when it is applied to the bathing medium (Harris-Warrick and Cohen, 1985). These results suggest that the frequency of the motor rhythm of the *in vitro* lamprey spinal cord, and thus the control of swimming speed, depend on the balance between the neuroactive substances present.

In the spinal dogfish preparation the rhythmical motor output appears immediately after the descending supraspinal pathways have been lesioned (Roberts, 1969) and remains fairly constant for several days. The frequency of the motor output seems to be determined by the intrinsic spinal cord properties and is similar in different individuals. The motor rhythm recorded from the paralysed preparation is more variable and usually has a longer mean cycle period than that recorded from the same animal when it is swimming. Consequently, we view the activity seen under paralysis as representing the 'free-running' activity of a self-sustaining oscillator with its own natural period. It differs between animals, presumably because of differences in the intrinsic properties of the spinal cord such as the strength of synaptic coupling between neurones and the concentrations of various neuroactive substances, but these differences between individuals only become obvious after the inputs from descending and sensory systems have been eliminated. The difference in frequency levels between the free running and the swimming rhythms in the same individual appears as soon as the body movement ceases after curare application (Williamson and Roberts, 1986), a finding that suggests that movement-related sensory feedback, which would be eliminated by the paralysis, plays a role in setting the cycle period of the rhythm.

8.5. **Sensory control of movement**

The importance of rhythmical sensory input to the spinal cord for the frequency of the motor rhythm can be seen when undulatory movements are imposed on to the body of swimming (Roberts, 1969) or paralysed (Grillner and Wallen, 1982; Williamson and Roberts, 1986) spinal dogfish (Fig. 8.6B). In response to an imposed movement, the motor output during fictive swimming synchronises one to one (entrainment) to frequencies of oscillation that are both higher and lower than the free-running rhythm (Fig. 8.6C). The entrainment is lost, however, at frequencies that extend beyond a particular 'capture range', which is itself dependent on the frequency of the intrinsic free-running rhythm. Over the capture range, the exogenous sensory signals arising from body movement dominate the endogenous oscillator and so determine the frequency of the rhythm of

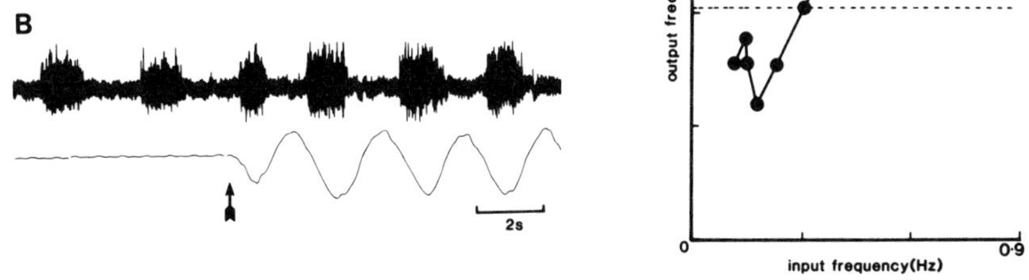

Fig. 8.6. Sensory entrainment of motor output in *Scyliorhinus*. (A,B) Top trace shows motor nerve activity during body movement recorded from (A) a swimming, and (B) a paralysed, spinal dogfish; bottom trace shows body movement measured by a mechanotransducer attached to the midline. Note in B the spontaneous motor output in the absence of any body movement and the synchronisation to the oscillation applied at the arrow. (C) Relationship between the frequency of an applied oscillation to the body of a paralysed, spinal dogfish and the motor output. The dotted line indicates the 'free-running' value of the output when no oscillation is applied. All data from Williamson and Roberts (1986).

the motor output. Such entrainment of central oscillators by rhythmical sensory input has been reported now for several rhythmical motor systems and is, we believe, important for their stability, preventing the motor output from changing unpredictably in response to inputs that are, in the context of the movement, trivial and irrelevant (see Roberts, 1984; Roberts and Ballintijn, 1988, for a fuller discussion).

8.6. **Descending control of locomotion**

Many basic features of locomotion are determined by the spinal cord circuitry. However, the stereotypical output of these circuits has to be modified continually so as to give the swimming animal a complete repertoire of movement, and sensorimotor analysis taking place in the brain is required for this. An interface between the higher brain centres and the pattern generators in the spinal cord is provided by brainstem neurones with axons descending into the spinal cord. The general arrangement of these loosely grouped cell masses is similar in fishes to that seen in other vertebrates, and comprises reticulospinal and vestibulospinal systems. Some of these neurones are involved in the initiation of rapid escape movements (see later), some participate in the control of fin movement (Roberts *et al.*, 1990), and others appear to regulate cruising locomotion. Thus, these descending interneurones appear to be involved in the selection of motor output. Not all descending pathways are essential for steady locomotion because goldfish swim satisfactorily when only 50% of the descending axons have regenerated following spinal cord transection (Bernstein and Gelderd, 1970).

8.6.1. *The action of descending neurones in continuous movement*

We know little about the action of brainstem supraspinal neurones in fishes except that many of them monosynaptically excite spinal cord motoneurones and interneurones (Rovainen, 1974; Roberts and Williamson, 1982). In the lamprey, stimulation of some Muller cells can modify the rate of the cord motor rhythm (Buchanan and Cohen, 1982), and the activity of certain reticulospinal neurones is modulated phasically during fictive locomotion (Kasicki *et al.*, 1989). These reticulospinal neurones discharge if the fictive output is strong, but otherwise their activity is subthreshold. As Kasicki *et al.* (1989) point out, it is somewhat surprising that so many reticulospinal neurones are rhythmically active because steady tonic driving of cord interneurones would seem sufficient to enable them to produce an appropriate motor programme. However, it is possible that a steady rhythmical input to the cord may have an important stabilising influence on cord circuits in a similar manner to the influence of movement-related sensory input. Certainly, rhythmical stimulation of

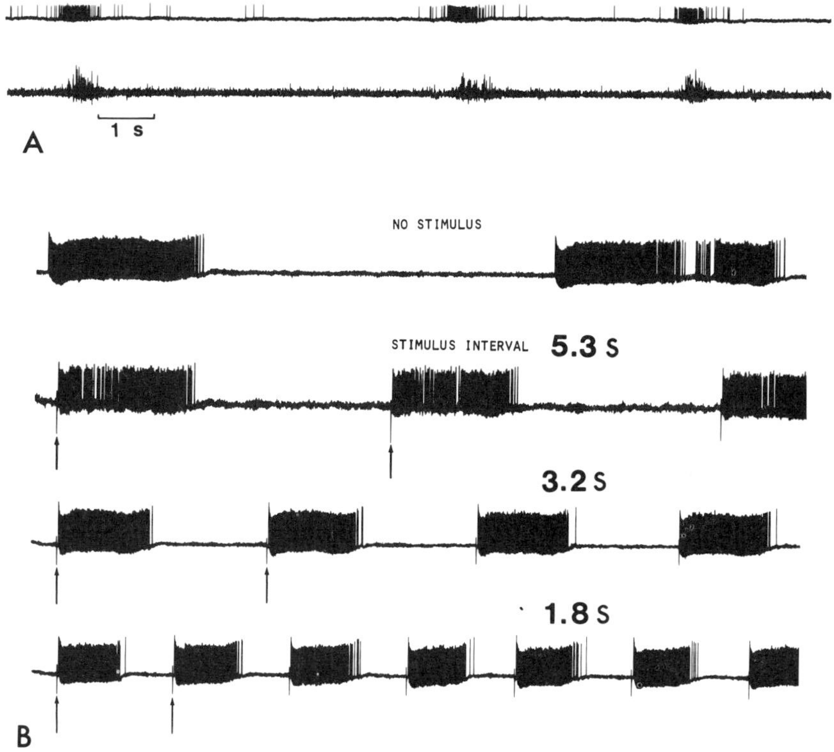

Fig. 8.7. Activity of reticulospinal neurones recorded from paralysed, decerebrate *Anguilla*. A: Top trace: reticulospinal neurone; bottom trace: recording from an ipsilateral spinal nerve in the region of the anus. B: Rhythmically active reticulospinal neurone entrained by rhythmical stimulus applied to ipsilateral trigeminal nerve. All data from Mos *et al.* (1986).

the descending pathways can entrain the motor output in the fictively swimming dogfish (Roberts and Williamson, 1983).

Rhythmic activity can be recorded from many reticulospinal and other brainstem neurones in decerebrate paralysed preparations of the eel, *Anguilla*, if a rhythmical motor output is being produced by the spinal cord (Fig. 8.7; Mos *et al.*, unpublished). However, when this output is irregular, the activity of reticulospinal neurones is also erratic and usually has no clear relationship to the spinal cord output. Many of the rhythmically active reticulospinal neurones are driven synaptically when the spinal cord is stimulated (Mos *et al.*, 1986), suggesting that there are reciprocal stabilising interactions between oscillating circuits in the brainstem and in the spinal cord. The activity of such neurones can be entrained by stimuli applied rhythmically to the trigeminal nerve (Fig. 8.7B).

8.6.2. *The action of descending neurones in transient movement*

A great deal is known about the neural circuitry that underlies the rapid escape movements made by fishes in response to certain types of stimuli. Most attention has been given to the so-called C-start manoeuvre which is seen when a fish, such as a goldfish, is given a sudden visual or vibratory stimulus. The eyes, jaws, operculum, body and fins move together in a rapid sequence that is aimed at propelling the fish away from the stimulus (Eaton *et al.*, 1977). The C-start has two stages. In the first, the fish rapidly bends its body to one side, so forming a letter C, and this becomes a propulsive stroke in the second stage which accelerates the fish through the water and which is usually followed by rapid, continuous swimming. This whole behavioural sequence depends on the activity of many neurones, but most research has been focused on the role of just one component, namely the Mauthner (M) cell, which is a paired giant interneurone in the rhombencephalon that has an axonal projection to the contralateral side of the spinal cord. The precise role of the M-cell and its associates in this command group of neurones has been much debated (see Nissarov and Eaton, 1989, for review) because the results of acute experiments using synchronous stimulation of the octavus nerve to activate the C-start indicate that a single M-cell action potential is sufficient, whereas work with more natural forms of stimulation supports the idea that the M-cell is just one component of a system, the combined activity of which is needed in order to obtain the full expression of the escape movement.

Within the spinal cord the M-cell axon has an intimate anatomical relationship with each primary motoneurone (Myers, 1985), the axon of which, as it passes from the cell body to the musculature, loops around the M-cell axon with which it makes synaptic contact (Fetcho and Faber, 1988). The combined morphological and physiological data obtained by Fetcho and Faber (1988) from the anaesthetised goldfish indicate that this is a chemical excitatory connection. Secondary motoneurones are also activated by the M-cell, but indirectly, via an interneurone whose axon descends the ipsilateral spinal cord. In each segment the M-cell also drives (electrotonically) one interneurone on each side that inhibits interneurones and primary motoneurones located contralaterally (Fig. 8.8). We have already seen for the zebrafish that the primary motoneurones exclusively innervate the white musculature and that every white muscle fibre is innervated by one of the three primary motoneurones that are found in each half segment of the spinal cord. Consequently the intimate relationship between the M-axon and these primary motoneurones should ensure that the mass of the white muscle fibres would be activated in concert when the M-cell discharges. This expectation has been confirmed in experiments on restrained zebrafish in which Liu and Westerfield (1988) were able to distinguish, in recordings made from the white

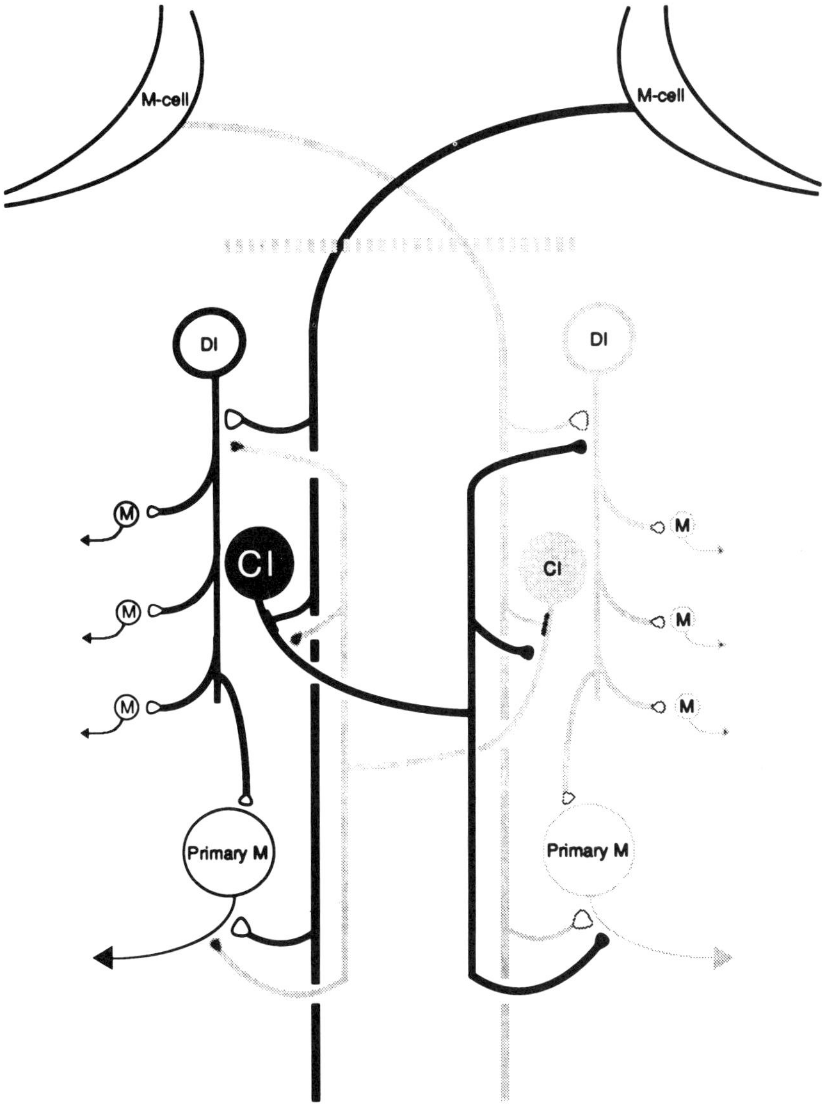

Fig. 8.8. Diagram of the spinal circuitry underlying the C-start movement in the goldfish. The Mauthner cell (M-cell) connections on the left side are drawn in solid lines; those on the right side are shaded. DI, Descending interneurones; CI, commissural interneurones; M, secondary motoneurones; Primary M, primary motoneurones. From Fetcho and Faber (1988).

muscle, the potentials that were evoked by activation of the primary motoneurones; the primary motoneurones always became active when a startle response was initiated (Fig. 8.3).

8.7. Control of descending systems

How the brainstem neurones that provide the supraspinal control for locomotion are themselves programmed and controlled is little understood, although some data are available about midbrain centres and the cerebellum.

8.7.1. *The 'mesencephalic locomotor region'*

Since the demonstration by Shik and Orlovsky (1976) that tonic stimulation in a restricted region of the midbrain could lead to regular locomotion in a decerebrate cat placed on a treadmill, it has been accepted that a 'mesencephalic locomotor region' (MLR) is involved in the supraspinal control of movement in mammals. This region does not project directly to the spinal cord but operates via relays that are most probably located in the reticulospinal system.

Kashin *et al.* (1974) showed that tonic mesencephalic stimulation could evoke regular movement in the carp, and they suggested that this region was equivalent to the MLR of mammals. In other species midbrain stimulation has also been reported to evoke locomotion (sting ray, Leonard *et al.*, 1979; lamprey, McClellan and Grillner,1984, dogfish, Grillner and Wallén, 1984). In the eel, *Anguilla*, for example, organised rhythmical motor activity can be established within the spinal cord by localised midbrain stimulation (Fig. 8.9). It remains to be seen, however, to what extent these 'MLRs' are homologous throughout vertebrates.

8.7.2. *The cerebellum*

In most fishes the cerebellum is large and has an internal circuitry similar to that of mammals; as in mammals, the Purkinje cells are the output neurones and they inhibit cerebellar efferent neurones which in turn project on to other brainstem systems (Paul and Roberts, 1981).

The activity of the Purkinje cells during locomotion has been studied in the dogfish *Scyliorhinus* (Paul and Roberts, 1984). The preparation used for these experiments was the decerebrate dogfish in which the medial longitudinal fasciculus and, thus, many supraspinal pathways, had been transected. Such a fish makes regular undulatory movements in a way similar to the spinal preparation, and its head can be held stationary with respect to the experimenter. About one-third of the Purkinje cells were found to be rhythmically active in time with these swimming movements.

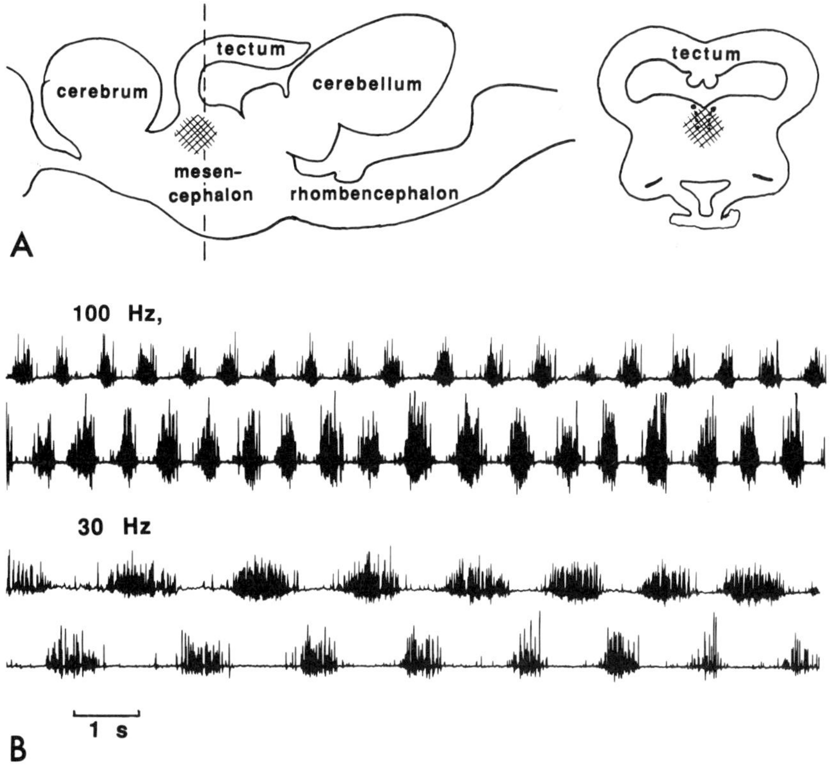

Fig. 8.9. Rhythmical spinal motoneuronal activity induced by steady mesencephalic stimulation. (A) Sagittal and tranvserse views of the location of the stimulating electrodes in the mesencephalon. (B) Recordings from spinal nerves on both sides of the same segment show the rhythmical output evoked by stimulation at 100 and 30 Hz.

The active Purkinje cells were distributed throughout the cerebellum, which suggests that the whole cerebellum participates in some way in the control of continuous locomotion. They did not discharge in unison but varied in their phase relationship to a defined point on the body (Fig. 8.10A). The source of the cerebellar rhythmicity probably lies in the spinal cord circuits that are involved in the generation of the locomotory rhythm because the rhythmical activity was still present in fish paralysed with curare. Further, the spinal cord neurones that project to the cerebellum only seem to be active when the spinal cord is generating an output (Paul and Roberts, 1981, 1984, 1990).

Unfortunately, it has not yet been possible to obtain recordings from Purkinje cells during movements of a range of tail beat frequencies. This information is crucial for our understanding of the role of the cerebellum

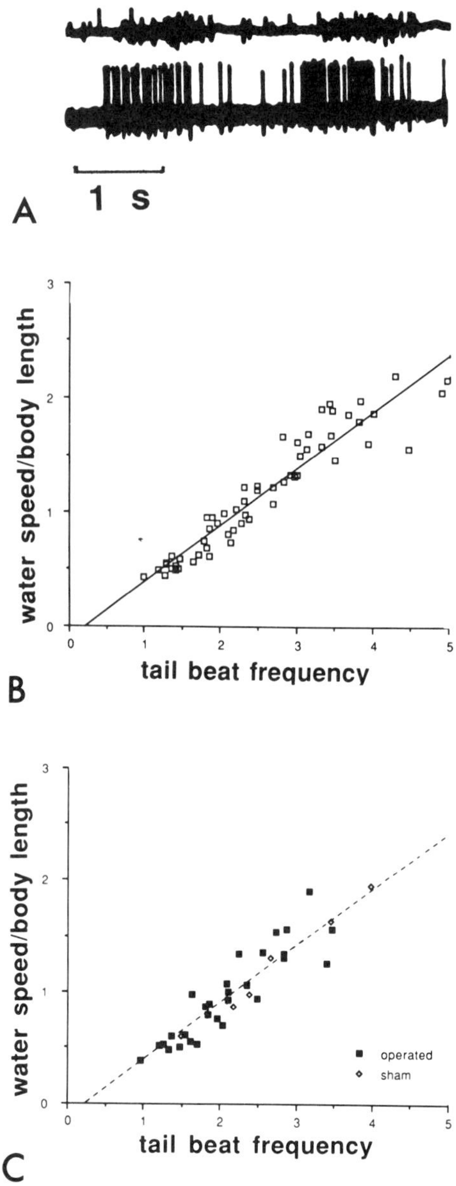

Fig. 8.10. The role of the cerebellum in locomotory control (A) Recordings from Purkinje cell in cerebellum (bottom trace) in swimming dogfish. The top trace is the electromyogram taken from the red musculature. (B,C) Measurements of tail beat frequency of trout swimming at different speeds in a water tunnel before (B) and after (C) cerebellectomy. Operated fish can no longer swim at speeds above 2 body lengths/s. (A) Unpublished data from Paul and Roberts; (B,C) from Roberts *et al.* (1989b).

for it is possible that the cerebellar output is highly nonlinear in relation to the fish's forward velocity and, for example, implements extra input to the brainstem at higher swimming speeds. Armstrong and Edgeley (1988) have made recordings from the Purkinje cells of walking cats and, surprisingly, found little change with changing speed. They have, however, no data for other gaits of locomotion.

The first attempts in fish to relate cerebellar function to locomotion were based on ablation and gave very confusing results but suggested that the cerebellum contributed little to movement control (reviewed in Healey, 1957). However, close examination reveals that there are changes in locomotory performance of a fish after much of its cerebellum has been ablated. In a trout, for example, the ability to swim continually at higher speeds is disrupted after cerebellar ablations (Roberts *et al.*, 1989b). As is well known, trout stay on 'station' in a flowing river, and in a water tunnel they will also retain a steady position as water speed changes by adjusting the tail beat frequency so as to match their forward speed to the water velocity (Fig. 8.10B). After cerebellectomy, a trout maintains its position in the tunnel at slow water speeds but is unable to do so at higher speeds (Fig. 8.10C) when it is swept backwards by the water stream. Maintaining a stable position at these water speeds requires more than three tail beats/s, which is the frequency at which white muscle fibres are recruited (Hudson, 1973). In the absence of the cerebellum, the transformation of sensory information about water speed into an appropriate tail-beat frequency is taking place satisfactorily but is not leading to the selection of the appropriate set of neurones needed for higher swimming speeds. This miscalculation could take place either in the sensory centres or in brainstem 'motor' regions; we favour the latter simply because in a previous study (Paul and Roberts, 1979) we showed that even reflexes mediated in the cord were markedly influenced by cerebellectomy, apparently because of changes in brainstem output.

Cerebellectomy in the trout has no effect on the execution of the C-start. It does, however, delay the onset of the contralateral contraction and thus the 'return flip', presumably because the cerebellum influences the non-Mauthner, reticulospinal circuits (Bosch and Roberts, unpublished).

8.7.3. *The role of the basal ganglia*

The striking changes in the locomotory performance of mammals that are observed in experiments in which midbrain dopaminergic (DA) neurones are stimulated or destroyed have indicated the importance of these DA-containing regions for locomotory coordination, but it is as yet unknown whether such dopaminergic systems are involved in the coordination of movement in other vertebrates. However, the recent demonstration of

DA-immunoreactive neurones in the midbrain of the elasmobranch fish *Raja* that can be homologised on the basis of their location and connections with the ventral tegmental area and substantia nigra of amniotes (Meredith and Smeets, 1987) implies that such systems may occur in anamniote vertebrates. Midbrain dopaminergic regions have not yet been observed in other fish (*Anguilla*, Roberts *et al.*, 1989a; *Gnathonemus*, Meek *et al.*, 1989), but a heavy dopaminergic innervation is present in the motor regions of the brainstem and spinal cord. The significance of this DA-innervation is unknown but the levels of noradrenaline and dopamine in the brain of *Anguilla* vary cyclically over a period of 24 h (Le Bras, 1979) and perhaps this catecholamine rhythm, accompanied by a diurnal variation in 5-HT levels (Fingerman, 1976; Genot *et al.*, 1976), may be important in determining the swimming activity of fishes.

8.8. **Concluding remarks**

The forward velocity of a swimming fish depends on its tail-beat frequency and hence on the cycle period of the segmental motor output. We have seen that this is determined by the generating circuits of the spinal cord in response to sensory and descending inputs. Steady movements can be considered as being the product of a single yet highly modifiable motor programme in which swimming speed is changed not only by changing the cycle length of regular contraction, but also by recruiting the white muscle system for faster and stronger contractions. This new recruitment results from activation of neurones projecting to the spinal cord from brainstem structures and may be under the control of the cerebellum. A quite distinct programme, using separate pathways, motoneurones and muscle fibres is implemented for very rapid, transient movements. These two programmes seem to operate independently in many fishes, but in higher teleosts they can be switched so as to achieve wide ranges of speed. At present the neural basis for this switching is unknown.

References

Altringham, J. D. and Johnston, I. A. (1988) Activation of multiply innervated fast and slow myotomal muscle fibres of the teleost *Myoxocephalus scorpius*. *J. exp. Biol.* **140**: 313–324.

Armstrong, D. M. and Edgeley, S. A. (1988) Discharges of interpositus and Purkinje cells of the cat cerebellum during locomotion under different conditions. *J. Physiol.* **400**: 425–445.

Bainbridge, R. (1958) The speed of swimming of fish as related to size and to the frequency and amplitude of the tail beat. *J. exp. Biol.* **35**: 109–133.

Bernstein, J. J. and Gelderd, J. B. (1970) Regenerative capacity of long spinal tracts in the goldfish. *Brain Res.* **19**: 21–26.

Bone, Q. (1966) On the function of the two types of myotomal muscle fibre in elasmobranch fish. *J. mar. biol. Ass. UK* **46**: 321–349.

Bone, Q. (1978) Locomotor muscle. In:*Fish Physiology* VII, W. S. Hoar and D. J. Randall, eds. Academic Press. New York, pp.361–424.

Bone, Q., Kiceniuk, J. and Jones, D. R. (1978) On the role of the different fibre types in fish myotomes at intermediate swimming speeds. *Fish. Bull.* **76**: 691–699.

Brodin, L. and Grillner, S. (1985) The role of putative excitatory amino acid neurotransmitters in the initiation of locomotion in the lamprey spinal cord. I. The effects of excitatory amino acid antagonists. *Brain Res.* **360**: 139–148.

Brodin, L., Grillner, S. and Rovainen, C. M. (1985) N-Methyl-d-aspartate (NMDA), kainate and quisqualate receptors and the generation of fictive locomotion in the lamprey spinal cord. *Brain Res.* **325**: 302–306.

Buchanan, J. T. and Cohen, A. H. (1982) Activities of identified interneurons, motoneurons, and muscle fibers during fictive swimming in the lamprey and effects of reticulospinal and dorsal cell stimulation. *J. Neurophysiol.* **47**: 948–960.

Cohen, A. H. and Wallén, P. (1980) The neuronal correlate of locomotion in fish. 'Fictive swimming' induced in an *in vitro* preparation of the lamprey spinal cord. *Exp. Brain Res.* **41**: 11–18.

Dale, N. (1986) Excitatory synaptic drive for swimming mediated by amino acid receptors in the lamprey. *J. Neurosci.* **6**: 2662–2675.

Dale, N. and Grillner, S. (1986) Dual-component synaptic potentials in the lamprey mediated by excitatory amino acid receptors. *J. Neurosci.* **6**: 2653–2661.

De Graaf, F. van Raamsdonk, W. van Asselt, E. and Diegenbach, P. C. (1990) Identification of motoneurons in the spinal cord of the zebrafish (*Brachydanio rerio*), with special reference to motoneurons innervating intermediate muscle fibres. *Anat. Embryol.* **182**: 93–102.

Droge, M. H. and Leonard, R. B. (1983). Swimming rhythm in decerebrated, paralyzed stingrays: normal and abnormal coupling. *J. Neurophysiol.* **50**: 178–191.

Eaton, R. C., Bombardieri, R. A. and Meyer, D. L. (1977) The Mauthner-initiated startle response in teleost fish. *J. exp. Biol.* **66**: 65–81.

Fetcho, J. R. (1986) The organization of the motoneurons innervating the axial musculature of vertebrates. I. Goldfish (*Carassius auratus*) and mudpuppies (*Necturus maculosus*). *J. Comp. Neurol.* **249**: 521–550.

Fetcho, J. R. and Faber, D. S. (1988) Identification of motoneurons and interneurons in the spinal network for escapes initiated by the Mauthner cell in goldfish. *J. Neurosci.* **8**: 4192–4213.

Fingerman, S. W. (1976) Circadian rhythms of brain 5-hydroxytryptamine and swimming activity in the teleost, *Fundulus grandis*. *Comp. Biochem. Physiol.* **54C**: 49–53.

Genot, G., Soulier, Ph. and Chèze, G. (1976) Relation entre le taux de sérotonine dans l'encéphale et l'activité motrice spontaneé de l'anguilla. *Jl de Physiol* **72** 42A.

Gray, J. and Sand, A. (1936) The locomotory rhythm of the dogfish (*Scyllium canicula*). *J. exp. Biol.* **13**: 200–209.

Grillner, S. (1974) On the generation of locomotion in the spinal dogfish. *Exp. Brain Res.* **20**: 459–470.

Grillner, S., Brodin, L., Sigvardt, K. and Dale, N. (1986) On the spinal network generating locomotion in lamprey: transmitters, membrane properties and circuitry. In: *Neurobiology of Vertebrate Locomotion*. S. Grillner, S. G. Stein, D. G. Stuart, H. Forssberg and R. M. Herman, eds. Macmillan, London, pp. 335–352.

Grillner, S., McClellan, A., Sigvardt, K., Wallén, P. and Wilen, M. (1981) Activation of NMDA-receptors elicits 'fictive locomotion' in lamprey spinal cord *in vitro*. *Acta Physiol. Scand.* **113**: 549–551.

Grillner, S., Perret, C. and Zangger, P. (1976) Central generation of locomotion in the spinal dogfish. *Brain Res.* **109**: 255–269.

Grillner, S. and Wallén, P. (1980) Does the central pattern generation for locomotion in lamprey depend on glycine inhibition? *Acta Physiol. Scand.* **110**: 103–105.

Grillner, S. and Wallén, P. (1982) On peripheral control mechanisms acting on the central pattern generators for swimming dogfish. *J. exp. Biol.* **98**: 1–22.

Grillner, S. and Wallén, P. (1984) How does the lamprey central nervous system make the lamprey swim? *J. exp. Biol.* **112**: 337–357.

Harris-Warrick, R. M. and Cohen, A. H. (1985) Serotonin modulates the central pattern generator for locomotion in the isolated lamprey spinal cord. *J. exp. Biol.* **116**: 27–46.

Healey, E. G. (1957) The nervous system. In: *The Physiology of Fishes II*, M. E. Brown, ed. Academic Press, New York, pp. 1–119.

Hidaka, T. and Miyahara, T. (1987) Excitatory and inhibitory neuromuscular transmission in fish red muscle. *Zool. Sci.* **4**: 819–823.

Hudson, R. C. L. (1969) Polyneural innervation of the fast muscles of the marine teleost *Cottus scorpius* L. *J. exp. Biol.* **50**: 47–67.

Hudson, R. C. L. (1973) On the function of the white muscles in teleosts at intermediate speeds. *J. exp. Biol.* **58**: 509–522.

Johnston, I. A., Davison, W. and Goldspink, G. (1977) Energy metabolism of carp swimming muscles. *J. Comp. Physiol.* **114**: 203–216.

Kashin, S. M., Feldman, A. G. and Orlovsky, G. N. (1974) Locomotion of fish evoked by electrical stimulation of the brain. *Brain Res.* **82**: 41–47.

Kasicki, S., Grillner, S., Ohta, Y., Dubuc, R. and Brodin, L. (1989) Phasic modulation of reticulospinal neurones during fictive locomotion and other types of spinal motor activity in lamprey. *Brain Res.* **484**: 203–216.

Le Bras, Y. M. (1979) Circadian rhythm in brain catecholamine concentrations in the teleost: *Anguilla anguilla* L. *Comp. Biochem. Physiol.* **62C**: 115–117.

Leonard, R. B., Rudomin, P., Droge, M. H., Grossman, A. E. and Willis, W. D. (1979) Locomotion in the decerebrate stingray. *Neurosci. Letts* **14**: 315–319.

Liu, D. W. and Westerfield, M. (1988) Function of identified motoneurons and co-ordination of primary and secondary motor systems during zebra fish swimming. *J. Physiol.* **403**: 73–89.

McClellan, A. D. and Grillner, S. (1984) Activation of 'fictive swimming' by electrical microstimulation of brainstem locomotor regions in an in vitro preparation of the lamprey central nervous system. *Brain Res.* **300**: 357–361.

Meek, J., Joosten, H. W. J. and Steinbusch, H. W. M. (1989) Distribution of dopamine immunoreactivity in the brain of the mormyrid teleost *Gnathonemus petersii*. *J. Comp. Neurol.* **281**: 362–383.

Meredith, G. E. and Smeets, W. J. A. J. (1987) Immunocytochemical analysis of the dopamine system in the forebrain and midbrain of *Raja radiata*: evidence for a substantia nigra and ventral tegmental area in cartilaginous fish. *J. Comp. Neurol.* **265**: 530–548.

Mos, W., de Schepper, F. V. J. H. and Roberts, B. L. (1986) Electrophysiological characterization of reticulospinal neurons in the eel. *Neurosci. Lett. Suppl.* **26**: S376.

Mos, W., Maslam, S. and Armée-Horvath, E. (1988) Changes in the distribution of synapses during growth: a quantitative morphological study of the neuromuscular system of fishes. *Neuroscience* **24**: 1061–1069.

Mos, W., Roberts, B. L. and Williamson, R. (1990a) Activity patterns of motoneurons in the spinal dogfish in relation to changing fictive locomotion. *Phil. Trans. R. Soc. Lond B* **330**: 329–339.

Mos, W., Roberts, B. L. and Williamson, R. (1990b) Interneuronal activity patterns during fictive locomotion of spinal dogfish. *Phil. Trans. R. Soc. Lond B* **330**: 341–349.

Mos, W. and Williamson, R. (1986) A quantitative analysis of the spinal motor pool and its target muscle in the growing dogfish, *Scyliorhinus canicula. J. Comp. Neurol.* **248**: 431–440.

Myers, P. Z. (1985) Spinal motoneurons of the larval zebrafish. *J. Comp. Neurol.* **236**: 555–561.

Nissarov, J. and Eaton, R. C. (1989) Reticulospinal control of rapid escape turning maneuvers in fishes. *Amer. Zool.* **29**: 103–121.

Paul, D. H. and Roberts, B. L. (1979) The significance of cerebellar function for a reflex movement of the dogfish. *J. Comp. Physiol.* **134**: 69–74.

Paul, D. H. and Roberts, B. L. (1981) The activity of cerebellar neurones of an elasmobranch fish (*Scyliorhinus canicula*) during a reflex movement of a fin. *J. Physiol.* **321**: 369–383.

Paul, D. H. and Roberts, B. L. (1984) The activity of cerebellar neurones of the decerebrate dogfish *Scyliorhinus* during spontaneous swimming movements. *J. Physiol.* **352**: 1–16.

Paul, D. H. and Roberts, B. L. (1990) Spinal neuronal activity during the pectoral fin reflex of the dogfish: pathways for reflex generation and cerebellar control. *J. exp. Biol.* **148**: 403–414.

Poon, M. L. T. (1980) Induction of swimming in lamprey by *L*-DOPA and amino acids. *J. Comp. Physiol.* **136**: 337–344.

Roberts, A., Soffe, S. R. and Dale, N. (1986) Spinal interneurons and swimming in frog embryos. In: *Neurobiology of Vertebrate Locomotion*. S. Grillner, S. G. Stein, D. G. Stuart, H. Forssberg and R. M. Herman, eds. Macmillan, London, pp. 279–306.

Roberts, B. L. (1969) The co-ordination of the rhythmical fin movements of dogfish. *J. Mar. Biol. Assn UK* **49**: 357–425.

Roberts, B. L. (1984) Role of sensory information in the control of locomotion in fishes. In: *Comparative Physiology of Sensory Systems*, R. D. Keynes and S. H. P. Maddrell, eds. Cambridge University Press, Cambridge, pp. 623–636.

Roberts, B. L. and Ballintijn, C. M. (1988) Sensory interaction with central 'generators' during respiration in the dogfish. *J. Comp. Physiol.* **162**: 695–704.

Roberts, B. L., Meredith, G. E. and Suharti Maslam (1989a) Immunocytochemical analysis of the dopamine system in the brain and spinal cord of the European eel, *Anguilla anguilla. Anat. Embryol.* **180**: 401–412.

Roberts, B. L., Rossem, A. van and Jager, S. de (1989b) The significance of the cerebellum for speed selection in swimming fish. *Eur. J. Neurosci.* Suppl. **2**: 132.

Roberts, B. L., Timerick, S. J. B. and Paul, D. H. (1991) Circuits for vestibular control of pectoral fin muscles in dogfish. In: *Locomotor neural mechanisms in arthropods and vertebrates*, D. M. Armstrong and B. M. H. Bush, eds. Manchester, University Press, Manchester, pp. 285–291.

Roberts, B. L. and Williamson, R (1982) Interactions between long descending pathways and cord neurons in the spinal dogfish. *J. Physiol.* **322**: 32P.

Rome, L. C., Loughna, P. T. and Goldspink, G. (1985) Temperature acclimation: improved sustained swimming performance in carp at low temperatures. *Science* **228**: 194–196.

Rovainen, C. M. (1974) Synaptic interactions of reticulospinal neurons and nerve cells in the spinal cord of the sea lamprey. *J. Comp. Neurol.* **154**: 207–224.

Shik, M. L. and Orlovsky, G. N. (1976) Neurophysiology of locomotor automatism. *Physiol. Rev.* **56**: 465–501.

Smit, W. A., Roberts, B. L. and Velzing, E. H. (1991) Changes in size and number of spinal motoneurons in relation to growth of the musculature in the eel, *Anguilla*. *Dev. Brain Res.* **58**: 73–80.

van Leeuwen, J. L., Lankheet, M. J. M., Akster, H. A. and Osse, J. W. M. (1990) Function of red axial muscles of carp (*Cyprinus carpio*): recruitment and normalized power output during swimming in different modes. *J. Zool. Lond.* **220**: 123–145.

van Raamsdonk, W., Veer, L., van't Veeken, K., Heyting, C. and Pool, C. W. (1982) Differentiation of muscle fiber types in the teleost *Brachydanio rerio*, the zebrafish. Posthatching development. *Anat. Embryol.* **164**: 51–62.

Videler, J. J. (1985) Fish swimming movements: a study of one element of behaviour. *Neth. J. Zool.* **35**: 170–185.

Videler, J. J. and Weihs, D. (1982) Energetic advantages of burst-and-coast swimming of fish at high speeds. *J. exp. Biol.* **97**: 169–178.

von Holst, E. (1935) Erregungsbildung und Erregungsleitung im Fischrückenmark. *Pflüg. Arch. ges. Physiol.* **235**: 345–359.

Wardle, C. S. (1975) Limit of fish swimming speed. *Nature* **255**: 725–727.

Williams, B. J., Droge, M. H., Hester, K. and Leonard, R. B. (1981) Induction of swimming in the high spinal stingray by *L*-DOPA. *Brain Res.* **220**: 208–213.

Williamson, R. M. and Roberts, B. L. (1980) The timing of motoneural activity in the swimming spinal dogfish. *Proc. R. Soc. Lond. B* **211**: 119–133.

Williamson, R. M. and Roberts, B. L. (1986) Sensory and motor interactions during movement in the spinal dogfish. *Proc. R. Soc. Lond. B* **227**: 103–119.

9 *Jenny Kien and Jennifer S. Altman*

Decision-making in the insect nervous system: a model for selection and maintenance of motor programmes

9.1. Introduction

Deciding what to do now and next is a central and continuous part of the neural control of behaviour in all animals: whether to keep doing the same thing or to do nothing at all, to groom or to eat, to walk straight or to turn. We still know very little about the mechanisms used to make such decisions: how is the behaviour appropriate to the input conditions of the moment selected? A widely held view, especially among invertebrate neurobiologists, is that these decisions are made in centres in the brain that dictate what the lower centres of the nerve cord should do. But this concept of a linear hierarchy of information processing and instructional levels cannot explain recent experimental findings in invertebrates (Kien, 1983, 1990a,b; Davis, 1976; this volume) and vertebrates (Eaton and DiDomenico, 1985; Alexander and Crutcher, Chapter 11, and Grobstein, Chapter 13, this volume).

Here we examine the data available for selection, initiation, maintenance and stopping of walking and flight in locusts. The complexity emerging from these studies emphasises that simple circuit diagrams are inadequate to describe the neural organisation of these behaviours and that a systems-based analysis is needed to provide a context for understanding the experimental observations. This need prompted us to develop a descriptive model that considers the nervous system as a whole and takes into account the flexibility of behaviour in intact animals which is often ignored in experimental analyses of motor control (Altman and Kien, 1987a, 1989).

In this model the selection of an appropriate output, which we term decision making, is a function distributed over the whole nervous system: a particular behaviour occurs as a result of the balance of activities in different parts of the nervous system at any instant. Further advances in understanding motor control systems will require formal models that can be tested mathematically and in computer simulations. Our model belongs to the Hopfield class of neural network models (Hopfield and Tank, 1986), for which there are formal analytical procedures, and we are now using these to make formal tests of our predictions.

9.2. Organisation of motor behaviour

A behaviour consists of sequences of motor activity each of which must be appropriately selected and initiated to produce a coherent and meaningful outcome. Once initiated a sequence is maintained until the aim of the sequence is achieved or until circumstances dictate a change of output. For example, cricket courtship behaviour, under laboratory conditions (see Rost and Honegger, 1988), is composed of a series of orientation, searching, stridulation and copulation or fighting sequences; the change from one sequence to the next is prompted by an appropriate response from the partner. Each sequence results from a basic pattern of motor activity which is continually tailored to fit the needs of the moment by sensory inputs and the internal state of the animal. This involves not only the generation of the specific effector pattern, e.g. wing movements in stridulation, but also the integration of the activity of the whole body to ensure coherent posture and movement.

In insects the specific motor patterns are generated predominantly in the segmental ganglia. Inputs from the brain and suboesophageal ganglion (SOG) are necessary for spontaneous expression of these patterns and for the plasticity seen in normal behaviour (Huber, 1960; Kien, 1983). The descending input from the head ganglia is usually referred to as 'higher motor control' but, for the reasons set out below, we consider that this input, rather than being instructive, co-operates interactively with the segmental ganglia in the production and regulation of the motor output.

Here we consider what is known about the nature of information descending from the brain during walking and the interaction between descending activity and thoracic neurones in the generation of flight.

9.3. Walking

9.3.1. *Spontaneous starting and stopping of walking*

Self-initiated behaviour, in insects as in vertebrates, is a complex process in which the change from quiescence to movements or from one behaviour to another can take several seconds to complete (Kien, 1990a). Spontaneous walking is preceded by a preparatory phase of up to 6 s in which small stepping 'intention' leg movements are made before walking starts with a fully developed step pattern and speed (Kien, 1990a).

The spontaneous start of walking is accompanied by complex changes in the firing patterns of descending interneurones (DINs) with axons that run from the brain and SOG to the thoracic ganglia. Figure 9.1 shows that spontaneous initiation of walking consists of at least three phases: before the preparatory phase (standing), the preparatory phase and the

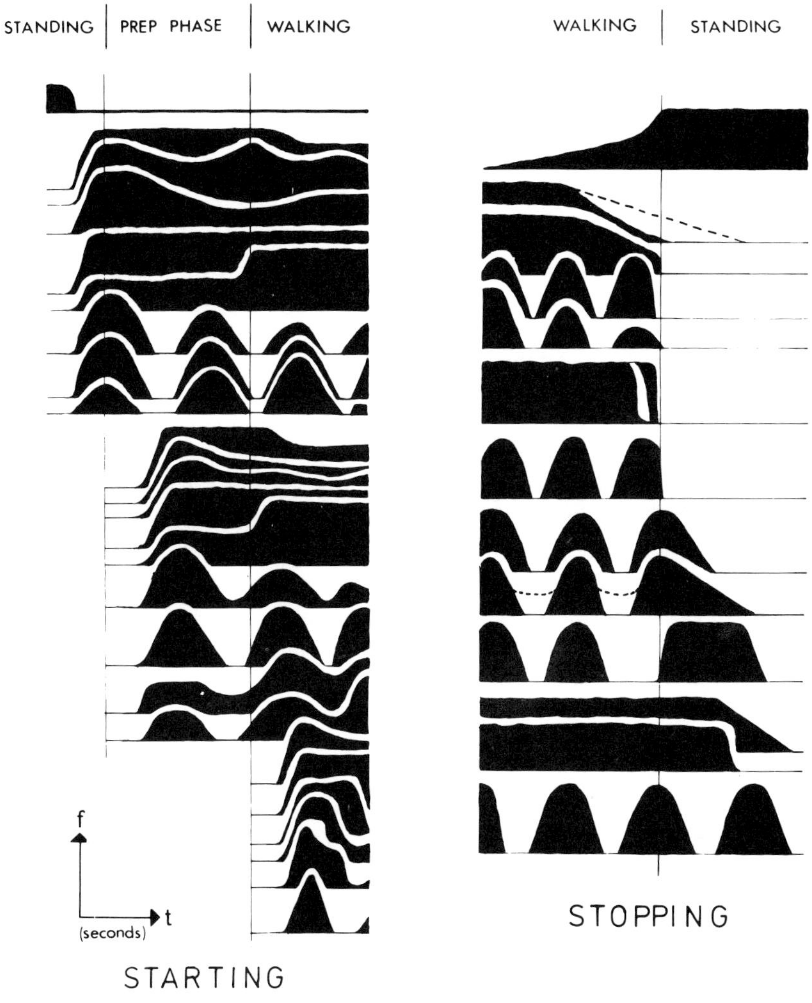

Fig. 9.1. The complexity of activity in the output from the head ganglia recorded during spontaneous starting and stopping of walking. One group of neurones starts to fire before the preparatory phase (standing), one during the preparatory (prep.) phase and a third when walking begins. The changes during stopping are equally complex, lasting from several seconds before stopping to several seconds afterwards. The patterns are schematised from the envelopes of instantaneous spike frequencies (f_{max} *c.*100 HZ) in single neurones recorded from many animals. Each pattern represents between one and six neurones. Dashed lines represent variations in a pattern. The activity shown here includes both brain and SOG DINs (after Kien, 1990a).

release of walking. These may be considered as corresponding to programme selection, organisation and performance.

Some neurones change their firing before the preparatory phase, others during the preparatory phase, and still others after the release of walking. As well as different times of onset, there are many different firing patterns in each phase (Fig. 9.1A). Reading down the columns of Fig. 9.1 gives an idea of the complexity of the input to the motor centres in the thorax during each phase. It is important to note that most neurones continue to fire throughout the following phases and for as long as the animal walks irrespective of the phase at which they become active.

The neuronal activity before the preparatory phase must be involved in selecting the walking programme because leg movements start during the preparatory phase. Before this can occur, ongoing behaviour, which includes standing still, must be stopped and indeed some neurones that fire during quiescence are inhibited before the preparatory phase begins. The neurones that start firing before the preparatory phase may also be involved in inhibiting other behaviours. For example, extracellular microstimulation at one site evoked walking and inhibited flight (Kien, 1983).

The walking programme seems to be organised in detail during the preparatory phase because, when walking begins, its form and speed are fully developed. When walking begins, yet more DINs become active (Fig. 9.1A). Once neurones start to fire, most remain active as long as the animal walks. The functions in each phase are thus initiated and maintained by both the newly recruited neurones and those previously active.

DINs in both brain and SOG fire before the preparatory phase and after the release of walking whereas the DINs that start firing during the preparatory phase are mostly in the SOG. This means that DINs in both the brain and SOG are involved in all three phases but the SOG DINs seem to contribute more than brain DINs to the preparatory phase, that is, to the organisation of the walking programme.

When walking stops spontaneously, the behavioural changes are as complex as during spontaneous starting. The transition from active walking to complete quiescence may take several seconds as the animal often slows down before stopping; after it stops, subthreshold rhythmical muscle activity may continue or muscle tone may remain elevated for some time. The patterns of neuronal activity in brain and SOG DINs change in equally complex ways (Fig. 9.1B); some neurones stop firing several seconds before, others at, and still others after, walking ceases. Stopping should therefore be considered as a deselection procedure; essentially it is equivalent to, or even part of, selection of a new behaviour, in this case standing still.

When walking is evoked by sensory stimulation, there is no preparatory phase and the complex sequence of DIN recruitment does not occur. Instead many neurones start firing at the same time and have a much more limited range of activity patterns during the stimulus and immediately after walking starts. The sensory input probably provides a rapid increase in excitation which replaces the complex sequence of recruitment of DINs before and during the preparatory phase of spontaneous walking.

9.3.2. *Role of brain and SOG output during walking*

Although the basic motor pattern for walking is organised in the thoracic ganglia (Huber, 1960; Kien, 1983; Bässler *et al.*, 1985), the brain and the SOG are necessary for the regulation of the direction, the speed and the details of the stepping pattern (Kien, 1983, 1990b). The activity recorded in the descending neurones from the brain and SOG during walking is as complex as that during starting or stopping (Fig. 9.2, cf. Fig. 9.1). Because neurones that fire before and during the preparatory phase continue to fire during walking, they may help to maintain walking by continually repeating the selection and organisation process. This function has also been suggested for some tonically active SOG flight DINs (Ramirez, 1988; see below).

Some neurones fire tonically, others in bursts. In some neurones the bursts do not correlate with any obvious features of the step cycle (Fig. 9.2); this patterned activity could regulate walking by influencing the timing of activity in the thoracic pattern generators. In other neurones the activity does correlate with stepping; the combination of correlations for each neurone is termed its 'motor field'. The motor fields each comprise a certain number of legs (Fig. 9.3). Most fields are independent of the direction of walking but nearly all correlate with particular phases of the step cycle, i.e. with the 'function' of a movement rather than with the movement itself. Only a few fields represent leg position or movement, and most of these include some movement functions. Thus, in the neuronal output from the head ganglia, the step cycle appears to be represented predominantly in terms of movement functions or sets of functions, rather than as the parameters of individual movements. Conversion from coding for function to coding for movement must occur in the local thoracic motor networks.

9.3.3. *Interactions between head and thoracic ganglia*

Information about walking does not flow simply from the head ganglia to the thorax but passes in both directions. In the SOG, some neuropiles provide a substrate for the exchange of information between neurons originating from and projecting to different levels of the nervous system because the branches of SOG interneurones mingle with those of through-running axons from other ganglia (Altman and Kien, 1987b,

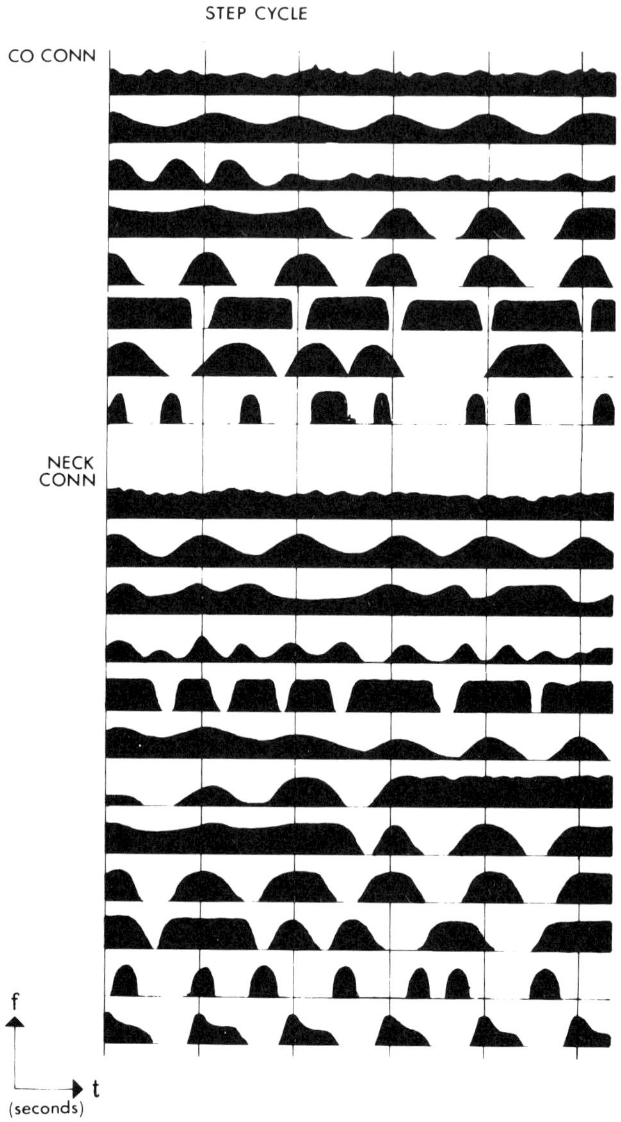

Fig. 9.2. Output from brain (CO conn) and brain and SOG (neck conn) during part of an established walking sequence (same conventions as in Fig. 9.1). The patterns of activity in the neurones are as complex as those during starting and stopping. The firing patterns of some neurones change during the sequence, whereas most are stable. Only some patterns correlate with the step cycle. (After Kien, 1990b.)

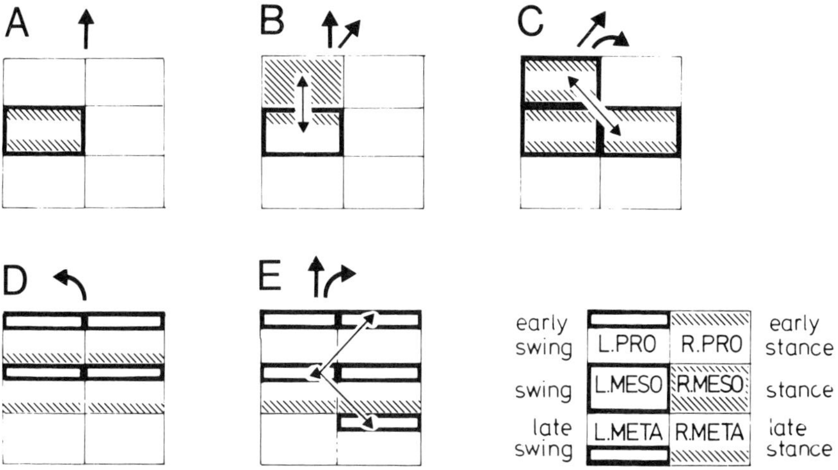

Fig. 9.3. Motor fields of descending neurones from the head ganglia during walking. These fields illustrate the involvement of the head ganglia in the step-by-step regulation of walking. They include parts of the step cycle regardless of the direction of walking, which suggests that functions of movements are represented in the head ganglia rather than the movements themselves. Examples of fields for one to five legs are given. A key to the schema is given in the figure. The six boxes in each represent the six legs of the locust; pro, prothoracic; meso, mesothoracic; meta, metathoracic legs; l, left; r, right. Stance and swing refer to the parts of the step cycle during which the neurones fired. Arrows above each schema give the direction of the walking from which the field was obtained. Arrows within the fields show which actions occurred at the same time. (After Kien, 1990b.)

Kien *et al.*, 1990). Intracellular recording shows that DINs receive sensory information from the body as well as from the head. For example, the brain DINs known as PI2(5) and O3 fire when the animal walks and PI2(5) also responds when the legs are moved passively (Kien and Altman, 1984). Some DINs in the SOG fire before spontaneous movements of the legs, as well as before and during walking, and respond when the legs are passively moved (Kien and Altman, 1984). Many SOG DINs have complex receptive fields with inputs from a wide variety of mechanoreceptors and proprioceptors all over the body (Ramirez, 1983; Kien and Altman, 1984; Kien, 1990a) and some also respond to very discrete tactile stimuli within their large fields (Ramirez, 1983). During ventilation in the cricket the ascending inputs produce patterned outputs in SOG DINs (Otto and Campan, 1978).

There are also interconnections between the brain and the SOG. Ascending interneurones (AINs) that originate in the SOG terminate in the brain (Boyan and Altman, 1985; Kien *et al.*, 1990). Some brain DINs which are active during walking have inputs to SOG DINs (Kien and Altman, 1984) and SOG DINs respond to visual and antennal inputs (Kien and Altman, 1984; Boyan and Altman, 1985).

9.3.4. *Organisation of walking*

We contend that these data, taken together, show that walking is organised by a distributed network of interconnections between the brain, the SOG and the segmental ganglia. The brain and SOG do not simply switch on walking but are involved in its organisation, maintenance and execution. This includes detailed coordination of the form and direction of stepping. The thoracic pattern generators alone can drive stepping movements but the highly patterned input from the head ganglia is required to organise steps into walking. The patterning of this input seems to be generated, at least in part, in the DINs in response to feedback from the thoracic ganglia and the sensory inflow from the body that results from movement (Kien and Altman, 1984). The system is thus organised in the form of loops operating in parallel, each loop with a particular combination of functions. The selection and maintenance of walking are emergent properties of this system.

9.4. **Flight**

9.4.1. *Networks involved in initiation of flight*

Most information about flight interneurones comes from intracellular recordings in reduced preparations so there are no detailed studies of neuronal activity in flight that has started spontaneourly, and little attention has been paid to neuronal activity during the start of flight evoked experimentally by wind. From the information available it seems that the flight rhythm generator is activated by at least fast-conducting wind-sensitive intersegmental interneurones and a distributed flight-selecting network, but the connections between the two systems are unknown.

Activity in the fast-conducting wind-sensitive interneurones may lead to flight; electrical stimulation of interneurones such as the paired tritocerebral commissure giant (TCG) neurones from the brain and cercal giant neurones GINs 2 and 4 from the terminal ganglion can evoke flight or rhythmical activity in the flight muscles (Bicker and Pearson, 1983; Boyan *et al.*, 1986; Boyan and Ball, 1989). When the TCG axons are cut, flight can still be evoked, but much less easily, indicating that other wind-sensitive pathways are also involved (Bicker and Pearson, 1983). GINs 1 and 4 make direct connections to neurones which are part of the flight pattern generator or which belong to the premotor interneurone pool (Boyan and Ball, 1989). There are also slower-conducting interneurones

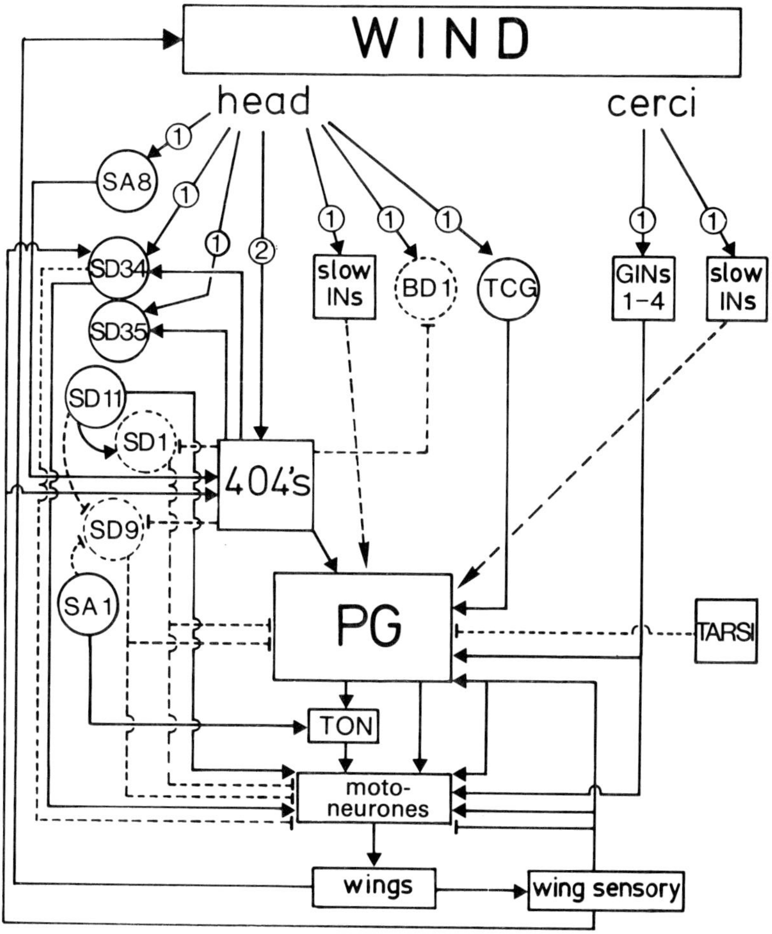

Fig. 9.4. Flight selection: a schematic representation of the information available on the neurones active during the initiation and maintenance of wind-evoked flight. Wind inputs excite fast-conducting intersegmental interneurones (right-hand side of figure) such as the TCG and cercal giants (GINs), and many neurones in the SOG and thoracic ganglia which are interconnected to form a distributed network (left-hand side of figure). The fast interneurones and the distributed network seem to work in parallel. Paths ending in arrows indicate excitatory connections; neurones with broken outlines and dashed pathways ending in bars are inhibitory. Paths with long dashes from slow INs are inferred, as is the inhibitory pathway from the tarsi. SD and SA are SOG descending and ascending interneurones (see Kien *et al.*, 1990); other neurone nomenclature as in the papers cited below. Boxes indicate groups of neurones, circles indicate individual neurones. 1 indicates a short latency response to wind, 2 a longer latency response. PG, pattern generator; TON, thoracic ocellar neurones. Data from Bacon and Möhl (1983), Bicker and Pearson (1983), Pearson *et al.* (1985), Reichert *et al.* (1985), Ramirez (1986, 1988) Boyan and Ball (1989).

from the terminal ganglion (Boyan *et al.*, 1989) which transmit wind information to the flight circuitry.

The distributed network includes a group of at least four neurones in the mesothoracic ganglion termed the 404s, which excite several SOG interneurones (Fig. 9.4). All these neurones are activated by wind on the head, although the response of the 404s is delayed with respect to that of the SOG interneurones, and they all evoke flight when stimulated electrically (Pearson *et al.*, 1985; Ramirez, 1988). They also inhibit flight-inhibitory neurones in the SOG (Ramirez, 1988). Several other neurones in the brain and SOG (not shown in Fig. 9.4) can also evoke flight activity when stimulated, but information about them is sketchy (Kien *et al.*, 1990).

Several neurones in the SOG seem to be involved in the wing opening at the release of flight. They are excited by wind, and fire at the beginning of flight, but they are inhibited during flight. Stimulating some of them evokes activity in the elevator motoneurones which help open the wings (Ramirez, 1986).

Both the wind-sensitive fast-conducting interneurones and the flight-selecting network are probably involved in maintaining flight, because all the flight-evoking neurones fire throughout flight. The TCGs reinforce the flight rhythm through their phase-locked responses to the turbulences caused by the beating of the wings (Bacon and Möhl, 1983). Although the 404 neurones are not responsible for the generation of the rhythmic motor pattern, their firing frequency seems to set the frequency of the wing beat cycle (Pearson *et al.*, 1985). The connections of the flight selection network to the flight rhythm generator are unknown.

9.4.2. *Networks involved in flight rhythm generation*

The flight rhythm generator includes neurones in the SOG, thoracic ganglia and fused abdominal ganglia. It has been suggested that the rhythm is generated by six interconnected bilateral groups of neurones, known as 201, 206, 301, 501, 503 and 504 (Fig. 9.5, see legend for details and references). This network alone could in principle produce an alternation of wing depression and elevation (Robertson and Pearson, 1985; Robertson, 1987). However, more neurones seem to be involved. For example, the 401 neurones feed delayed excitation both to the 301 neurones and to the depressor premotor pool. The 520 and 202 neurones receive delayed excitation from the 301 and 401 neurones respectively, but their output connections and hence their roles are not known. Many other neurones fire in phase with either depressor or elevator motoneurones (Fig. 9.5), and from the few interconnections published it seems that they could be included in flight rhythm generation networks.

Even with these additional neurones, the description of the pattern

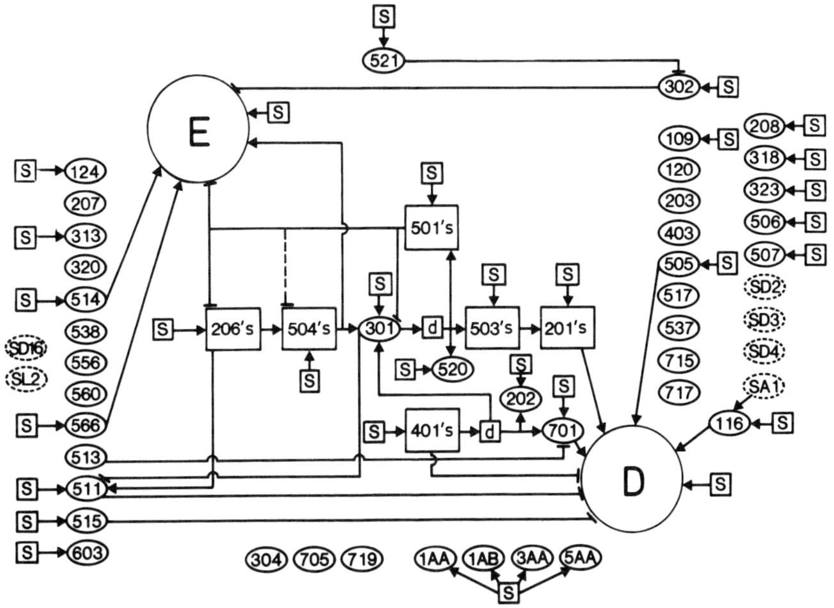

Fig. 9.5. Flight rhythm generation: a schematic representation showing that the rhythm-generating neurones also form a complex highly interconnected network. Excitation in the 206 neurones is transmitted via the 504s both to elevator motoneurones and to the paired 301 s. These produce a delayed excitation (d) in the 501 s, which feed back to inhibit the 206s, and in neurones which excite depressor premotor neurones (503 s to 201 s). This circuit is thought to be responsible for the alternating elevator and depressor motor output (Robertson and Pearson, 1985). A large number of other interneurones have been found which are rhythmically active during flight but whose connections are not known. They are lined up in columns near the motoneurone pool with which they fire in phase. The AA neurones below receive flight sensory inputs. The other neurones below are rhythmically active but no further information has been given. Ovals represent bilateral pairs of neurones, and squares represent groups. Cells with broken outlines are in the SOG; all others are thoracic. Inhibitory pathways are indicated by bars at the end. The dashed pathway is inferred. E, D represent elevator and depressor motoneurone pools. s, Sensory input from stretch receptor and tegula. Neurone nomenclature and data from Robertson and Pearson (1983, 1984, 1985), Robertson (1987), Pearson and Robertson (1987), Pearson and Wolf (1988), Reye and Pearson (1987), Ramirez (1986), Ramirez and Pearson (1988), Reichert and Rowell (1985), Boyan (1985), Boyan and Ball (1989), Elson (1987).

generator in Fig. 9.5 is probably incomplete. Ronacher *et al.* (1988) have shown that transecting ganglia or cutting connectives produces less disruption of the flight pattern than would be predicted from Fig. 9.5. The central rhythm generator may thus include subunits in each hemiganglion coordinated by the neurones in Fig. 9.5.

Figure 9.5 is also simplified because only some of the many sensory inputs to flight interneurones are included (stretch receptor: Reye and Pearson, 1987; tegula: Pearson and Wolf, 1988). Sensory afferents also make monosynaptic connections with motoneurones (Peters *et al.*, 1985; Pearson and Wolf, 1988). During flight, sensory input to the flight motoneurones dominates and masks the inputs from the rhythm generator (Wolf and Pearson, 1989). Thus the sensory inputs themselves must be regarded as part of the pattern generator (Altman, 1982; see also Möhl, 1985).

9.4.3. *Organisation within the flight system*

Figures 9.4 and 9.5 make it clear that the operations of the flight system, like those of the walking system, cannot be described in terms of the activity of single neurones. Rather, flight is initiated and maintained by a large constellation of neurones distributed between the brain, SOG and thoracic ganglia. There are also more interconnections than are shown in

Table 9.1. SOG neurones regulate various aspects of flight. Reading down the columns shows that each of these functions is subserved by many neurones, and reading across the columns shows that each neurone is involved in more than one function. Neurones included under 'Prohibit' are those that inhibit flight and so must be inhibited before flight can occur. Neurone numbering after Kien *et al.* (1990). Data from Ramirez (1986, 1988), Kien *et al.* (1990).

Select	*Release*	*Prohibit*	*Maintain*	*Pattern/execute*
SD1	SD1	SD1		
SD10	SD10	SD10		
SD11	SD11	SD11		
SD7		SD7		
SD9		SD9		
SA8			SA8	
SD12	SD12		SD12	
SD18	SD18		SD18	
SD25	SD25		SD25	
SD34	SD34		SD34	
SD35	SD35		SD35	
	SD2		SD2	SD2
			SD3	SD3
			SD4	SD4
			SD5	SD5
			SD16	SD16
			ST6	ST6
			ST15	ST16
			SA1	SA1
			SL2	SL2

Figs 9.4 and 9.5: several of the prothoracic flight interneurones are known to project at least to the SOG (Ramirez, 1986), some even to the brain (Göring, unpublished data). SOG interneurones that terminate in the brain exert strong effects on thoracic interneurones (Ramirez, 1988) and brain DINs receive ascending information (Kirn and Altman, 1984; Hensler, 1989). Many neurones have more than one behavioural function in flight, and each function is carried out by many neurones (Table 9.1). There are thus several parallel but overlapping networks, each involved in the regulation of a particular aspect of the flight motor output.

9.5. General principles which can be derived from walking and flight

These examples underline the concept that the neuronal substrates of walking and flight, two 'relatively simple' behaviours, cannot be understood merely through connection diagrams. The circuits shown in Figs 9.4 and 9.5 are already so complex that complete diagrams would probably be incomprehensible, particularly as many parts of such circuits operate simultaneously and not in linear sequences.

Some principles underlying the organisation of both these systems have been established (Altman and Kien, 1987a, 1989): first, any function results from the concerted activity of many neurones, which can be described as an across-fibre pattern (AFP); secondly, the output of a network depends on a consensus between the AFP of its inputs and its ongoing activity — that is, networks are not passively driven by commands from other centres; and, thirdly, the neurones are highly interconnected in a number of parallel loops, each of which regulates a different set of functions. These principles are the basis of a general descriptive model developed for the insect motor system (Altman and Kien, 1987a, 1989).

9.5.1. *Across-fibre patterns and neuronal ensembles*

An AFP is the spatio-temporal pattern of activity in the inputs to or the output from a group or ensemble of neurones that are highly interconnected and working together to produce a discrete function at a particular time. The concept of an AFP can be applied to any level in the system, from single neurones to large populations. It can be applied equally well to all types of function, from sensory inputs (the application for which the concept was first developed, Erickson, 1963), to the descending activity from the brain or the output of motoneurones innervating a group of muscles. Units in an AFP may be active or silent, excitatory, inhibitory or neuromodulatory, tonic or phasic. It is important to recognise that the units making up an AFP are not necessarily all equally effective inputs to their target neurones.

A good example of ensembles and AFPs is the motoneurones that drive the thoracic muscles in locusts. Some of these muscles are bifunctional, moving both leg and wing (Wilson, 1962). The motoneurones innervating them are active in both flight and walking and so participate in different ensembles (sensory neurones, interneurones and motoneurones) in the two behaviours. The muscles that are antagonists in flight are, however, synergists in walking; their motoneurones are active in different combinations in the two behaviours so that there are two alternative patterns of activity or AFPs in the nerves innervating the thoracic muscles. Large changes in the AFP may result in a shift from one behaviour to the other, whereas small changes in the AFP may lead to small adjustments in the current motor output producing, for example, changes in wing angle necessary for flight stability and steering (Möhl and Zarnack, 1977) or in leg positioning during walking.

In walking, the output from the head ganglia forms an extremely complex AFP to the thoracic motor centres. Reading down the columns in Figs 9.1 and 9.2 shows that, when walking starts spontaneously, there are three successive AFPs, one associated with deciding to walk, one with preparing to walk, and the third with the start of walking. Within each phase, previously silent fibres become active and continue to fire for as long as the animal walks. When walking is evoked by sensory stimulation, there is a sudden shift from the standing AFP to a walking AFP as the three groups of fibres all start to fire at roughly the same time.

In the flight system, the input to the flight motoneurones can be described by the AFPs of activity in the elevator and depressor premotor interneurones (Fig. 9.5). The activity in each motoneurone group cannot be predicted from activity in any one premotor interneurone, and changes in the activity of only one such interneurone are unlikely to produce large changes in the motoneurone output. But, owing to interconnections within the network, large changes in the activity of one neurone are likely to alter the firing of many others and it is the activity of this group which alters the motor output.

AFPs can also be usefully applied to describing the activity in the rhythm-generating circuits. Most of the elements in the circuit of Fig. 9.5 consist of groups rather than single neurones and the output of each group forms an AFP, designated by the arrows in the figure, which is part of the input to the next element. Indeed Table 9.1 shows that describing the activity of single neurones in isolation is completely inadequate to explain how any function in the flight system is carried out.

9.5.2. *Consensus*

The outputs of a network such as the flight rhythm generator are not passively driven by instructions from higher levels but result from the balance between the activity in the network and in all the inputs to it. These include local sensory inflow as well as inputs from the brain and SOG.

Several examples of such a consensus can be found in walking and flight. Walking cannot reliably be evoked by focal electrical stimulation in the CNS in a locust that is active in other ways or not in a 'walking mood' (Kien, 1983). Locusts and stick insects will not walk unless certain sense organs are stimulated by loading the legs; otherwise they make only searching movements (Bässler, 1983; Macmillan and Kien, 1983). A locust will start to fly when a stream of air is blown over the head but usually only when the feet are not in contact with the ground (Pond, 1972). In course correction during flight, the response to tilting of the horizon when the animal starts to roll involves a consensus between the descending neurones carrying the sensory information from the head and the flight neurones in the thoracic ganglia that generate the motor output; the descending input is effective only when the flight pattern generator is active. Even then, the information reaches the motoneurones only when the pattern generator is in a particular phase (Reichert *et al.*, 1985).

9.5.3. *Loops*

In walking and flight, both individual neurones and groups of neurones receive inputs from other neurones to which, either directly or indirectly, they send outputs. That is, these connections form loops in which the activities of the components or ensemble are continually modified by each other. The output from such a loop is thus the result of the activity in the connections between the components rather than that of individual neurones or groups of neurones. The concept of loops emphasises that the origin of information has no significance for the neurones within an ensemble, which see only their own input AFP at each instant.

Loops are well illustrated by the sensory inputs to the flight system (Figs 9.4 and 9.5). The wing movements stimulate wing sense organs that form part of the input to both the flight motoneurones and the flight pattern generator; this loop provides cycle-by-cycle regulation of the wing beat. At the same time, the wing movements themselves phasically modulate the flow of air over the head and body which results in phase-locked firing in the TCG interneurones that convey the information back to the thoracic ganglia, where the motor output to the wings is generated (Fig. 9.4). This loop contributes both to the maintenance of flight and to course correction (Bacon and Möhl, 1983). It also influences the firing of other neurones involved in flight maintenance, for example the 404 group and SD34. The two loops converge on the flight interneurones and motoneurones, and so reinforce each other.

Thus every loop has a particular function. Those between adjacent ganglia regulate the local coordination of effectors, as in the locust thorax, where both direct afferent connections and interneurones between neighbouring ganglia are involved in local coordination of legs of wings, as well as in motor pattern generation. Loops involving both the head and segmental ganglia ensure coordinated responses of the whole body through neurones that have collaterals in most segments (Altman and Kien, 1985). By acting in consensus with the local loops, the longer loops prevent different parts of the body from expressing incompatible actions and ensure that all segments respond appropriately to diverse stimuli. Some loops may be more important than others but all are required for the system to function efficiently in a biologically meaningful way.

The connections between head and segmental ganglia form more than one loop as brain DINs and SOG DINs project separately to the segmental ganglia from which they both receive inputs (Fig. 9.6; Kien and Altman, 1984; Altman and Kien, 1987a). One operation that could be carried out by these parallel loops is to compare the inputs from the brain, which give the animal's orientation in space, with proprioceptive information from the whole body and with corollary motor discharge which is integrated in the SOG (Fig. 9.6). The result of this comparison would continually modulate the motor output generators to achieve the desired output. In walking, the loops between the SOG and thoracic ganglia and between the brain and thoracic ganglia are involved in the whole range of functions from selection of walking to execution of details of the motor output. These functions overlap and cannot be clearly assigned to specific loops. Although we have previously suggested that the brain–thorax loop has a 'what movement' or selection function and the SOG–thorax loop a 'how to do it' or organisation and execution function, it now seems more likely that each can peform both functions but to different extents: in the brain–thorax loop 'what' is the stronger component whereas the SOG–thorax loop has a larger 'how' component.

9.6. **The model**

The model (Fig. 9.7A) considers the motor system as several interconnected stations which approximate to brain, SOG and segmental ganglia in the locust. Each station contains the networks that generate the output of the station. The three basic principles — AFPs, consensus and loops — operate at every station in the system. The output of any station at a given time is derived from the consensus between the total AFP of activity in the inputs to its local networks and the ongoing activity in the local networks. The inputs to all stations include information about novel events and self-generated changes in the environment, about changes in the internal state such as arousal, and about the effects of learning and memory and hormonal changes, as well as the outputs from other

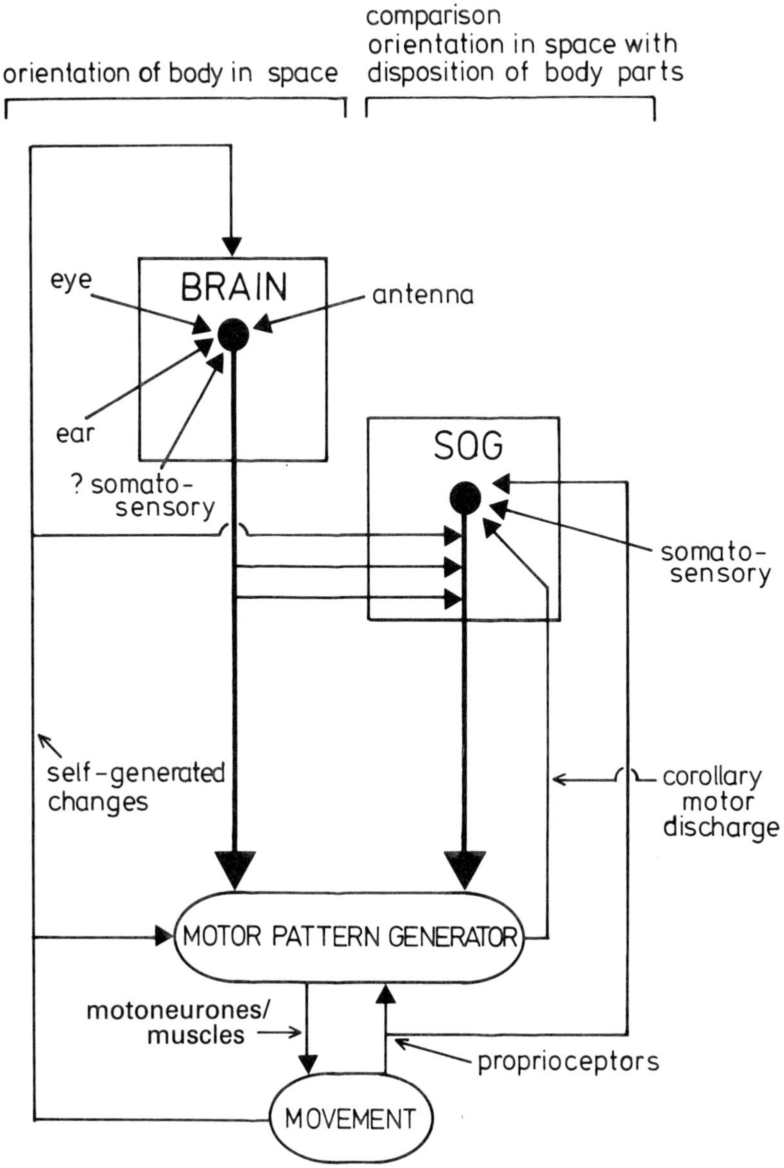

Fig. 9.6. A scheme illustrating how activity in the two loops connecting the brain and SOG to the segmental ganglia could result in a comparison between orientation of the body in space and the disposition of the body parts. Brain DINs respond more to orientation of the body in space; SOG DINs receive this information as well as detailed proprioceptive inputs. Thus these two inputs are compared in the SOG. Both sets of DINs send inputs to the thoracic motor system where there is a further comparison. The activity arising from comparisons between orientation in space and disposition of body parts can interact with the pattern generators so that these can produce a motor output which achieves the desired behaviour. (This comparison is discussed in more detail in Altman and Kien, 1987a.)

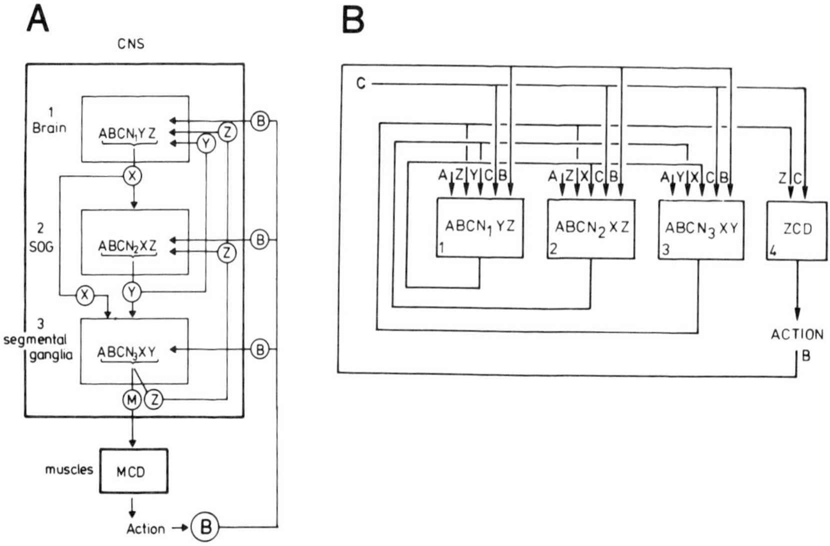

Fig. 9.7. (A) a decision-making model composed of several stations (1–3) representing different levels of functional organisation and different local circuits (N1–3). In the locust the three levels are brain, SOG and segmental ganglia. The output of each station (X, Y, Z) is determined through operation of the principles of AFPs and consensus, and provides one component of the input to the other stations so that all stations are kept informed about the state of the others. Each station also receives information about the external world (A, novel input, B, self-generated changes) and internal state of the animal (C). Each station therefore combines a different set of inputs. An input may achieve different weighting if it is re-represented in the outputs of several stations. The model operates as several loops, each of which makes a different contribution to the selection and execution of the final output (M) to the effectors. The performance of the effectors is determined by this output in consensus with internal state (C) and mechanical constraints (D). This performance (B) in turn feeds back to all stations (B). The model redrawn in Hopfield and Tank (1986) notation. (From Altman and Kien, 1989.)

stations. The system as a whole thus consists of several loops acting in parallel. We propose that it is the consensus of the activity in these loops that regulates behaviour and provides the basis for decision making.

The network is not homogeneous with everything connected to everything else. First, although the same information is represented in the inputs to each station, it does not necessarily have the same weight or significance at every level. For example, the local circuits of the thoracic ganglia require detailed information about joint angles and muscle tension in the legs to control stepping, but in the brain and suboesophageal ganglion information on the overall leg movement, such as its phase in stepping, is needed to regulate the speed, direction and pattern of walking. Secondly, not all the neurones in a station necessarily receive the same inputs. Thirdly, the local circuits in each station are unique, so the output of each station will be different and hence each station also receives a unique combination of inputs from the others.

The stations also all receive signals about the internal state of the animal which include modulatory inputs that alter the responsiveness of local circuits. Many of these effects are mediated by neuromodulatory neurones which are not directly involved in the generation of coordinated motor outputs, although there is increasing evidence that they play an important part in the selection and coordination of behavioural routines (see Menzel and Bicker, 1987). By acting at several points in the system simultaneously, they form a diffuse loop acting in concert with the loops directly concerned with the generation of outputs. As an element of the input to each station, their effects will be governed by the same principles of AFP and consensus. Similarly, they should receive the same range of inputs as the output-generating circuits, and their activity will likewise be determined by the AFPs of these inputs.

The model in Fig. 9.7A is qualitative and descriptive but lends itself to formalisation because it can be easily redrawn in the Hopfield and Tank (1986) notation for neuronal networks (Fig. 9.7B; Altman and Kien, 1989). Gelperin (1986) and Hopfield and Tank (1986) have shown how this type of model is useful in analysing the functions of various molluscan networks and is able to simulate learning and memory; it provides a powerful tool for the formal modelling of motor function that is now essential for a deeper understanding of the operating principles involved in motor control.

A feature of Hopfield and Tank networks is that, under certain conditions, they become stable, i.e. they produce a constant output pattern for a given set of inputs. If the inputs are changed, the system is perturbed and will then settle into a stable state again. If the perturbation is small the system may move back exactly to or close to its original stable state; if large it may move to a new state. For example, irregularities in the ground may produce a temporary change in stepping pattern but the advent of a predator may change the output from walking to flight. In the Hopfield and Tank version of our model (Fig. 9.7B), the process of reaching a stable state *is* decision making and the output of this state *is* the decision.

Using the Hopfield and Tank notation emphasises our thesis that selection of behaviour and decision making are distributed functions of the system and that no higher order control unit is needed to decide what the rest of the system will do. In our model (Fig. 9.7A) decisions are thus an emergent property of the whole system; they are the outcome of the total activity in all the loops at any time, where each loop regulates different aspects of motor output and hence of behaviour.

9.7. Conclusion

The 'holistic' approach we describe here does not exclude or replace reductionist methods that dissect the operations of the system. Rather it provides a conceptual frame of reference that enables the results of such dissections to be understood in the context of the whole system. In a system as complex as the nervous system, where components interact at many levels, it is necessary not to lose sight of the requirements and limitations of the whole system when studying its parts.

A sound conceptual framework becomes even more essential as experimental methods become more demanding. Techniques such as intracellular recording from single identified neurones are powerful analytical tools, but on their own can give results which can be very misleading. A clearer understanding of the neuronal control of behaviour may be achieved by combining the single-unit approach with methods that can give more information about spatio-temporal patterns of activity, such as multiunit recording and perhaps voltage-sensitive dyes (see also Chapter 6). But results from such experiments will be difficult to interpret without good models to guide the formulation of questions as well as to provide an independent test of results (e.g. Lockery *et al.*, 1989).

Although our discussion has been limited mainly to insects, the evidence presented elsewhere in this volume indicates that the principles we propose here are generally applicable to the study of motor systems in a wide range of animals from molluscs to mammals, and perhaps to sensory processing as well. All these studies emphasise not only the distributed nature of decision making but also the circular operation of the nervous system. The three basic principles that determine our model — AFPs, consensus and interconnections in loops — offer tools for analysing the operations of such circular systems, taking the dynamics and flexibility of behaviour into account.

Acknowledgement

Supported by SFB-4 project H2 of the Deutsche Forschungsgemeinschaft.

References

Alexander, G. E. and Crutcher, M. D. (1991) Control of goal-directed limb movements in primates: neurobiological evidence for parallel, distributed motor processing. In: *Neurobiology of Motor Programme Selection: New Approaches to the Study of Behavioural Choice*, J. Kien, C. R. McCrohan and W. Winlow, eds. Manchester University Press, Manchester, pp. 203–224.

Altman, J. S. (1982) Sensory inputs and the generation of the locust flight pattern: from the past to the future. In: *Biona Report* 2, W. Nachtigall, ed. Fischer Verlag, Stuttgart, pp. 127–136.

Altman, J. S. and Kien, J. (1985) The anatomical basis for intersegmental and bilateral co-ordination in locusts. In: *Co-ordination of Motor Behaviour*, B. M. H. Bush and F. Clarac eds., Cambridge University Press, Cambridge, pp. 92–119.

Altman, J. S. and Kien, J. (1987a) A model for decision making in the insect nervous system. In: *Nervous Systems in Invertebrates*, M. Ali, ed. Plenum Press, New York, pp. 621–643.

Altman, J. S. and Kien, J. (1987b) Functional organisation of the suboesophageal ganglion in insects and other arthropods. In: *Arthropod Brain: its Evolution, Development, Structure and Function*, A. P. Gupta, ed. John Wiley and Sons, New York, pp. 265–301.

Altman, J. and Kien, J. (1989) New models for motor control. *Neural Computation* **1**: 173–183.

Bacon, J. and Möhl, B. (1983) The tritocerebral commissure giant (TCG) wind-sensitive interneuron in the locust I. *J. Comp. Physiol.* **150**: 439–452.

Bässler, U. (1983) *Physiology of Stick Insect Behaviour. Neuroethology of a Simply Organised Insect*, Springer-Verlag, Berlin.

Bässler, U., Foth, E. and Breutel, C. (1985) The inherent walking directions differ for the prothoracic and mesothoracic legs of stick insects. *J. exp. Biol* **116**: 301–311.

Bicker, G. and Pearson, K. G. (1983) Initiation of flight by an identified wind sensitive neurone (TCG) in the locust. *J. exp. Biol.* **104**: 289–293.

Boyan, G. S. (1985) Auditory input to the flight system of the locust. *J. Comp. Physiol.* **156**: 79–91.

Boyan, G. S. and Altman, J. S. (1985) The suboesophageal ganglion: A 'missing link' in the auditory pathway of the locust. *J. Comp. Physiol A* **156**: 413–428.

Boyan, G. S., Ashman, S. and Ball, E. E. (1986) Initiation and modulation of flight by a single giant interneuron in the cercal system of the locust. *Naturwiss.* **73**: 272–274.

Boyan, G. S. and Ball, E. E. (1989) The wind-sensitive cercal receptor/giant interneurone system of the locust, *Locusta migratoria*. III. *J. Comp. Physiol. A* **165** 523–538.

Boyan, G. S., Williams, J. L. D. and Ball, E. E. (1989) The wind-sensitive cercal receptor/giant interneurone system of the locust. IV. *J. Comp. Physiol. A* **165**: 538–552.

Davis, W. J. (1976) Organisational concepts in the central motor networks of invertebrates. *Adv. Behav. Biol.* **18**: 265–292.

Eaton, R. C. and DiDomenico, R. (1985) Command and the neural causation of behaviour: a theoretical analysis of the necessity and sufficiency paradigm. *Brain Behav. Evol.* **27**: 132–164.

Elson, R. C. (1987) Integration of wing proprioceptive and descending exteroceptive sensory inputs by thoracic interneurones of the locust. *J. exp. Biol.* **128**: 193–217.

Erickson, R. P. (1963) Sensory neural patterns and gustation. In: *Olfaction and Taste*, Y. Zotterman, ed. Pergamon Press, New York, pp. 205–213.

Gelperin, A. (1986) Complex associative learning in small neural networks. *Trends Neurosci.* **9**: 323–328.

Grobstein, P. (1991) Directed movement in the frog: spatial representation, motor choice, free will? In: *Neurobiology of Motor Programme Selection: New Approaches to the Study of Behavioural Choice*, J. Kien, C. R. McCrohan and W. Winlow, eds. Manchester University Press, Manchester, pp. 250–279.

Hensler, K. (1989) The pars intercerebralis neurone PI(2)5 of locusts: convergent processing of inputs reporting head movements and deviations from straight flight. *J. exp. Biol.* **140**: 511–533.

Hopfield, J. J. and Tank, D. W. (1986) Computing with neural circuits: a model. *Science* **233**: 625–633.

Huber, F. (1960) Untersuchungen über die Funktion des Zentralnervensystems and insbesondere des Gehirns bei der Fortbewegung und der Lauterzeugung der Grillen. *Z vergl Physiol.* **44**: 60–132.

Kien, J. (1983) The initiation and maintenance of walking in the locust: an alternative to the command concept. *Proc. Roy. Soc. Lond. B* **219** 137–174.

Kien, J. (1990a) Neuronal activity during spontaneous walking. I. Starting and stopping. *Comp. Biochem. Physiol.* **95**A: 607–621.

Kien, J. (1990b) Neuronal activity during spontaneous walking. II. Correlation with stepping. *Comp. Biochem. Physiol.* **95A** 623–638.

Kien, J. and Altman, J. S. (1984) Descending interneurons from the brain and suboesophageal ganglia and their role in the control of locust behaviour. *J. Insect Physiol.* **30**: 59–72.

Kien, J., Fletcher, W. A., Altman, J. S., Ramirez, J. -M. and Roth, U. (1990) Anatomical organisation of intersegmental interneurons in the suboesophageal ganglion of *Schistocerca gregaria* (Forskal) and *Locusta migratoria migratorioides* (Reiche and Fairmaire) (Acrididae, Orthoptera). *Int. J. Ins. Embryol. Morphol.* **19**: 35–60.

Lockery, S. R., Wittenberg, G., Kristan, W. B. Jr and Cottrell, G. A. (1989) Function of identified interneurons in the leech elucidated using neural networks trained by back-propagation. *Nature* **340**: 468–471.

Macmillan, D. M. and Kien, J. (1983) Intra- and intersegmental pathways active during walking in the locust. *Proc. Roy. Soc. Lond.* **218**: 287–308.

Menzel, R. and Bicker, G. (1987) Plasticity in neuronal circuits and assemblies of invertebrates. In: *Neural and Molecular Bases of Learning*, J.-P. Changeaux and M. Konishi, eds. Wiley and Sons, Chichester pp. 433–471.

Möhl, B. (1985) Sensory aspects of flight pattern generation in the locust. In: *Insect Locomotion*, M. Gewecke and G. Wendler, eds. Parey, Berlin, pp. 139–148.

Möhl, B. and Zarnack, W. (1977) Activity of the direct down-stroke flight muscles of *Locusta migratoria* during steering behaviour in flight. II. *J. Comp. Physiol.* **118**: 235–247.

Otto, D. and Campan, R. (1978) Descending interneurons from the cricket suboesophageal ganglion. *Naturwiss.* **65**: 49–50.

Pearson, K. G., Reye, D. N., Parsons, D. W. and Bicker, G. (1985) Flight-initiating interneurons in the locust. *J. Neurophysiol.* **53**: 910–925.

Pearson, K. G. and Robertson, R. M. (1987) Structure predicts synaptic function of two classes of interneurons in the thoracic ganglia of *Locusta migratoria. Cell Tiss. Res.* **250**: 105–114.

Pearson, K. G. and Wolf, H. (1988) Connections of hindwing tegulae with flight neurones in the locust, *Locusta migratoria. J. exp. Biol.* **135**: 381–409.

Peters, B. H., Altman, J. S. and Tyrer, N. M. (1985) Synaptic connections between the hindwing stretch receptor and flight motor neurones in the locust revealed by double cobalt labelling for electron microscopy. *J. Comp. Neurol.* **233**: 269–284.

Pond, C. (1972) The initiation of flight in unrestrained locusts, *Schistocerca gregaria. J. Comp. Physiol.* **80**: 163–178.

Ramirez, J.-M. (1983) Untersuchung der sensorischen Eingänge von multimodalen Neuronen im Unterschlundganglion der Heuschrecke (*Schistocerca gregaria*). Diplom Thesis, Universität Regensburg, Regensburg.

Ramirez, J.-M.(1986) Interneuronal control of flight. Doctoral Thesis, Universität Regensburg, Regensburg.

Ramirez, J.-M.(1988) Interneurons in the suboesophageal ganglion of the locust associated with flight initiation. *J. Comp. Physiol.* **162**: 669–685.

Ramirez, J.-M.and Pearson, K. G. (1988) Generation of motor patterns for walking and flight in motoneurons supplying bifunctional muscles in the locust. *J. Neurobiol.* **19**: 257–282.

Reichert, H. and Rowell, C. H. F. (1985) Integration of nonphase-locked exteroceptive information in the control of rhythmic flight in the locust. *J. Neurophysiol.* **53**: 1201–1218.

Reichert, H., Rowell, C. H. F. and Griss, C. (1985) Course correction circuitry translates feature detection into behavioural action in locusts. *Nature* **315**: 142–144.

Reye, D. N. and Pearson, K. G. (1987) Projections of the wing stretch receptors to central flight neurons in the locust. *J. Neurosci.* **7**: 2476–2487.

Robertson, R. M. (1987) Neuronal circuits controlling flight in the locust: central generation of the rhythm. *Trends Neurosci.* **9**: 278–280.

Robertson, R. M. and Pearson, K. G. (1983) Interneurons in the flight system of the locust: distribution, connections and resetting properties. *J. Comp. Neurol.* **215**: 33–50.

Robertson, R. M. and Pearson, K. G. (1984) Interneuronal organisation in the flight system of the locust. *J. Ins. Physiol.* **30**: 95–101.

Robertson, R. M. and Pearson, K. G. (1985) Neural circuits in the flight system of the locust. *J. Neurophysiol.* **53**: 110–128.

Ronacher, B., Wolf, H. and Reichert, H. (1988) Locust flight behaviour after hemisection of individual thoracic ganglia: evidence for hemiganglionic premotor centres. *J. Comp. Physiol.* **163**: 749–759.

Rost, R. and Honegger, H.-W. (1988) The timing of premating and mating behaviour in a field population of the cricket *Gryllus campestris* L. *Behav. Ecol. Sociobiol.* **21**: 279–289.

Wilson, D. M. (1962) Bifunctional muscles in the thorax of grasshoppers. *J. exp. Biol.* **39** 669–677.

Wolf, H. and Pearson, K. G. (1989) Comparison of motor patterns in the intact and deafferented flight system in the locust. III. Patterns of interneuronal activity. *J. Comp. Physiol. A* **165**: 61–74.

10 *William B. Kristan, Jr, Shawn R. Lockery, George Wittenberg and David Brody*

Making behavioural choices with interneurones in a distributed system

10.1. Introduction

Animals must constantly make a choice among many possible behaviours: to fight or to flee; to investigate or to ignore; to run or to remain motionless; to court or to be aggressive. We make such choices so routinely and so naturally that we are often not even aware that we are doing it. Now that many behaviours are beginning to be understood at a neuronal level (Selverston, 1985; Cohen *et al.*, 1988; Carew and Kelley, 1989), we can start to ask how the choices between behaviours are made among circuits of neurones. Sherrington (1906) was the first to point out that it is necessary to make such choices among interneurones because the same 'final common path' (i.e. the same motoneuronal pool) is activated in different behaviours. McFarland and Sibly (1975), in writing about how animals make such choices in their natural environment, have termed these interneuronal circuits the 'behavioural final common path'. Interneuronal control of behavioural choice is apparent at all levels of phylogenetic complexity, from rat feeding and courtship behaviour (Toates and Halliday, 1980) to leech swimming (Friesen, 1988).

Part of the motivation for studying behavioural choice is to explain the variation in the response of animals to seemingly identical stimuli (Kristan *et al.*, 1982; Drewes, 1984). Such variation may reflect subtle differences in the delivery of the stimulus, in the posture of the animal, or in the animal's internal state, or it may reflect an inherent variability — the 'noise' — in the underlying neuronal circuits. In addition, response variation may be advantageous to an animal (Toates 1980), allowing it to optimise its performance in a varying environment by periodically testing whether a given behaviour has become appropriate. Studying behavioural choice has therefore become largely a study of behavioural variation, usually in one of two ways. The first way is to give a stimulus which sometimes produces one behaviour and at other times produces a different behaviour; this is the 'one-stimulus, two possible response' regime. For instance, touching the mouth area of a sea slug with a rod dipped in food (liquefied squid) can elicit either withdrawal or feeding depending upon the concentration of the squid juice and the state of hunger of the slug; at intermediate levels, the same stimulus can produce one behaviour or the other (Davis, 1979). The second way behavioural choice has been studied is to present two different stimuli simultaneously, each of which elicits an incompatibly different behaviour; this is the 'two stimulus, one possible response' regime. For instance, sea slugs placed on their back will right themselves; if given food while hungry, they will eat the food; if they are turned on their back and given food while on their back, they will either eat the food or right themselves, but not both (Davis, 1979; Croll and Davis, 1987).

Analyses of physiological recordings have generated three general proposals for how neurones make choices among behaviours. The first might be termed *competing circuits*. The competition is usually envisaged as inhibitory interactions between 'decision' or 'command' interneurones (Kovac and Davis, 1980; Croll and Davis, 1987; Krasne and Lee, 1988), but it might involve more complex interactions among sensory, interneuronal and motor networks (Altman and Kien, 1987, 1989). Long-term modifications, such as reproductive state or learning, might change the strength of the interactions but the choice between behaviours is made by synaptic connections among circuits with individual functions. The second proposed mechanism for behavioural choice is *neuromodulation*. In this scheme, each network or collection of networks is very flexible so that each one can produce two or more qualitatively different behaviours; which pattern is expressed is determined by the action of neuromodulators on the cellular and synaptic properties of neurones in the network (Harris-Warwick, 1988; Davis *et al.*, 1989; Dickinson, 1989; Marder and Nusbaum, 1989). A third proposal, *stimulus selection*, proposes that a stimulus could activate either all or just part of a neuronal circuit, producing one behaviour if the whole circuit is activated and an entirely different behaviour if only part of the circuit is activated (Getting and Dekin, 1985). Stimulus selection could be implemented by a different neuronal mechanism than is used in the other two — for instance, by a fluctuating threshold in a critical interneuronal pool — but it might also be implemented by competition among interneurones or by neuromodulation. Therefore, the distinctions among the three types might blur at the level of neuronal mechanism, and they are clearly not mutually exclusive.

We approached the question of behavioural choice in the medicinal leech with these different mechanisms in mind, and we were fooled — twice. The major reason we were fooled is because we thought that each component interneurone would have a very specific function, closely linked to a particular behaviour or part of a behaviour. We have come to call such neurones *committed* to a particular function. Feature detectors, which are activated only by a very restricted set of sensory stimuli, are one sort of committed neurone. A second sort are those neurones that are active only during one type of behaviour (Cohen *et al.*, 1988).

('Command neurones', which trigger a behaviour by themselves or in concert with similar neurones, are a subset of the latter.) Instead of these types, the interneurones we found are activated by many patterns of sensory input and can produce motor patterns that are not very similar to any known behaviour. We speak of such interneurones as subserving *distributed* functions. Since these interneurones contribute to many different behaviours, sometimes in an antagonistic way, it is hard to predict the behaviour of networks composed of them. Because our results have caused us to restructure how we think about behavioural choice, we thought it would be instructive to follow through the logic of our approach and show what the inclusion of distributed neurones does to the notion of behavioural choice. To date we have studied two behaviours — local bending and shortening — with behavioural choice in mind. We will describe each behaviour, show the motoneurone pattern underlying each one, then present something of the interneuronal network responsible for each. In order to understand how distributed networks function, we have generated many model networks, using the back-propagation algorithm on simplified neurone-like units. We have also developed measures to assess the degree of committedness of an interneurone and of a network. We will then discuss these modelling studies and the measures of committedness, to indicate how they helped us to think about behavioural choice.

10.2. **Local bending in the medicinal leech**

10.2.1. *The behaviour in terms of sensory and motoneurones*

Leeches are annelid worms, composed of a head, a tail, and 21 nearly identical midbody segments (Muller *et al.*, 1981). If you push hard enough on the skin of a leech to indent it slightly, the part of the body pushed will actively bend away from the site of touch, whether the touch site was on the dorsal, ventral or lateral body wall (Fig. 10.1A). This bend is local in that it is limited to the segment being touched and possibly to the next segment fore and aft. In response to progressively stronger mechanical stimuli, the leech produces more vigorous, whole-body responses, such as shortening or swimming, but local bending predominates at the site of the stimulus. This behaviour is produced by a reciprocal coordination of the longitudinal muscles on either side of the animal: those on the side touched contract whereas those on the opposite side relax (Fig. 10.1B). From previous studies (Kristan, 1982) we knew that, to a first approximation, local bending is elicited by the activation of P ('pressure') mechanosensory neurones, and the bends are produced by motoneurones that excite or inhibit the dorsal and ventral longitudinal muscles; these are the DE, DI, VE and VI motoneurones (Ort *et al.*,

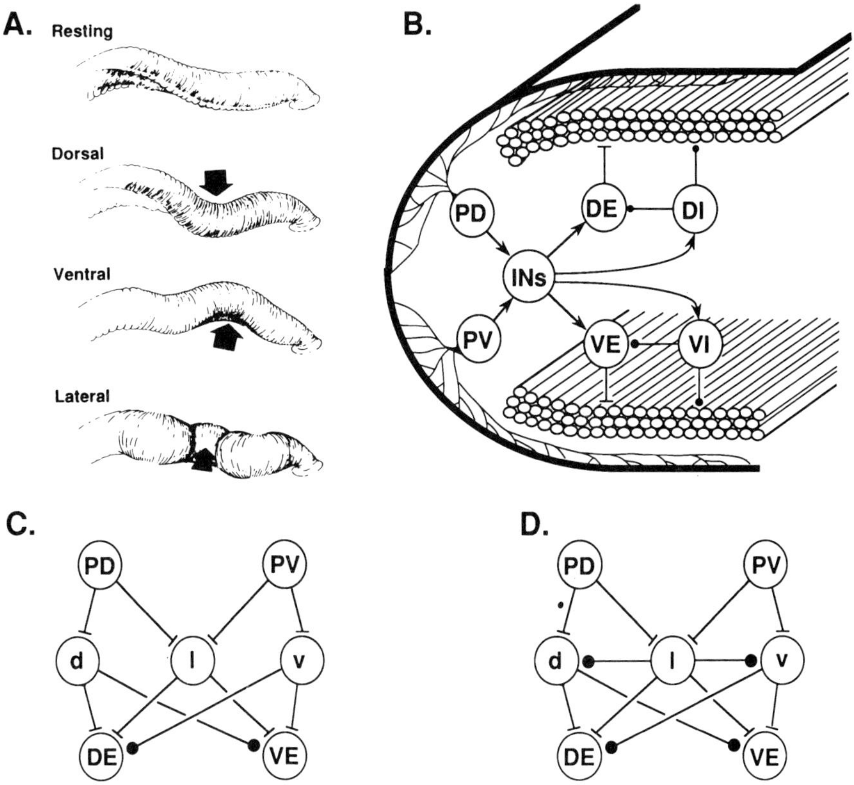

Fig. 10.1. Neurones involved in the local bending response of the medicinal leech. (A) The local bending behaviour of a leech to the application of pressure to the dorsal, ventral and lateral surfaces of its skin. (B) A schematic version of the neurones involved in the local bending response. Shown are the skin, the dorsal and ventral longitudinal muscles and neurones in one side of one segment in the middle of the leech. PD = dorsal P mechanosensory neurone; PV = ventral P-cell; INs = interneurones involved in local bending; DE = dorsal excitor (i.e. excitor of dorsal longitudinal muscles); DI = dorsal inhibitor; VE = ventral excitor; VI = ventral inhibitor. The somata of all the neurones are within the segmental ganglion, which is located on the inside surface of the ventral body wall. (C, D) Two types of neuronal circuits that we expected to be controlling the local bending response, both involving interneurones that are activated only by dorsal (d), ventral (v) or lateral (l) P-cell stimulation. T-junctions represent excitatory synaptic connections, and closed circles represent inhibitory connections. (C) A parallel circuit. (D) A circuit with inhibitory connections from interneurone l to both d and v.

1974). These earlier studies concluded that there were interneurones between the sensory and motoneurones; because the synaptic delays from sensory neurone spikes to the postsynaptic potentials in the motoneurones were short, there appeared to be only one or two layers of interneurones in the pathway.

10.2.2. *Possible interneuronal pathways*

One of us (SLR) decided to find all the interneurones controlling local bending. This is a reasonable goal in the leech nervous system because there are only 400 segmentally iterated neurones in each ganglion, of which most are bilaterally paired, and many of them had already been identified morphologically and physiologically (Muller *et al.*, 1981) Before the search was begun, we anticipated what the interneurones must be like; the sorts of network we imagined are shown in Figs 10.1C and 10.1D. The simplest possibility involved interneurones that respond only to input from dorsal P-cells (the d interneurones in Fig. 10.1C) and other interneurones which receive input only from ventral P-cells (the v interneurones). We anticipated that each of these interneurones would have output connections appropriate for their sensory input: d interneurones would excite DEs and inhibit VEs, and v interneurones would do the opposite. We recognised immediately that these two kinds of interneurone, by themselves, would not produce a *lateral* bend. Assuming that dorsal and ventral bends are of equal strength, activating both the dorsal and ventral P-cells on one side (a stimulus that produces lateral bending in the animal) would produce an equal excitation and inhibition of all the motoneurones and, rather than a lateral bend, this network would produce paralysis in response to a lateral touch. To make this network produce all three behaviours, we imagined a third type of interneurone, the l ('lateral') interneurone, which is activated only by co-stimulation of the dorsal and ventral P-cells on one side and in turn excites the DE and VE on the same side and inhibits the DE and VE on the other side. This system would work if it were arranged as in Fig. 10.1C, because the output of the l interneurone would sum with the mutually nullifying input from the other two types of interneurone to produce the appropriate output for a lateral bend. A somewhat more elegant solution is shown in Fig. 10.1D, in which the battle for the motoneurones is at the interneuronal level: the l interneurones inhibit the d and v interneurones on the same side, so that the only input the motoneurones receive is from the appropriate interneurone. Note that the network in Fig. 10.1C is a *parallel system* from the sensory input to the motor output, whereas the network in Fig. 10.1D depends upon lateral interactions and is therefore partly *serial* in function. We could imagine more complex sorts of interactions, but all of them required more neurones and we did not think the leech nervous system would squander its limited number of interneurones on such a simple set of behaviours. Also, we felt comfortable because the two types of networks made easily distinguishable predictions: one had connections among the interneurones and the other did not.

There are two more underlying assumptions in these auguments about the two types of network. The first is that, at the neuronal level, lateral bending is a behaviour incompatible with dorsal or ventral bending; for local bending to occur, the l interneurone must either override or inhibit the other interneurones. The second assumption is that, in both behaviours, the interneurones are committed to a particular behaviour. That is, both the input and output of all the interneurones are appropriate for just one kind of behaviour: dorsal, lateral or ventral local bending. In the end, when we found the connections among the real leech neurones, all of these predictions and assumptions turned out to be wrong.

10.2.3. *The actual interneurones*

In order to start searching for interneurones it was necessary first to characterise the various local bending behaviours in neuronal terms; i.e. to determine the responses of all the motoneurones to all possible sensory neurone combinations. The four P-cells in a single ganglion were stimulated individually and in pairs to simulate dorsal (left and right PD), ventral (left and right PV) and lateral (PD plus PV on either the left or the right) touches. At the same time, recordings were made from representatives of the eight kinds of longitudinal motoneurone (Lockery and Kristan, 1990a). The averaged responses of the motoneurones were plotted as shown in Fig. 10.2A. The leftmost plot shows that stimulating the dorsal P-cell on the left (PD(L)) excites the dorsal excitor (DE) and inhibits the dorsal inhibitor (DI) on that side. The diagonally opposite side relaxes, both because the ventral excitors (VEs) are inhibited and because the ventral inhibitors (VIs) are excited. The right dorsal and the left ventral quadrants show intermediate responses. The middle plot shows that stimulating the PV(L) mechanosensory neurone produces a ventral bend: the VE on the left is excited and the DE on the right is inhibited, with the inhibitors having responses that are appropriately opposite to those of the excitors. Comparing these two panels, the effects of stimulating PD and PV singly are nearly equal and opposite, although there are some differences. For instance, there is a bias towards dorsal bending in that stimulating a PD excites both DEs and inhibits both VEs, whereas stimulating a PV inhibits the contralateral VE as well as both DEs. However, it is clear that both kinds of stimulation produce a bend away from the site of stimulation.

The rightmost plot in Fig. 10.2A shows the motoneurone response to stimulation of both P-cells on one side, the sort of stimulation that would

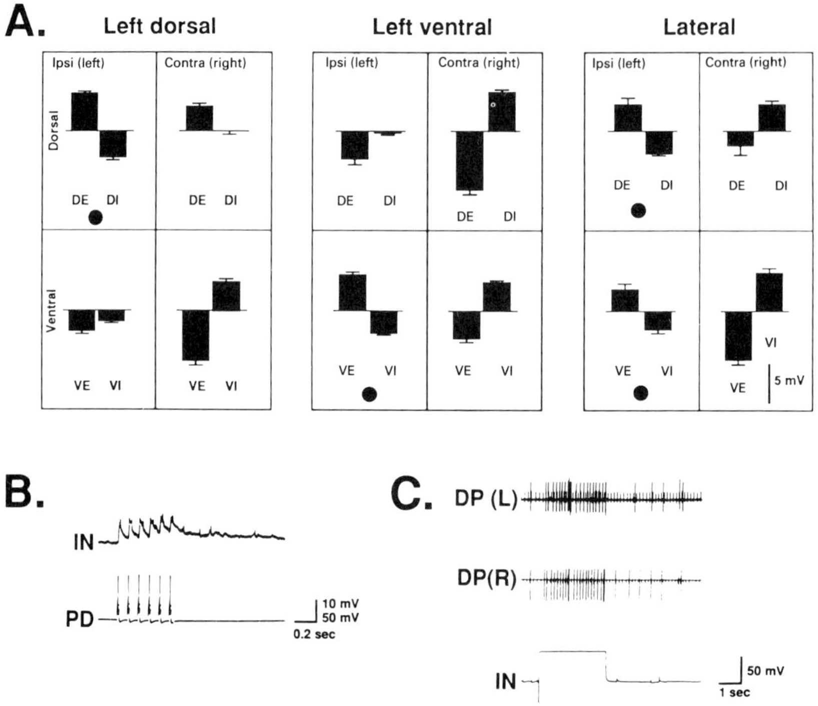

Fig. 10.2. Characterisation of the local bending response in neuronal terms. (A) Graphs of the averaged responses of the eight types of motoneurone (left and right DE, DI, VE and VI motoneurones, as described in the previous figure) which produce contractions and relaxations of the longitudinal muscles, in response to stimulation of individual P-cells which innervate the dorsal skin on the left (left dorsal panel), the ventral skin on the left (left ventral panel), and in response to stimulation of both the dorsal and ventral P-cells on the left (lateral panel) which mimics touching the lateral edge of the leech's body. The four boxes in each cluster represent the four quadrants of the body wall in one leech segment. See legend to Fig. 10.3 for explanation of data presentation. The calibration in the lower right of the lateral panel applies to all the graphs. The large dot indicates which quadrant's P-cell was stimulated to produce the response shown in that panel. DE: Dorsal excitor; DI: dorsal inhibitor; VE: ventral excitor; VI: ventral inhibitor. (B) Intracellular recordings from a dorsal P-cell (PD) and an interneurone (IN), showing the time-locked e.p.s.p.s in the interneurone after each P-cell action potential. (C) Intracellular recordings from the same interneurone (IN) and extracellular recordings from the left and right DP nerves from the same ganglion. The large spikes in the nerve recordings are from the same DE motoneurone (cell 3) whose intracellular recordings are the basis for the panels in part A. The upward deflection in the IN recording indicates when the interneurone was strongly depolarised by current injection.

produce a lateral bend in the animal. Again, the motor response is appropriate: the DE and VE on the side stimulated are excited and the DE and VE on the contralateral side are inhibited. In all cases, the sign of the response to the stimulation of the two P-cells is what is predicted from the summation of the responses to stimulation of the individual P-cells. For instance, the excitation in the left DE produced by the coactivation of PD(L) and PV(L) is predicted by summing the strong excitation from PD(L) with the moderate inhibition from PV(L), and, likewise, the inhibition of the left DI is not surprising because both sensory cells individually inhibit this DI neurone. In some cases these summations are not linear: the response to coactivation is smaller or larger than predicted by simply adding together the responses to the individual P-cells. (It is of some interest that all such nonlinearities are in a direction to strengthen the behavioural responses (Lockery, 1989), but this will not be considered here.)

With these motoneurone responses in hand, the search for local bend interneurones was begun. For a variety of technical reasons, it was decided to start the search by looking for 'dorsal local bend interneurones'. The search strategy was to dissect out a single ganglion, carefully cut away the connective tissue sheath on one or both surfaces to facilitate recording from all the neurones, then systematically search for interneurones that both were excited by PD cells and also activated the DE motoneurones (Lockery and Kristan, 1990b). As shown in Figs 10.2B and 10.2C, the input from PD was assessed by recording from the PD and the interneurone simultaneously whereas the output from the interneurone was determined by monitoring DE motoneurones extracellularly while stimulating the interneurone intracellularly. When this kind of interneurone was found, its position was noted and the search was continued for other neurones. At the end of the search, one or two of the interneurones were re-impaled and injected with the dye Lucifer Yellow to determine whether it was indeed an interneurone (because of the electrical connections among motoneurones, all DE motoneurones would test positive by the criteria used) and to use morphological features to distinguish among interneurones. In a single preparation, as many as 150 neurones were sampled. The conclusion, after many such exhaustive searches, was that there are 17 such 'dorsal interneurones' — eight bilateral pairs and one unpaired interneurone — distinguishable by their sensory input (Fig. 10.3A), the strength of their motor output (Fig. 10.3B), and differences in the branching patterns of their major neurites.

Once identified, the interneurones were further characterised in terms of their total output connections to all the longitudinal motoneurones. (Because the motoneurone somata are on the dorsal surface and only two of the interneurone pairs are on the same surface, the motor output of only these four interneurones could be determined rigorously.) The upper

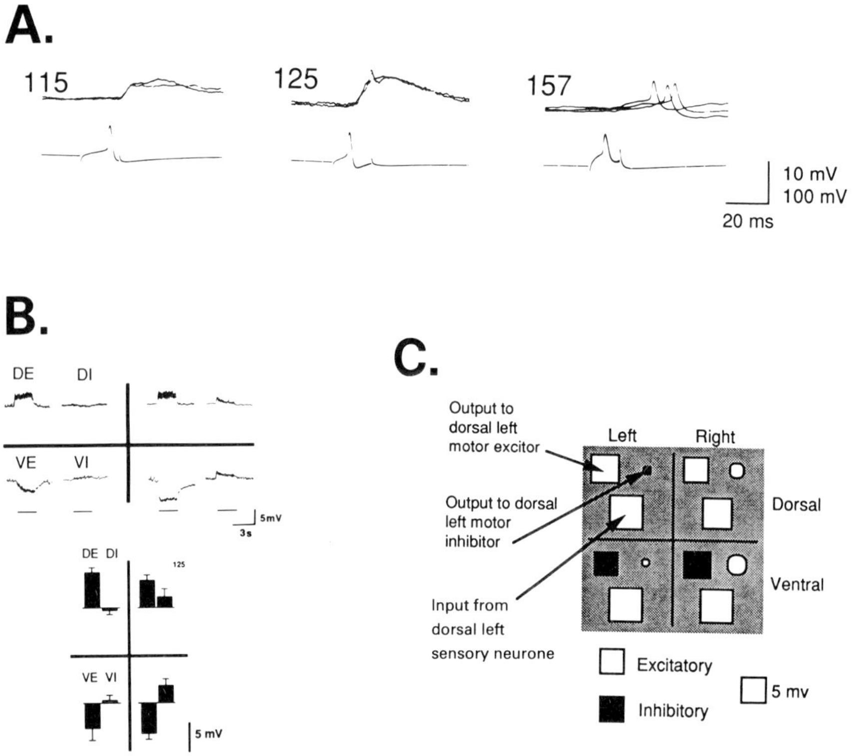

Fig. 10.3. Characterisation of the inputs and outputs of interneurones involved in dorsal local bending. (A) The e.p.s.p.s elicited in three interneurones (cells 115, 125 and 157) by action potentials in a P-cell (lower trace in all three examples). Each panel consists of three overlapping traces triggered by the P-cell action potentials. The upper voltage calibration applies to the interneuronal recording and the lower calibration applies to the P-cell recording. (B) Motoneurone responses to the activation of one interneurone, cell 125. The upper traces are intracellular recordings from the eight types of motoneurone responsible for the local bending response, as in Fig. 10.2A. In each case, the same cell 125 was depolarised at the time indicated by the bar below the bottom trace. The graphs below are averaged maximal responses of these same motoneurones from several different ganglia, indicating the degree of variation in these measures across preparations. (C) The representation of the inputs and outputs of an interneurone in a 'Hinton diagram' (Ackley *et al.*, 1985). Each square or circle represents either an input or output connection, with the size of the square or circle representing the amplitude of the response and the shading representing the quality of the response; white squares represent excitatory connections and black squares represent inhibitory connections.

part of Fig. 10.3B shows the response of each of the eight types of motoneurones to a 2.5-s. intracellular depolarisation of one interneurone, cell 125. These recordings show that this interneurone has connections largely appropriate for the dorsal bending response: it excites the DEs and inhibits the VEs. The bar graph at the bottom of Fig. 10.3B shows the averaged peak response of the motoneurones to cell 125 stimulation in several preparations. Comparing these graphs with those in Fig. 10.2A makes a strong case that this interneurone is related to dorsal bending: it excites and inhibits the excitatory motoneurones in essentially the same pattern as does stimulation of the left dorsal P-cell. In fact, the only qualitative differences between the motor responses to PD(L) and to cell 125 stimulation are in the responses of the right DI and left VI, which are quite small or variable.

We have adopted a graphic representation, called 'Hinton diagrams' to represent input and output connections (Ackley *et al.*, 1985), to represent the connections to and from the interneurones. An example is shown in Fig 10.3C, in which the eight bar graphs at the bottom of Fig. 3B have been redrawn as squares whose area is proportional to the length of the bars together with boxes representing sensory inputs; with white squares representing excitatory connections and black representing inhibitory connections. In addition, we have rounded the corners of the boxes representing outputs to the inhibitors to emphasise that inhibitors and excitors have opposite behavioural effects.

From its motoneurone connections, cell 125 was a prime candidate for being a d interneurone (see Fig. 10.1C,D), so we expected that its input would be exclusively from one or both of the PD cells. It was a major surprise, however, when we found that this neurone received strong input from all four P-cells (Fig. 10.3C). This observation means that cell 125 is strongly activated by touch to the ventral surface, a behaviour for which its output connections are nearly entirely inappropriate. We looked for connections among interneurones that would silence cell 125 or remove its influence from the motoneurones during ventral P-cell stimulation, but we found very little evidence for connections among the interneurones and no evidence for inhibitory connections (Lockery and Kristan, 1990b). Because of the lack of connections among the interneurones, the network controlling local bending appears to be nearly completely *parallel*, just as drawn in Fig. 10.1C. This means that other interneurones activated by the PV cells must make motoneurone connections that counteract all the inappropriate effects of cell 125 during ventral bending. Furthermore, cell 125 was not unique in receiving input from seemingly contradictory sets of P-cells; most interneurones received inputs from both ventral and dorsal P-cells. Therefore, the interneurones responsible for local bending are not committed to one of the types of local bending. Instead, each behaviour is controlled by the activation of most of the interneurones, each one contributing both appropriate and inappropriate influences. This kind of connectivity pattern is very *distributed*: the behaviour produced is a blending of the activity of many interneurones, no one of which produces a motor pattern that is equivalent to any of the individual behaviours or even a subset of any of them.

It is difficult to think about distributed networks because of the complexity of the interactions. We were not sure at first that networks were even capable of producing all three kinds of local bends, given the limited range of interneurone types we had found. Using the back-propagation algorithm to generate model neural networks, we convinced ourselves that there are many ways to connect networks of this complexity to produce these behaviours. In addition, we have used the same techniques to give us a feeling for what such distributed networks are capable of doing. Therefore, we will now take a substantial diversion into how these networks are generated and interpreted, so that we can consider how such parallel distributed networks contribute to behavioural choice.

10.3. **Neural network models of local bending**

10.3.1. *Use of back-propagation to generate networks capable of local bending*

A common network architecture trained by back-propagation contains three types of simplified neurone-like units: *input units*, which are activated in some sort of pattern; *output units*, whose pattern of activity is monitored and compared with the desired 'target' pattern; and, in between, a layer of *hidden units* (Rumelhart and McClelland, 1986; Rumelhart *et al.*, 1986). In its simplest representation the network is strictly 'feed forward': input units connect to hidden units and hidden units connect to output units, with no lateral connections (e.g. no connections among hidden units) and no back-connections (e.g. output units do not affect hidden units). The network is started by making all possible connections, each having a randomly selected small strength or 'weight'. To adjust the connection weights, the network is stimulated with one of the set of input patterns and the output is compared with the appropriate target output. The 'error' of the network is calculated by taking the difference between the actual output and the target output. The size and sign of the error determines how the connections are changed. (The algorithm is called 'back-propagation of error' because the changes are made at the hidden-output layer first, then at the sensory-hidden layer, so that the changes due to the error propagate backwards through the network.) Next, a second pattern is given, the output error is calculated for this pattern, and the connection weights are again changed. This procedure is repeated, typically 1000 to 50 000 times, for each pattern in the set or until the output error is reduced to some preset criterion level.

This algorithm is an efficient and powerful way to produce networks for which particular input and output patterns are known but whose synaptic connection strengths are unknown or known incompletely. It is therefore an efficient tool for constructing neural circuit models in advance of connectivity data. As such, the back-propagation algorithm is really little more than a form of multi-dimensional curve fitting or parameter estimation, in which the connection weights are the parameters to be adjusted and the input–output relations are the functions to be fitted. Although it is common to refer to back-propagation as a 'training' procedure, it is as a curve-fitting technique that we have employed this algorithm because we do not imagine that back-propagation is a model of learning (Crick, 1989), nor that connections in the leech are established by training.

There were several features of the neuronal circuit for local bending that made it ideally suited for the back-propagation algorithm (Fig. 10.4A). First, we knew in detail the activity pattern of output units (the motoneurones) in response to 10 different activity patterns in the input units (the P-neurones). Secondly, the lack of significant connections among leech interneurones justified modelling the interneurones as a single layer of hidden units. Thirdly, the question we were asking (what kinds of connectivity patterns will produce a particular set of motor patterns?) suited the curve-fitting strength of the algorithm. A general weakness of the back-propagation algorithm currently available is that the units are so simplified that they are very poor models of neurones. This is a problem which we keep in mind with each application of the algorithm. For instance, if back-propagation fails to find a network to produce a particular output, it may be because the units sum inputs linearly, or because they do not represent spatial interactions, or because they look only at maximal responses rather than at the full time course of synaptic events. Efforts are being made to make the model units more like neurones (Lockery *et al.*, 1990), but in some ways their very simplicity makes these units valuable, making it much easier to understand what is going on in a network. One can test networks with the simplest possible units, then add in the complexities one at a time to see what they add to the capabilities of the network. Using this strategy, one can sometimes parse a network's function into connectivity-dependent and unit-dependent properties.

Our simulation of the leech network had units representing what we knew of the local bend circuit: it had four input units representing the

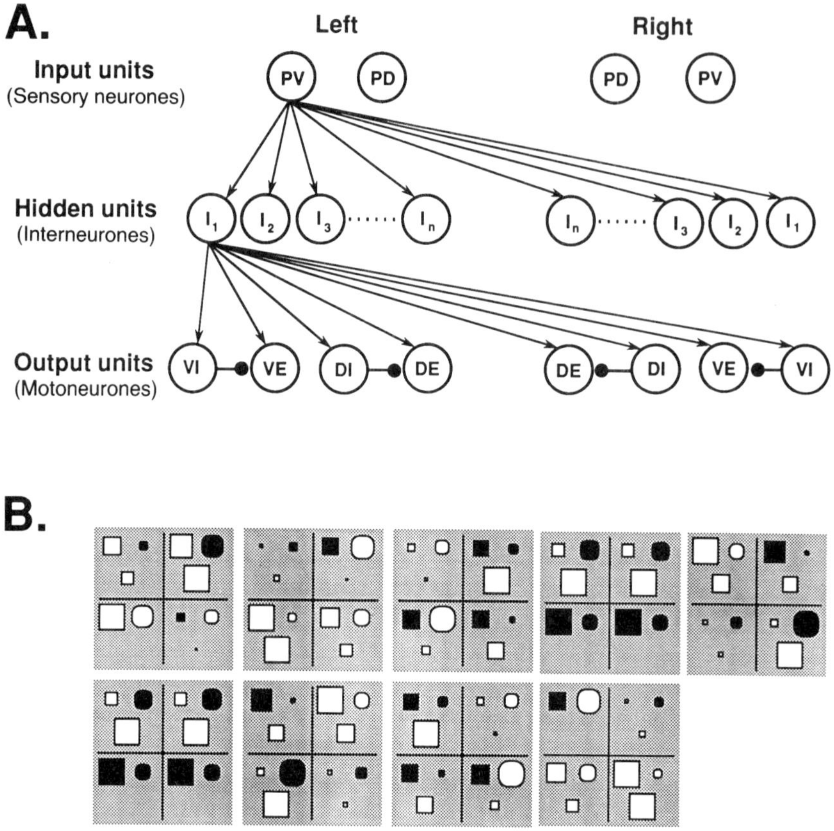

Fig. 10.4. The generation of local bending networks by back-propagation. (A) The elements and initial connections in the typical back-propagation network. At the start, all input units, simulating the P-cells are connected to all hidden units ('interneurones') by weak, excitatory connections (because only excitatory sensory-to-interneurone connections were found in the leech ganglia). All hidden units are connected to all output units ('motoneurones') by weak connections that could be either excitatory or inhibitory. The connections made by the inhibitors (DI and VI) on to the excitors (DE and VE) were strong and not modifiable, simulating the known connections between these neurones in leech ganglia (Granzow *et al.*, 1985). In all networks simulated, the same four P-cells and eight motoneurones were used, with from 4 to 18 hidden units. (These were varied in pairs to simulate the fact that most leech interneurones occur in pairs and their connections are mirror-image identical.) For most simulations, the training set consisted of the responses of the eight motoneurones to the ten different patterns of stimulation of the P-cells (each of the four singly and all six possible pairs). The weights of the connections were changed after each presentation of a training pattern, and the networks were run for 1000 to 50 000 iterations of the training set to produce a match to the target output pattern. (B) The input and output connections (using the Hinton diagram representation explained in Fig. 10.3C) for one network trained on the complete set of 10 different P-cell stimuli. This network was trained for 10 000 iterations and had nine pairs of hidden units. Each grey square represents one of the nine different kinds of hidden units generated; the other nine hidden units were mirror images of these nine units. This network duplicated all the responses of the eight motoneurones with less than a 0.1% error.

four P-cells, eight output units representing the different kinds of longitudinal muscle motoneurones, and 18 hidden units representing the number of known interneurones in each ganglion. (Eighteen were represented rather than the real number of 17 because of simplicity in implementing the third constraint mentioned below. As a test, networks with 16 hidden units were also tested, to simulate ablation of the unpaired interneurone, and they performed essentially the same as those with 18 hidden units.) Networks were initialised and run in a standard way, with all the connections — both those from input to hidden units as well as those from hidden to output units — assigned random, small weights. From what we knew of the neuronal circuit, we constrained the simulated network in three ways:

(1) The connections from the input layer could not be inhibitory, because all interneurones contributing to dorsal local bending had either excitatory input or no input from P-cells (Ort *et al.*, 1974; Lockery and Kristan, 1990b).
(2) There were fixed, strong, inhibitory connections from the DIs on to DEs and from the VIs on to VEs (Granzow *et al.*, 1985).
(3) The 18 hidden units were modelled as nine pairs; the synaptic weights of inputs to and outputs from each member of the pair were adjusted in concert so that all the inputs and outputs of both members of a pair were the same (Lockery and Kristan, 1990b).

Figure 10.4B shows the nine different kinds of hidden units generated from 10000 iterations of one such network. The 'target' output was the motoneurone responses in the eight motoneurones to 10 different P-cell inputs (stimulation of the four P-cells individually and the six possible pairs of P-cells). This network produced a very good representation of the local bend training set: it matched the 80 responses (eight motoneurone outputs in 10 input patterns) to a cumulative error of 4 mV. This amounts to an average error of about 0.002 mV per synapse per behaviour, a level of error that would certainly be undetectable in physiological experiments.

It was easy to show that this network was in no way unique: by starting the back-propagation algorithm to work on a different randomly connected network, a different network was generated that was every bit as good at performing the ten kinds of local bending (Kristan *et al.*, 1989; Lockery *et al.*, 1989). It was difficult to determine just how many different kinds of network would work equally well, because it is difficult to set rigid criteria for calling two hidden units different from one another, but our impression is that there are at least several dozen, different, equally effective networks. The next section outlines on way we have developed to analyse these simulated networks, so that we can start to make such comparisons among them.

10.3.2. *Committed vs distributed system*

From our previous considerations the interneurones responsible for any behaviour must fall somewhere along a spectrum of commitment, with *committed interneurones* being active during only a single behaviour and helping to shape that behaviour selectively, whereas *distributed interneurones* would participate in many behaviours. We have already seen how misleading this concept can be, however, in considering the function of cell 125: from its output connections, cell 125 would appear to be highly selective for dorsal bending, but, judged by its inputs, its function is distributed over all the local bending behaviours fairly equally. Therefore, in order to examine the properties of distributed systems more closely, one must define what it means for both a neurone and the whole system to be committed or distributed. To do this, it is useful to consider what it means for a system to be composed of totally committed or totally distributed units, and to consider what happens when one stimulates or ablates individual units in each type of network. A completely committed network is one in which each unit produces output that is appropriate to one and only one task, when and only when this unit receives the proper stimulus. Removing such a unit from a network composed only of such units would decrease or completely abolish the appropriate behaviour but would not change any other behaviour. In a completely distributed system, the output of each unit would contribute equally to all behaviours performed by the network, and this unit would respond to all stimuli that activate the network. When such a unit is ablated, all of the behaviours would be affected equally, to a relatively small extent. In an effort to develop a quantitative assessment of commitment, we have developed a scale based on such stimulation and ablation experiments. For this we have used simulations, because repeated stimulation and ablation experiments are much more readily performed on simulated networks than on real neuronal circuits.

10.3.2.1. *Committedness measured by stimulation of single units: match.* The first measure of committedness compares the motor pattern produced by stimulating a given hidden unit with each of the behavioural motor patterns; the best *match* for a hidden unit is the local bending pattern most similar to the motor pattern that this unit produces on its own. These comparisons are made using a vectorial method found to be useful for other modelling tasks (Jordan, 1986; Lockery *et al.*, 1989). First, the set of eight output unit activations (four excitors and four inhibitors) for each motor pattern is expressed as an eight-dimensional vector, with the amplitude of each output unit's response measured along one of the eight axes. Next, similar vectors are determined for the output patterns produced by activating each of the hidden units individually. Then, each hidden unit output pattern is compared with each behavioural output pattern by determining the distance between the ends of their vectors in 8-dimensional space; this is measured as the cosine of the angle between the vectors. The actual calculation is the dot product of the two vectors, normalised to the average response of the system:

$$\text{match}_{up} = \frac{\overrightarrow{\text{response}_{up}} \cdot \overrightarrow{\text{target}_p}}{\overrightarrow{|\text{response}_{up}|}\,\overrightarrow{|\text{target}_p|}}$$

where: u = hidden unit number, and p = behavioural pattern type.

This metric gives a convenient measure of similarity. Its values lie between +1.0 (identical motor activation patterns) and −1.0 (opposite patterns); if half the motoneurones were to have identical responses and the other half had opposite responses, a value near 0 would be found. The hidden unit with the 'best match' (i.e. the highest positive match value with a given behavioural output pattern) would produce, if activated on its own, a closer approximation of the behavioural output than would any other hidden unit.

10.3.2.2. *Committedness measured by ablation of single units: relative damage.* Although the match of hidden units to simulated behaviours was useful, we found that this metric failed to capture two types of feature which we felt were important. The first is exemplified by considering a hidden unit which activates exactly the same motor pattern as one of the behavioural patterns, but has very small connection weights, so that activating it produces almost no motor output. This unit would have a perfect match value of 1, but it would contribute almost nothing to the behaviour. Such a result would be expected in a very redundant network, in which this unit was just one of dozens or even hundreds of units doing essentially the same thing, but it was not what we expected in our relatively sparse networks. A second problem is a hidden unit with connections similar to those of cell 125: its output connections are appropriate for one behaviour (e.g. dorsal bending) but it is activated by sensory input that produces the opposite behaviour (e.g. ventral bending). This cell scores high for dorsal bending (match = 0.95), but this measure does not take into account the fact that, in ventral bending, its output connections are entirely inappropriate. Both of these problems are resolved by a second measure which is based on the behavioural effects of ablating each hidden unit. This is done by training a network to perform all the local bending behaviours, then comparing the performance of each behaviour by the intact network with its performance after removing one of the hidden units. (The values compared are actually the squares of the differences of each output unit's activation, as is standand for back-propagation algorithms (Rumelhart and McClelland, 1986)). The total damage sustained is less informative than is the *relative damage* caused, i.e. the amount of damage to a behaviour caused by removing one hidden unit divided by the sum of the damage caused by removing each of the other units. This provides a convenient number between 0 (when removal of a unit causes no damage) and 1 (when removal of a unit eliminates the behaviour and no other unit has any effect). Formally:

$$\text{relative damage}_{up} = \frac{\sum_{o=1}^{8} (\text{target}_{po} - \text{output}_{upo})^2}{\sum_{u=1}^{n} \left(\sum_{o=1}^{8} (\text{target}_{po} - \text{output}_{upo})^2 \right)}$$

where: u = hidden unit, p = behavioural pattern, o = output unit, and n = number of hidden units.

10.3.2.3. *A combined measure of commitment: involvement.* In order to take into account the information gained about commitment from both the stimulation and ablation experiments, we use a combined metric called *involvement*, which is the product of match and relative damage for a given unit in a given behaviour, with some minor arithmetic manipulation to produce a convenient range of possible values:

$$\text{involvement}_{up} = \frac{(\text{relative damage}_{up})(\text{match}_{up} + 1)}{2}$$

where: u = hidden unit, and p = behavioural pattern.

We add 1 to match values so that hidden units that produce motor outputs almost opposite to the desired pattern have match values near 0 and those producing nearly perfect matches have values close to 2. Because the range of relative damage values is also 0 to 1, the involvement values must also lie between 0 and 1. To have a high involvement value for a particular behaviour, therefore, stimulation of a hidden unit must produce an output pattern very similar to the behavioural motor pattern *and* ablation of the hidden unit must severely impair the same behaviour. The problems encountered in using either relative damage or match individually now disappear. For instance, a unit that produces a negligibly small output pattern that is very similar to a behavioural response (the first problem discussed in section 10.3.2.2) can score a nearly perfect 1 for match but near 0 for relative damage, so that the involvement value is small. On the other hand, a unit which produces a response opposite to the desired one will cause a great deal of relative damage when removed (because the compensating connections of the other hidden units are, by themselves, inappropriate for producing the behaviour — see section 10.3.2.2), but the second term has a value near 0, so that the involvement value is again very low.

10.3.2.4. *A measure of unit commitment.* Involvement is a convenient measure, but it does not take into account the specificity of a hidden unit for a behaviour. For this we need to compare a hidden unit's participation in one behaviour with its participation in all the others. To do this, we compare its 'best pattern' (i.e. its highest involvement value) with its match to the other motor patterns (i.e. its other involvement values). The individual differences are bounded by 0 (if the highest involvement value equals the other value) and 1 (if the highest involvement value is at 1 and the other values are 0). This measure is normalised by dividing the sum of the differences by the number of comparisons, so that the values of this metric are also between 0 and 1. Formally:

$$\text{Unit commitment} = \frac{\left(\sum_{p=1}^{np} (\text{involvement}_{ua} - \text{involvement}_{up})\right)}{np - 1}$$

where: u = unit, p = behavioural pattern, a = best pattern for unit u, and np = number of behavioural patterns.

This measure captures all the features that we feel are part of 'commitment': a unit must not only be involved heavily in the production of one behaviour, it must also be minimally involved in all the others.

10.3.2.5. *Measure of network commitment.* We can also use the involvement measures to estimate whether a behaviour is caused primarily by committed units or distributed ones. This is done by comparing the involvement value of the best unit for a given behaviour with all the other units' involvement values for the same behaviour. Again, this measure is normalised by dividing the summed differences by the number of other hidden units, thus making the *network commitment* equal to the average of the differences between the highest involvement value and all the others. With this normalisation, network commitment also ranges from 0 to 1. Formally:

$$\text{Network commitment} = \frac{\left(\sum_{u=1}^{nu} (\text{involvement}_{bp} - \text{involvement}_{up})\right)}{nu - 1}$$

where: b = best unit for behavioural pattern p, nu = number of units, p = behavioural pattern, and u = unit. This definition matches our qualitative judgement of what it means to have a committed network, because network commitment can be low in two ways: if the best involvement value for any hidden unit (i.e. involvement_{bp}) is low, or if no single hidden unit is much more involved than any other unit.

10.3.3. *Examples from simulated networks*

In order to test the utility of these measures of commitment, we have applied them to hidden units in network simulations. We started simply, with small networks such as the one shown in Fig. 10.5, which has only four pairs of interneurones. Despite having fewer units than the number of interneurones found in real leeches (Lockery and Kristan, 1990b), this network performs all ten local bending behaviours quite well. (The cumulative error averages out to less than 0.25 mV per output connection.) Figure 10.5A shows the connection strengths for one such network. It is not apparent from looking at these diagrams that any of the units are committed to any particular behaviour, but the new measures help to assess the level of commitment. In order to show how the commitment measures are applied, the performance of this network was analysed for all 10 types of P-cell stimulation; Figs 10.5B–D show the effects of three such stimuli. (In these three diagrams, the size of the boxes represents the activation level of the output units, which is the connection weights

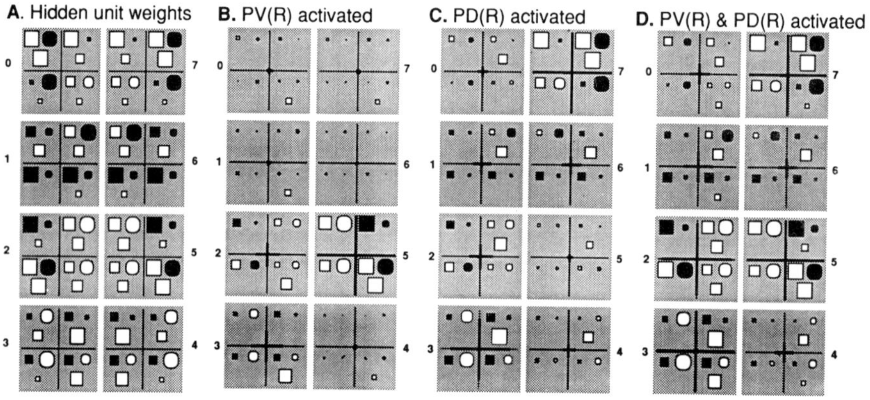

Fig. 10.5. Committed and distributed units in a network generated by back-propagation to produce local bending responses. This network consists of four pairs of hidden units; it was trained for 10 000 iterations and duplicated the output unit responses to the full set of ten different P-cell stimuli to an error of less than 1%. (A) Input and output weights for the four pairs of hidden units (labelled 0 to 7) in Hinton diagram representation. Note that each of the units in the left column is a mirror image of the units in the right column. (B, C and D) Responses of the same network to stimulation of the right ventral (B), right dorsal (C) and both right (D) P-cells. Instead of representing the input and output connection weights, the sizes of the squares and circles for these Hinton diagrams represent the *activation* of the inputs and outputs, which is the weight times the level of activity in each connection. The size of the darker cross in the middle of each unit's diagram represents the sum of the input activations on to the hidden unit produced by that stimulus.

times the activation level of the hidden unit rather than just the connection weights as is represented in Fig. 10.5A and all previous Hinton diagrams.) In response to stimulation of the ventral right P-cell (Fig. 10.5B), only one unit, number 5, is strongly activated and two others, numbers 2 and 3, are moderately activated. A different unit, number 7, is strongly activated when the dorsal P-cell is stimulated (Fig. 10.5C), and several units number (1, 2, 3 and 6) are weakly activated. When both P-cells are stimulated at the same time (Fig. 10.5D), four units are strongly activated and all the others are moderately activated. Qualitatively, therefore, we would say that ventral-right bending is produced by a unit (number 5) which is fairly committed, that dorsal-right bending is produced by a more distributed activation pattern, with one unit (number 7) predominating, and that right-lateral bending is produced by a very distributed pattern.

Application of the commitment measures bears out most of these impressions. For instance, the match for unit 5 to the ventral-right bend is 0.95, which is nearly perfect. When unit 5 is removed from this network, the relative damage is 0.65. Thus unit 5 is highly involved in the ventral-right bending behaviour (involvement = 0.63; unit commitment = 0.55). For purposes of comparison, unit 0 has little involvement in ventral-right bending; its output looks very little like the behaviour (match = −0.25), and its removal does not hinder the behaviour very much (relative damage = 0.004). A more telling comparison is with unit 3, which is the next most involved hidden unit after unit 5. The match of unit 3 to the ventral-right bend is 0.55 and causes a relative damage of 0.21 when removed; its involvement is 0.25, less than half that of unit 5. Intuitively, the activity of unit 3 can be thought to modify the response produced by unit 5 towards the correct output pattern. This modification is significant but not tremendously important, as indicated by the fact that the network commitment value for ventral-right bending is 0.60. This means that ventral-right bending is produced primarily by units with high involvement values, and unit 5 is the only unit with a high value.

Dorsal-right bending (Fig. 10.5C) is primarily accomplished by unit 7 (match = 0.73, relative damage = 0.43) whose involvement in this behaviour is 0.37, much less than the best unit for ventral-right bending. Also, the next most involved unit, number 3, has an involvement of 0.25, close to that of unit 7; therefore unit 3 significantly modulates the effect of unit 7. Because several units contribute to dorsal-right bending, and because the maximal involvement for this behaviour is lower than for ventral-right bending, the network commitment value for this behaviour is only 0.33. This is in line with the impression that dorsal-right bending is accomplished in a more distributed manner than is ventral-right bending.

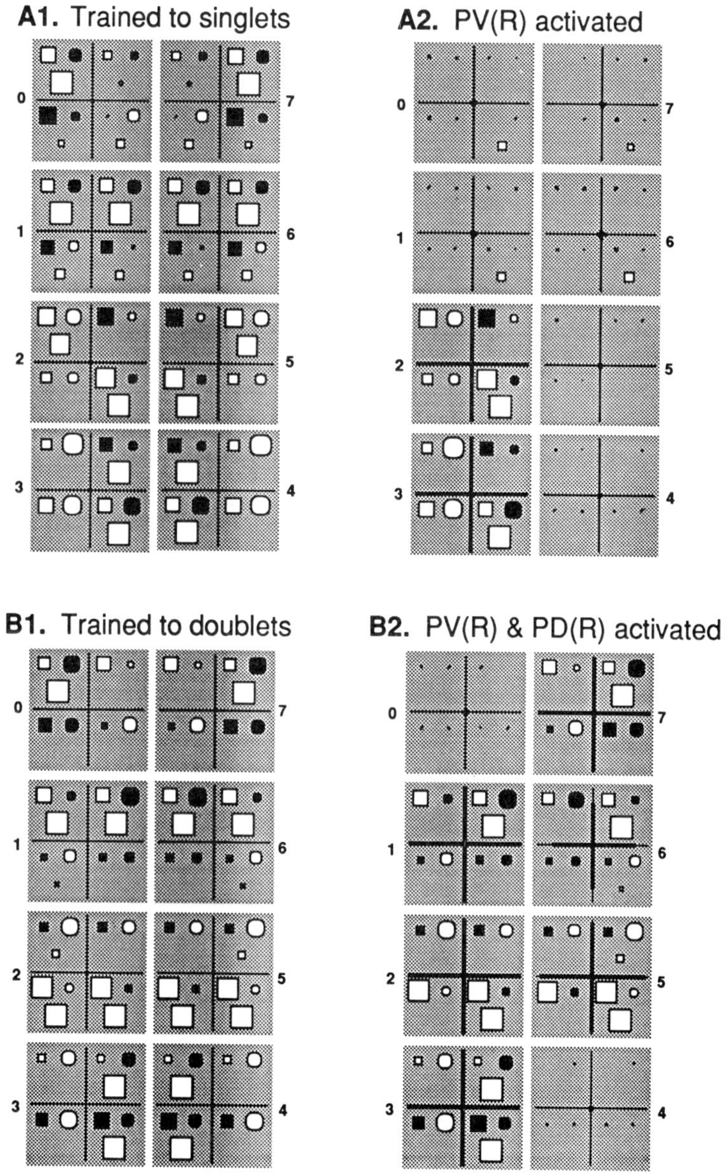

Fig. 10.6. Comparison of connection weights for two networks trained to perform two different sets of tasks. The two networks had the same number of hidden units, four pairs, with identical starting weights. (A1) Connection strengths for the network trained to respond to the activation of the four P-cells singly. This means that the training set consisted of the responses to four stimuli: PD left and right, and PV left and right. (A2) The response of the network shown in A to the stimulation of the right PV cell. (B1) Connection strengths for a network trained to respond to the activation of the four possible adjacent pairs of P-cells. This means that this training set also consisted of four stimulus conditions: both dorsal P-cells, both ventral P-cells, both P-cells on the left, and both P-cells on the right. (B2) The response of the network shown in C to the activation of the two right P-cells.

As mentioned previously, Fig. 10.5D shows that stimulating the two P-cell inputs together gives a very much more distributed response than does stimulation of either one alone. The most involved unit is number 3, but not by much: the involvement of unit 3 is 0.25 compared with 0.20 and 0.12 for units 2 and 7. The low degree of involvement of all the hidden units is reflected in the low network commitment value of 0.19.

Comparing Fig. 10.5D with Figs 10.5B and 10.5C suggests that behaviours triggered by input from more than one stimulus source are done in a more distributed fashion than behaviours triggered by a single input. To test this suggestion we trained two networks from identical starting weights under the same training conditions except that one network was trained to respond to the activation of each of the four P-cells individually (Fig. 10.6A1) and the other network was trained to respond to the activation of the four possible pairs of adjacent P-cells (Fig. 10.6B1). The response of the first network to the right-ventral P-cell is shown in Fig. 10.6A2; it is a highly committed one: only two units are activated and their outputs are very similar to the target pattern (#2 match = 0.52; #3 match = 0.97), and the network commitment for this behaviour is 0.72. The response of the second network, trained to respond to pairs of P-cell activity, is done in a highly distributed fashion. For example, stimulation of both P-cells on the right side (Fig. 10.6B2) strongly activates six of the eight units involved, none of which produces output patterns resembling rightward bending, the target behaviour. In fact, units 5 and 6, which are strongly activated, are mismatched for this behaviour (their match values are −0.25 and −0.38). Only in concert do the hidden units produce the desired behaviour, and no individual unit is essential to the performance of the network. This is reflected in a relatively low network commitment of 0.21. These examples were representative of all the behaviours produced for these two networks: the average network commitment value for all the networks trained on individual P-cell stimulation was 0.44, whereas the average network commitment value for the networks trained on paired P-cell stimulation was 0.18.

These results were in line with the suggestion that networks trained to respond to paired P-cell stimuli were more distributed than networks trained to single P-cell stimuli. However, the possibility remained that, by chance, the randomly chosen starting weights for the two networks in Fig. 10.6 produced the difference in commitment. This was tested by training three other pairs of networks from different starting weights, and the results were qualitatively similar: the network commitment averages for networks trained on single P-cell stimuli ranged from 0.42 to 0.65, whereas networks trained on paired P-cell stimuli had network commitment averages of 0.16 to 0.30. Thus it appears that the nature of the stimulus patterns used to train the network can strongly influence the style of performance — committed vs distributed — of the network.

10.3.4. *Conclusions about commitment and behavioural choice*

Whenever a unit or a network appeared to be committed, it scored high on the commitment scale, and vice versa. Such coherence gives us some confidence that the factors we judged to be important — the match of the units output with the behaviour, and the damage sustained when the unit was removed — are the relevant factors for deciding the degree of commitment of a unit. These measures will have to be elaborated and generalised in order to make them useful for networks in which all the inputs and outputs cannot be fully defined — that is, most neuronal networks — but for now these measures provide a useful way to assess whether a network uses primarily committed or distributed units to solve a given task.

From the modelling studies to date, we are beginning to get a feeling for how simulated networks perform specific behaviours. For instance, when a network is trained to respond to individual inputs, it uses fairly committed units, and when presented with a training set of paired input patterns, the network uses more distributed units. Why do the networks seem to use committed units for simple tasks and more distributed units for more complex ones? So far, we have identified two factors that lead to the production of committed units.

(1) When there is no overlap in the input units activated in the various patterns of the training set, the hidden units tend to be more committed than when there is overlap in the input pattern. For instance, the training set consisting of single P-cell stimuli has no input overlap, and the networks generated are composed of fairly committed units (Fig. 10.6A). The paired P-cell stimuli clearly do overlap (e.g. the right PD is activated with the left PD in one part of the training set and with the right PV in another), and the networks they generate are more distributed (Fig. 10.6B). In other simulations, we found that networks trained on paired but non-overlapping P-cell stimuli (i.e. the two diagonal P-cell pairs), produce very committed networks. In training sets that simulated both single and paired P-cell stimuli (Fig. 10.5), the single stimuli tended to activate more committed hidden units and the paired stimuli activated these same units plus others, so that the network commitment for each single task was higher than for the paired tasks. In a sense, the committed units produce the separate-stimulus tasks and the more distributed units modify the effect of co-activating groups of committed hidden units, to produce the appropriate paired-stimulus response.

(2) For a network to be committed, the number of tasks in the training set had to be nearly equal to the number of hidden units. One could imagine that when the number of tasks equals the number of hidden units, each stimulus could activate just one hidden unit, and the output of that unit would match the task exactly. However, even when there

was such a 1:1 match between tasks and hidden units, the networks generated by back-propagation were never completely committed. In fact, the larger the number of hidden units in a network (from 4 to 18), the more distributed the units became; also, the more behavioural patterns a network was trained to perform (from 2 to 10), the more distributed the networks became. It will be interesting to see whether these tendencies hold for other behaviours, other networks, and other simulation algorithms.

These considerations lead to the prediction that leech behavioural networks would have interneurones that are committed to responding to single P-cells and other interneurones that modulate the responses of these interneurones when more than one P-cell is stimulated. Given the number of interneurones present (17 that produce strong output to dorsal excitors) and the large number of behaviours they control, our experience with simulation suggests that the networks controlling this system would be quite distributed. The characterisation of the interneurones to date, although incomplete, suggests that they are distributed (Lockery and Kristan, 1990b). Now that we have the metrics to quantify and compare the degree of commitment of these units, we can approach the issue of committed vs distributed in a real neuronal circuit in a more substantial way.

We must keep in mind that coordinating the performance of local bending behaviours, however complex it turns out to be, is but one of many functional constraints faced by actual interneurones. For instance, interneuronal networks must function reliably; they must be able to grade the amplitude of the behaviour to different stimulus intensities; and they may not be able to make certain patterns of connections because of limitations imposed by the mechanisms of forming synaptic connections during embryogenesis. Any or all of these might promote or prevent certain types of connectivity patterns a neurone can establish. One intriguing possibility is that interneurones may be involved in many behaviours, even seemingly quite different ones (Getting and Dekin, 1985; Kristan *et al.*, 1988) One such case involves one of the neurones, cell 115, already considered in relation to local bending, which is also involved in swimming (Friesen, 1988) and shortening (Wittenberg *et al.*, 1989). In the following section, we will consider the performance of interneurones involved in both local bending and shortening.

10.4. **The shortening response**

10.4.1. *The physiology of shortening*

Shortening is a graded, whole-body response to mechanosensory stimulation (Kristan *et al.*, 1982). Activation of the same mechanosensory

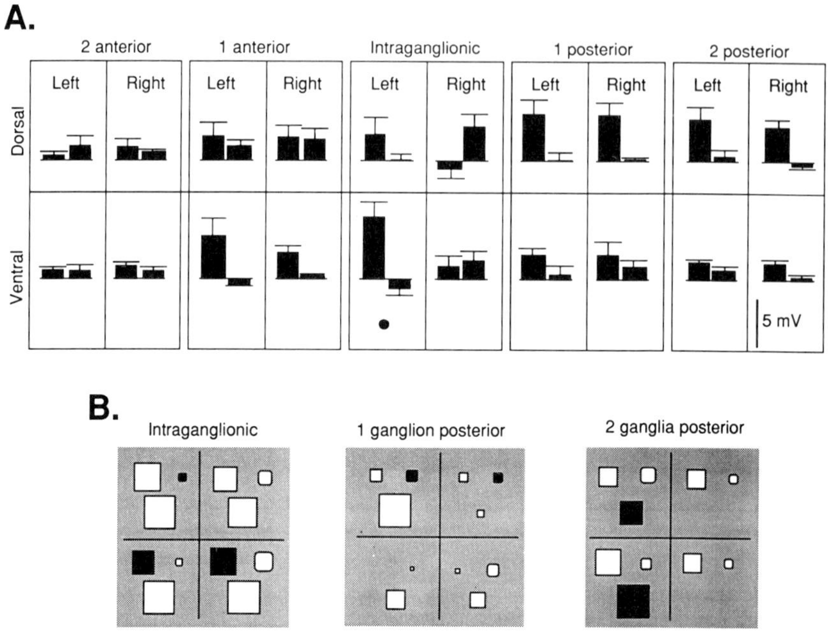

Fig. 10.7. Neuronal characterisation of the shortening response. (A) Averaged motoneurone responses for the same eight types of longitudinal motoneurones (i.e. DE, DI, VE and VI on the left and right sides) in response to stimulation of a left ventral P-cell. The same conventions are used here as for Figs 10.2A and 10.3B, except that the responses are shown for motoneurones in five successive ganglia: the one in which the stimulus was given (the middle cluster of four boxes), the two ganglia more anterior (to the left) and the two ganglia more posterior (to the right). Note that this stimulation produces a left ventral bend in its own ganglion (the left VE is excited and the right DE is inhibited), whereas it produces shortening in the most distant ganglia (both DEs and both VEs are excited in the end ganglia) and an intermediate response in the adjacent ganglia. (B) The connections of cell 125, one of the interneurones involved in local bending, to motoneurones in its own and in adjacent ganglia. As pointed out previously, the output connection within its own ganglion is very similar to dorsal local bending. Its connectivity pattern in adjacent segments is more similar to shortening, however.

neurones — the P-cells — that elicit local bending will also cause contraction of all the longitudinal muscles in segments remote from the site of stimulation. The stronger the activation of the P-cells, the more vigorous is the shortening and the further the shortening spreads from the site of stimulation. The motor output pattern underlying these movements was studied by intracellular stimulation and recording in isolated nerve cords (Wittenberg and Kristan, 1988). The output pattern observed for single ventral P-cell stimulation is shown in Fig. 10.7A. The shortening movement is caused by excitation of both DE and VE motoneurones in ganglia one or more removed from the site of stimulation. In the ganglion in which the P-cell was stimulated, however, the output pattern is that of local bending. It was known that all of the dorsal local bending interneurones had interganglionic projections (Lockery and Kristan, 1990b) and thus that they could potentially contribute to shortening. However, if they made the same connections in remote ganglia, they would produce local bending there as well. Therefore, in order to produce shortening, either the output connections of known interneurones in remote ganglia must differ from their connections in their own ganglion, or there must be additional interneurones activated during shortening whose connections add to the effects of the local bend interneurones in remote ganglia to produce shortening. So far, no such additional, shortening-specific interneurones have been discovered, but local bending interneurones do have different effects on motoneurones locally and remotely (Wittenberg *et al.*, 1989). For instance, the output connection of cell 125 promotes local bending within its own segment but shortening (i.e. co-contraction of dorsal and ventral longitudinal muscles) two segments posterior (Fig. 10.7B).

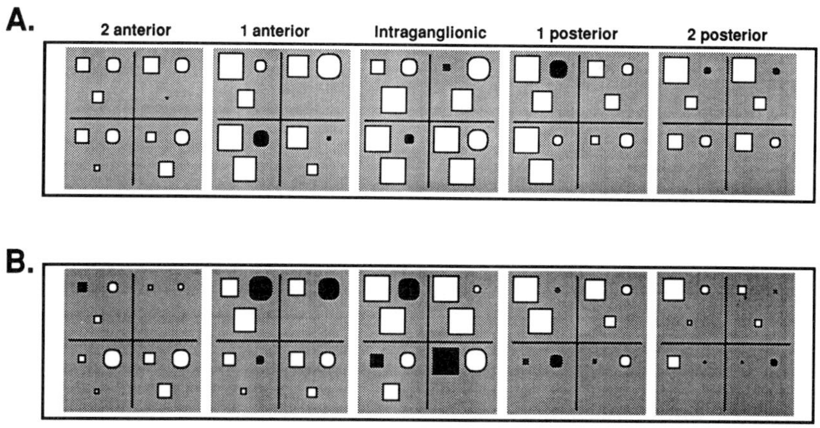

Fig. 10.8. The connections of two hidden units in a network trained by back-propagation to produce local bending in its own ganglion and shortening in each of the two ganglia anterior and posterior. The network contained nine pairs of hidden units. The target output connections were similar to those shown in Fig. 10.7A. The network produced a good match to both behaviours in just 65 iterations. (A) A hidden unit which has qualitatively similar connections in remote ganglia to those in its own ganglion. (B) A hidden unit which has quite different connections in remote ganglia, particularly the more anterior ganglion, from the connections in its own ganglion.

10.4.2. *Network model of shortening*

We have constructed network models capable of shortening, using physiological data like those shown in Fig. 10.7A as the training set (Wittenberg *et al.*, 1989). The target response of the output units was the average response of each of the motoneurones during the 1-s interval following stimulation of a single P-cell. The input and output patterns of two hidden units from one such model network are shown in Fig. 10.8, typifying the two types of hidden unit we have found to date. Neurones of the first type have very similar connections from one segment to the next (Fig. 10.8A), whereas those of the second type are more like cell 125, with different connections in their own segment from those in adjacent ganglia (Fig. 10.8B). Some of these units have one input–output pattern in their own segment, and the nearly opposite pattern that is repeated in each of the other segments. This 'centre/surround' organisation may serve to sharpen segmental differences in motor output in a way similar to the sharpening of contrast by lateral inhibition in sensory systems. We are encouraged by the fact that hidden units show up in the simulations which are so much like the identified interneurone, cell 125. The remaining hidden units will be used as predictors for as yet unidentified interneurones that participate in shortening.

10.5. **Conclusions about behavioural choice in a distributed neuronal circuit**

We started our studies by assuming that the different sorts of local bending were different behaviours, controlled by different interneurones which compete with each other for the control of the motoneurones. We thought that, once we knew how interneurones caused local bending, we would find the additional interneurones responsible for shortening, then see how the interneuronal circuits interacted. We now know that this way of looking at the two systems appears to be doubly wrong. First, local bending is produced by a blending of the actions of interneurones with very distributed actions. For any given behaviour, each interneurone produces some appropriate and some apparently inappropriate effects. It is not as though local bending is an assembly line, with each interneurone tightening down its two bolts and twisting a flange into place, either in series or in parallel, with the co-worker interneurones. Instead, it is as though each interneurone tightens one bolt, loosens another, and pulls the flange part-way to its proper position; other interneurones must retighten the bolt and complete the flange adjustment. It is harder to think about this type of processing, but that appears to be the way leeches perform local bending — and it may be the way molluscs withdraw their gills (Frost *et al.*, 1988; Zecevic *et al.*, 1989), and insects coordinate their behaviours (Altman and Kien, 1987, 1989), and maybe even how vertebrates sense and move (Georgopoulos *et al.*, 1988; Lehky and Sejnowski, 1988; Zipser and Andersen, 1988; Anastasio and Robinson, 1989; Grobstein, 1989).

Our second mistake was in thinking that shortening was composed of a separate interneuronal network. From what we know now, many of the same interneurones used for local bending are also used in shortening, and no interneurones activated during shortening are inactive during local bending. From our modelling studies, we know that it is possible for networks with the number of interneurones used in local bending to produce both shortening and all the local bends. In retrospect, it is not so surprising that these two behaviours might not have separate networks controlling them because shortening and local bending are fully compatible behaviours.

Does this mean that all behaviours triggered by P-cell stimulation are produced by the same set of distributed interneurones? This is very unlikely. For instance, the interneuronal network responsible for leech swimming (Friesen, 1988) has at least three levels of interneuronal processing between the sensory neurones (among them, the same P-cells) and the motoneurones (the same longitudinal muscle motoneurones), with extensive interconnections — largely inhibitory — among the interneurones; the interneuronal network for local bending has no such interconnections. Also, in the search for local bend interneurones, 15 of the 17 interneurones found were different from the ones identified as swimming-generating interneurones. It is intriguing, however, that one of the local bend interneurones — cell 115 — is also part of the swimming pattern-generating circuit. It will be of interest to determine whether any of the other local bend interneurones also participates in swimming. It may well be that the leech nervous system chooses swimming in the way we expected the other behaviours to interact, i.e. by shutting down the other circuits and allowing only the 'swim interneurones' to be active. In this scheme, cell 115 would be like the DSI and VSI interneurones in the sea slug *Tritonia*; both of these molluscan interneurones are active both during reflex withdrawal and during swimming (Getting and Dekin, 1985). We were, however, fooled before by the capabilities of distributed networks, so we will keep an open mind about how the shortening vs swimming choice is made.

In one sense, local bending and shortening are never being chosen between; they are compatible behaviours which are being coordinated. In another sense, however, the lack of choice was apparent only after we analysed the behaviours in terms of the details of their motor patterns and their interneuronal connections. One lesson from these findings is that behavioural choice may turn out to be quite different at a neuronal level from how it appears at a behavioural level, especially if the interneuronal network is quite distributed in its function. It will not be surprising if all the proposed mechanisms for behavioural choice turn out to be used for different behaviours. It is good to keep in mind, however, that the types of mechanism used in a distributed system are likely to be quite different from those used in a committed network.

References

Ackley, D. H., Hinton, G. E. and Sejnowski, T. J. (1985) A learning algorithm for Boltzman machines. *Cog. Sci.* **9**: 147–155.

Altman, J. S. and Kien, J. (1987) A model for decision making in the insect nervous system. In: *Nervous Systems in Invertebrates*, M. A. Ali, ed. Plenum, London, pp. 621–643.

Altman, J. S. and Kien, J. (1989) New models for motor control. *Neural Comput.* **1**: 173–183.

Anastasio, T. and Robinson, D. A. (1989) The distributed representation of vestibulo-oculomotor signals by brain-stem neurons. *Neural Comput.* **1**: 230–241.

Carew, T. J. and Kelley, D. B. (1989) *Perspectives in Neural Systems and Behavior*, Liss, New York, pp. 73–92.

Cohen, A. H., Rossignol, S. and Grillner, S. (1988) *Neural Control of Rhythmic Movements in Vertebrates*, Wiley, New York, pp. 285–332.

Crick, F. (1989) The recent excitement about neural networks. *Nature* **337**: 129–132.

Croll, R. P. and Davis, W. J. (1987) Neuronal mechanisms of motor program switching in *Pleurobranchaea*. In: *Higher Brain Functions: Recent Explorations of the Brain's Emergent Properties*, S. P. Wyse, ed. Wiley, New York, pp. 157–179.

Davis, M., Jacobs, B. L. and Schoenfeld, R. I. (1989) *Modulation of Defined Vertebrate Neural Circuits*, Vol. 563, Ann, N Y Acad. Sci, Washington, DC.

Davis, W. J. (1979) Behavioral hierarchies. *Trends Neurosci.* **2**: 5–7.

Dickinson, P. (1989) Modulation of simple motor patterns. *Seminars Neurosci.* **1**:15–24.

Drewes, C. D. (1984) Escape reflexes in earthworms and other annelids. In: *Neural Mechanisms of the Startle Response* R. C. Eaton, ed. Plenum, New York, pp. 43–91.

Friesen, W. O. (1988) Neuronal control of leech swimming movements. In: *Neuronal and Cellular Oscillators*, J. Jacklet, ed. Marcel Dekker, New York, pp. 269–316.

Frost, W. N., Clark, G. A. and Kandel, E. R. (1988) Parallel processing of short-term memory for sensitization in *Aplysia*. *J. Neurobiol.* **19**: 297–334.

Georgopoulos, A. P., Kettner, R. E. and Schwartz, A. B. (1988) Primate motor cortex and free arm movements to visual targets in three-dimensional space. II. Coding of the direction of movement by a neuronal population. *J. Neurosci.* **8**: 2928–2937.

Getting, P. A. and Dekin, M. S. (1985) *Tritonia* swimming: a model system for integration within rhythmic motor systems. In: *Model Neural Networks and Behavior*, A. I. Selverston, ed. Plenum, New York, pp. 3–20.

Granzow, B., Friesen, W. O. and Kristan, W. B., Jr (1985) Physiological and morphological analysis of synaptic transmission between leech neurons. *J. Neurosci.* **5**: 2035–2050.

Grobstein, P. (1989) Organization in the sensorimotor interface: a case study with increased resolution. In: *Visuomotor Coordination*, J. P. Ewert and M. A. Arbib, ed. Plenum, New York, pp. 537–568.

Harris-Warrick, R. M. (1988) Chemical modulation of central pattern generators. In: *Neural Control of Rhythmic Movements*, A. H. Cohen, S. Rossignol and S. Grillner, eds. Wiley, New York, pp. 285–332.

Jordan, M. I. (1986) An introduction to linear algebra in parallel distributed processing. In: *Parallel Distributed Processing*, D. E. Rumelhart, J. L. McClelland and PDP Research Group, eds. MIT Press, Cambridge, MA, pp. 365–422.

Kovac, M. P. and Davis, W. J. (1980) Neural mechanism underlying behavioral choice in *Pleurobranchaea*. *J. Neurophysiol.* **43**: 469–505.

Krasne, F. B. and Lee, S. C. (1988) Response-dedicated trigger neurons as control points for behavioral actions: selective inhibition of lateral giant command neurons during feeding in crayfish. *J. Neurosci.* **8**: 3703–3712.

Kristan, W. B., Jr (1982) Sensory and motor neurones responsible for the local bending response in leeches. *J. exp. Biol.* **96**: 161–180.

Kristan, W. B., Jr, Lockery, S. R., Wittenberg, G. and Cottrell, G. W. (1989) Behavioral choice — in theory and in practice. In: *The Computing Neuron*, R. Durbin, C. Miall and G. Mitchison, eds. Addison-Wesley, New York, pp. 180–204.

Kristan, W. B., Jr, McGirr, S. J. and Simpson, G. V. (1982) Behavioural and mechanosensory neurone responses to skin stimulation in leeches. *J. exp. Biol.* **96**: 143–160.

Kristan, W. B., Jr, Nusbaum, M. P. and Stern-Tomlinson, W. (1988) Multifunctional interneurons in behavioral circuits of the medicinal leech. In: *Invertebrate Neuroethology*, J. Camhi, ed. *Experientia* **44**: 383–389.

Lehky, S. R. and Sejnowski, T. J. (1988) Network model of shape-from-shading: neural function arises from both receptive and projective fields. *Nature* **333**: 452–454.

Lockery, S. R. (1989) Distributed processing of sensory information in the leech. PhD dissertation, University of California San Diego.

Lockery, S. R., Fang, Y. and Sejnowski, T. J. (1990) Neural network analysis of distributed representations of dynamical sensory-motor transformations in the leech. In: *Advances in Neural Information Processing Systems*, D. Touretzky, ed. Morgan Kaufmann, San Mateo, CA.

Lockery, S. R. and Kristan, W. B., Jr (1990a) Distributed processing of sensory information in the leech. I. Input–output relations of the local bending reflex. *J. Neurosci.* **10**: 1811–1815.

Lockery, S. R. and Kristan, W. B., Jr (1990b) Distributed processing of sensory information in the leech. II. Identification of interneurons contributing to the local bending reflex. *J. Neurosci.* **10**: 1816–1829.

Lockery, S. R., Wittenberg, G., Kristan, W. B. Jr and Cottrell, G. W. (1989) Function of identified interneurons in the leech elucidated using neural networks trained by back-propagation. *Nature* **340**: 468–471.

Marder, E. and Nusbaum, M. P. (1989) Peptidergic modulation of the motor pattern generators in the stomatogastric ganglion. In: *Perspectives in Neural Systems and Behavior*, T. J. Carew and D. B. Kelley, eds. Liss, New York, pp. 73–92.

McFarland, D. J. and Sibly, R. M. (1975) The behavioral final common path. *Trans. Roy. Soc. London B* **270**: 265–293.

Muller, K. J., Nicholls, J. G. and Stent, G. S. (1981) *Neurobiology of the Leech*, Cold Spring Harbor Laboratories, Cold Spring Harbor, NY.

Ort, C. A., Kristan, W. B., Jr and Stent, G. S. (1974) Neuronal control of swimming in the medicinal leech. II. Identification and connections of motor neurons. *J. Comp. Physiol.* **94**: 121–154.

Rumelhart, D. E., Hinton, G. E. and Williams, R. (1986) Learning internal representations by error propagation. *Nature* **323**: 533–536.

Rumelhart, D. E. and McClelland, J. L. (1986) *Parallel Distributed Processing*, 2 vols, MIT Press; Cambridge, MA.

Selverston, A. I. (1985) *Model Neural Networks and Behavior*, Plenum, London.

Sherrington, C. (1906) *The Integrative Action of the Nervous System*, Yale University, Press, New Haven, CT.

Toates, F. M. (1980) *Animal Behavior: A Systems Approach*, Wiley, New York.

Toates, F. M. and Halliday, T. R. (1980) *Analysis of Motivational Processes*, Academic Press, London.

Wittenberg, G. and Kristan, W. B. Jr (1988) Behavioral choice in the leech; characterization of the shortening motor pattern. *Soc. Neurosci. Abs.* **14**: 690.

Wittenberg, G., Lockery, S. R. and Kristan, W. B., Jr (1989) Intersegmental interneurons contribute to multiple behaviors in the leech. *Soc. Neurosci. Abs.* **15**: 1119.

Zecevic, D., Wu, J.-Y., Cohen, L. B., London, J. A., Hopp, H. P. and Falk, C. X. (1989) Hundreds of neurons in the *Aplysia* abdominal ganglion are active during the gill-withdrawal reflex. *J. Neurosci.* **9**: 3681–3689.

Zipser, D. and Andersen, R. A. (1988) A back-propagation programmed network that simulates response properties of a subset of posterior parietal neurons. *Nature* **331**: 679–684.

Part III

Selection of directed movements

11 *Garrett E. Alexander and Michael D. Crutcher*

Control of goal-directed limb movements in primates: neurobiological evidence for parallel, distributed motor processing

11.1. The problem of motor control

The control of goal-directed limb movements requires that the target or goal of each movement be translated quickly and efficiently into an appropriate set of muscle activation patterns that will carry the limb along the required trajectory. From an analytical perspective (Fig. 11.1), this basic problem of motor control is enormously complex, requiring sequential computations of target location, limb trajectory, inverse joint kinematics, inverse joint dynamics, and muscle activation patterns (Paul, 1981; Hollerbach, 1982). These computations also require the reconciliation of multiple spatial frames of reference, including those of external space (which must be translated from retinotopic, to craniotopic, to body-centred coordinates), joint space, and muscle space (Saltzman, 1979; Pellionisz and Llinas, 1985). The problem is of sufficient magnitude that even within the highly constrained environments in which robots are expected to operate, robotics engineers generally arrange for these computations to be carried out during a planning phase that precedes the initiation of any new goal-directed movement of an artificial limb, rather than 'on-line' during the course of movement execution (An *et al.*, 1988).

But the luxury of a substantive planning phase is often not available to animals making goal-directed movements, nor are they generally afforded the stable conditions of a highly controlled environment. Thus, it is difficult to overestimate the complexity of the problem faced by primates when they attempt to make targeted limb movements in three-dimensional space (Bernstein, 1984). The primate motor system represents an evolutionary solution to this problem whose success has yet to be approximated by human engineering. And since the neural mechanisms that underlie the evolutionary solution remain largely unknown, primates and most other mammals continue to be vastly superior to robots in the performance of motor tasks that require flexible interactions with a changing environment.

Though often employed in robotics, the serial/analytic approach to motor processing seems a relatively implausible account of how the brain might control movement, because of the extremely slow (by semi-conductor standards), noisy and stochastic operating characteristics of its

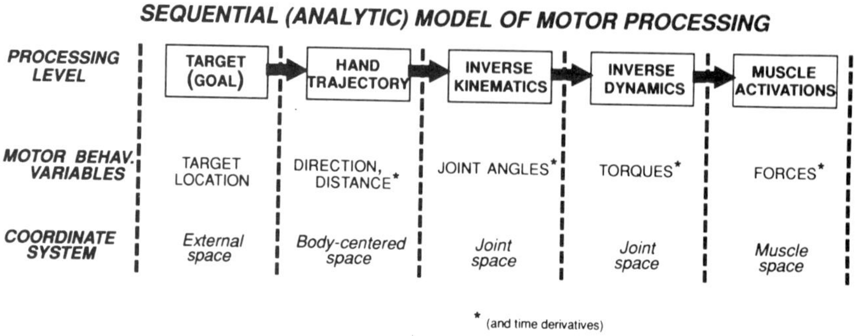

Fig. 11.1. Schematic representation of the theoretical levels of motor processing that would be required if the brain used a sequentical, analytical approach to motor processing. From Alexander and Crutcher (1990b).

component neurones (Loeb, 1983). Even if computational short cuts were employed (Hinton, 1984), the length and complexity of the calculations required to compute analytical solutions for each new movement would seem to be incompatible with the manifest speed and agility of primates and other mammals, especially in view of the notorious inefficiency of brains as compared with computers in the performance of arithmetic operations (Rumelhart and McClelland, 1986).

11.2. **Concept of motor programmes**

To many theorists, the concept of the *motor programme* has seemed to offer an attractive, alternative account of motor processing because it suggests that the brain may be able to circumvent, or at least to minimise, the seemingly prohibitive complexity and temporal demands of the serial, analytical approach by using a limited set of generic, prepackaged motor commands to control a large repertoire of goal-directed movements (Schmidt *et al.*, 1979; Rosenbaum, 1985; Pew, 1989). The notion has also been motivated by the fact that some movements can be carried out in the absence of sensory feedback, suggesting that the motor system must contain some types of internal representation that are used to control movements of this sort (Keele, 1981). Over the past two decades the idea that movements are controlled by motor programmes has become quite popular, and this view is now inherent, in one form or another, in most contemporary theories of motor control.

In its most general form the concept of the motor programme is relatively unassailable. It simply refers to a theoretical (but unspecified) mechanism that makes it possible for a subject to execute a class of related movements in an efficient and reproducible manner. When used in this fashion, without implied insight into the neurobiology of the underlying mechanism, the concept provides a useful summation of one of the central tenets of motor psychophysics: namely, that most movements represent coherent patterns of behaviour whose reproducibility indicates that the information required for their control must be both stored and later retrieved by some mechanism residing within the motor system.

The motor programming concept is especially attractive when applied to simpler organisms with supposedly limited motor behavioural repertoires, where there may be considerable explanatory power in the notion of a finite set of 'commands' that can be combined in various ways to control an assortment of hard-wired motor pattern generators. However, even in these preparations, such ideas are now open to question (see Winlow *et al.*, Chapter 4, and Kien and Altman, Chapter 9), and when the concept has been extended to animals with more complex behavioural repertoires, including primates making goal-directed limb movements, its explanatory value has been rather limited. Though frequently offered as an explanation for certain invariances observed in studies of human motor psychophysics (e.g. constancies in the geometric features of limb trajectories or in the linked kinematics of adjacent joints during multijoint movements), the notion that the brain uses motor programmes to control movements has done little in itself to illuminate the mechanisms that might underlie such phenomena. Most likely this is due to the fact that motor programmes are generally construed in terms of a standard computer metaphor, despite overwhelming evidence that the functional architecture of the vertebrate brain is fundamentally different from that of a digital computer.

The traditional concept of a motor programme implies a firm distinction between the motor system's hardware (sets of interconnected neurones) and software (presumably, the motor programmes themselves, each consisting of a specific set of motor 'commands'). But such a model presents insuperable difficulties when one tries to imagine how motor programmes might actually be represented within the brain's neural architecture. What, for example, would constitute the 'software' in such a model? The conventional answer is likely to be 'patterns of neuronal discharge' among the neurones that make up the hardware of the motor system. But this fails to address such questions as where and how such software might be stored while it was not being 'run', and, more importantly, how might the algorithms and software that constitute these motor programmes be created in the first place. We are not aware of anyone having put forward a detailed model of a motor programme that addresses these issues.

In attempting to look beyond behavioural phenomenology and suggest more plausible neurobiological substrates for the acquisition, retention and reproducible execution of goal-directed movements, it seems increasingly worthwhile to consider a group of models that differ radically from those based either on the traditional notion of motor programmes or on the serial/analytical approach.

11.3. **Parallel distributed processing**

The emerging concept of *parallel distributed processing* (PDP) (Rumelhart and McClelland, 1986) suggests that it may be feasible to account for even the complex and highly adaptive behaviour of primates with models of motor processing that are both meaningful and plausible from a neurobiological perspective. In recent years, PDP models have been developed to account for a variety of behavioural processes, including vision (Linsker, 1986), somaesthesia (Grajski and Merzenich, 1990), vestibulo-ocular function (Anastasio and Robinson, 1990) and motor control (Kuperstein, 1988; Grossberg and Kuperstein, 1989). While it would be inappropriate here to attempt to review the characteristic features of PDP models in any detail (for reviews of this topic see Rumelhart and McClelland (1986); Schwartz, 1990), it does seem worthwhile to consider some of the implications of such models with respect to the issue of motor processing in general and, more particularly, the control of goal-directed limb movements.

The essence of PDP models of brain function is that they are self-organising, massively parallel networks of neurones in which information is stored not in discrete locations (e.g. as bytes of information) but rather in the overall pattern of variable-strength connections among neurones. Thus, there is no fundamental distinction between hardware and software in PDP models, and no mystery about the manner in which information determining the network's response properties is stored while the network is not actively engaged in processing new inputs. Motor processing, according to such schemes, consists of two distinct phases, a learning phase and an execution phase. In PDP models, solutions to immediate or prospective motor control problems are achieved during the learning phase, in which the strengths or 'weights' of the network's connections are adjusted through a process of trial-and-error and repetition. In models with purported relevance to neurobiology, the weight adjustments are determined by purely local, activity-dependent processes operating at the level of each synapse, usually according to a Hebbian (Bear *et al.*, 1987) learning rule. The execution phase of parallel, distributed motor processing consists of the trained motor network's response to massively parallel inputs, either from the periphery or from other networks with which the motor system is connected. From a PDP perspective, the motor system's response during the execution phase is predetermined by the current pattern of synaptic weights (which has resulted from the cumulative effects of all previous episodes of both trial-and-error learning and repetition of learned responses through practice and/or habit) and thus does not really require the computation of a new solution for the motor problem at hand.

It is important to realise that there is a crucial distinction between the nature of information processing in PDP networks vs 'serial/analytical' or traditional 'motor programming' models. In PDP networks, the processing of information is essentially '*non-algorithmic*', i.e. it is not based on a well defined set of logical or mathematical operations. Rather, the manner in which information is processed within a PDP network is determined by the current overall pattern of synaptic weights, which has resulted in turn from an essentially *random* process of distributed, trial-and-error learning that selects for the successful solutions (i.e. the optimal weight configurations) simply because they are successful and, as such, tend to be reinforced through repetitive behaviour (either through habit or practice) on the part of the organism.

According to the general definition of a motor programme as a mechanism that accounts for the efficient reproducibility of motor behaviour, it might be argued that motor programmes are both meaningful and useful constructs within the framework of PDP, as they can be conceived as being represented by the patterns of synaptic weights within the motor system. It should be noted, however, that in PDP models of motor control all movements that a network has been trained to execute will be represented by the same overall pattern of synaptic weights throughout that network. Thus, from the PDP perspective, motor programmes could not be considered to exist as discrete, separately identifiable entities within a trained but inactive network. On the other hand, when executed, each movement would be associated with a unique *pattern of activation* of different elements within the network, and to this extent it could be said that the 'running' of each motor programme would indeed be expected to have a unique physiological signature.

Regardless of whether the motor programme concept is considered useful from the standpoint of PDP models of motor processing, the models themselves are attractive not only because of their biological plausibility but because they seem to account for a number of psychophysical observations that present serious difficulties for other theories of motor control. Using distributed processing to achieve non-algorithmic solutions to the problem of motor control, PDP models show the same type of fault tolerance that is manifested in biological systems (Rumelhart and McClelland, 1986; Schwartz, 1990), whereas analytical/algorithmic approaches have been shown to be extremely sensitive to small computational errors (Hogan *et al.*, 1987). While generally absent in other theories of motor control, the property of self-organisation is an additional feature of PDP models that appears to be shared by natural motor systems (Kuperstein, 1988; Grossberg and Kuperstein, 1989). PDP models may also provide an explanation for one of the most important yet least understood aspects of goal-directed limb movements in primates: the phenomenon of 'motor equivalence'.

11.4. **Motor equivalence**

One of the salient features of human motor behaviour is the ease with which learned movements, even those with complicated trajectories such as drawing or writing, can be readily scaled in time and space, transferred between extremities, and adjusted for the use of hand-held tools, all with little or no practice. This uniquely biological phenomenon of 'motor equivalence' is well known (Soechting and Lacquaniti, 1981; Viviani and Terzuolo, 1982; Bernstein, 1984), but its neural basis is poorly understood. Such a capacity for ready adaptability to changing task demands is, by comparison, virtually absent in conventional robots (Paul, 1981; An *et al.*, 1988). One explanation for the ease with which subjects can modify previously learned movements is that motor control in humans and other primates may be organised primarily in terms of the spatial aspects of the task, rather than the kinematics and dynamics of the joined limb (Morasso, 1981; Bernstein, 1984; Hogan *et al.*, 1987).

In natural reaching towards objects, the hand's trajectory tends to be straight and its velocity profile smooth and bell-shaped, as would be expected if such movements were planned and controlled in terms of the coordinate system of external space (Morasso 1981; Abend *et al.*, 1982). Because of the complex relations between multi-joint (wrist, elbow, shoulder) kinematics and hand trajectory, however, if target-directed movements were controlled primarily in terms of joint kinematics even small degrees of incoordination among the joints involved could produce irregular and more curvilinear hand trajectories (Hogan *et al.*, 1987). Observations such as these have prompted several motor theorists to propose that goal-directed limb movements may be organised in terms of the target or goal of the movement, rather than the kinematic or dynamic features of the movement itself (Morasso, 1981; Bernstein, 1984).

One possibility is that the motor system may receive information from other brain areas that are concerned with identifying the target or goal of forthcoming movements, with the motor system in turn translating that target or goal into an appropriate set of muscle activation patterns by means of non-algorithmic PDP mechanisms that include the emergent properties of fault tolerance and pattern generalisation (Rumelhart and McClelland, 1986; Kuperstein, 1988; Grossberg and Kuperstein, 1989). These latter properties could readily account for the phenomenon of motor equivalence. It is important to recognise, however, that although PDP models do employ non-algorithmic mechanisms to solve problems of motor control, they do not necessarily predict an absence of neural representations for those levels of motor processing considered, from an analytical/algorithmic perspective (see Fig. 11.1), to be 'lower' than that of the target or goal of the movement. In parallel distributed motor processing, neural representations of the 'lower' processing levels might well be expected to emerge incidentally during the course of motor learning, even though the processing within PDP networks is not strictly organised according to a logically prescribed sequence of computations (see, for example, Linsker, 1986; Kuperstein, 1988).

11.5. **Neural representations of movement direction vs muscle pattern**

In an attempt to address some of the above-outlined issues regarding the neural substrates of goal-directed limb movements in primates, we recently carried out a series of chronic neurophysiological studies in awake, behaving monkeys (Alexander and Crutcher, 1990a,b; Crutcher and Alexander, 1990). The studies were based on the assumption that, if any or all of the analytically defined levels of processing were represented within the motor system, there are at least two ways in which they might be organised. One possibility, consistent with the serial/analytical model, is that the different processing levels (and their associated motor variables) might be represented in separate, *serially* connected motor structures, each of which would be functionally specialised to deal with a particular aspect of motor control. The motor system might also be organised *in parallel*, however, with individual motor structures participating simultaneously in several levels of processing, and with each processing level being distributed over multiple structures.

We also sought to determine which, if any, of the analytically defined levels of motor processing were represented during the *preparation* for movement, and which during movement *execution*. Previous studies have shown that neurones in multiple motor areas, including the precentral motor fields and their projection zones within the basal ganglia, show striking changes in discharge rate following presentation of an instructional stimulus that specifies the direction of an impending, stimulus-triggered arm movement (Tanji and Evarts, 1976; Thach, 1978; Weinrich and Wise, 1982; Alexander 1987). These directionally specific, instruction-dependent changes in activity are characteristically sustained until the occurrence of the movement-triggering stimulus, and generally subside as the pre-planned movement is initiated. This type of 'preparatory' activity is generally thought to represent the neural correlate of 'motor set', the state of being 'ready' or 'prepared' to execute a particular movement (Gibson, 1941; Ryan, 1970). Motor set appears to be directly analogous to directed attention or 'perceptual set' in the sensory realm (Evarts *et al.*, 1984).

We compared neuronal activity associated with the planning and execution of visually guided limb movements in three motor areas of the monkey: the supplementary motor area (SMA), primary motor cortex (MC), and the putamen. All are components of the basal ganglia-thalamocortical 'motor circuit' (Alexander *et al.*, 1986), with the SMA and MC sharing reciprocal connections and projecting in turn to the putamen, which returns its own influences to the SMA and rostral MC via intermediate connections in the globus pallidus and ventrolateral thalamus (for a recent review of the functional organisation of this circuit see Alexander *et al.* (1990)).

Rhesus monkeys were trained to perform a set of motor tasks that permitted dissociation of a number of the motor behavioural variables involved in making visually guided limb movements to capture targets in space (Fig. 11.2). Each of the tasks was a delayed visuomotor step-tracking paradigm in which elbow movements were made, both with and without prior instruction about the direction of the forthcoming movement. In this way, the tasks made it possible to differentiate between those processes involved in movement preparation and those related to movement execution. In addition, the tasks dissociated three different levels of motor processing (see Fig. 11.1). To dissociate the direction of limb movement (and preparatory set) from the task-related patterns of tonic and phasic muscular activation, some trials included the application of a constant torque load that either assisted or opposed the movements required by the behavioural paradigm. This effectively dissociated *trajectory/kinematics*-level variables from *dynamics/muscle*-level variables. Also, during some blocks of trials, the cursor and forearm moved in the same direction, but on others there was an inverse relationship between the directions of movement of the forearm and cursor. Together, these two types of trial dissociated the spatial features of the target or goal of the movement from those of the movement itself, that is, *target*-level variables from *trajectory/kinematics*-level variables.

While each monkey performed the tasks, the activity of single neurones from the arm regions of the SMA, MC and putamen contralateral to the working limb was recorded extracellularly. A total of 741 task-related neurones were studied, including 222 within the SMA, 202 within the MC, and 317 within the putamen. In the MC and putamen, the majority of task-related cells showed changes in activity *only* during *movement execution* (63% and 67%, respectively), while less than half of those in the SMA (45%) showed this pattern of activity. Each area also contained substantial proportions of neurones that manifested *preparatory* activity, i.e. cells that showed task-related changes in discharge rate during the post-instruction (preparatory) interval when the monkey knew the direction of the impending movement. The largest proportion of cells that showed *only* preparatory activity was seen in the SMA (32%) and the smallest in the MC (11%), with the putamen having an intermediate

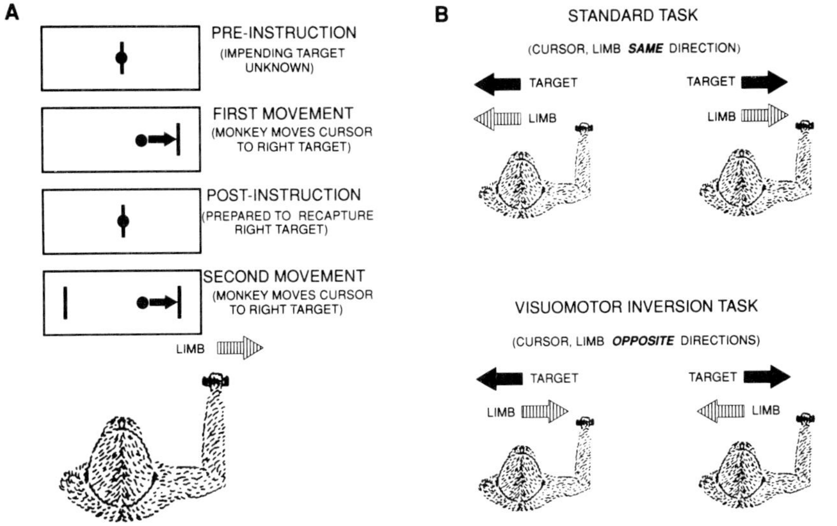

Fig. 11.2. (A) Schematic representation of the display viewed by the monkey during performance of the visuomotor step-tracking task. The four rectangles show the CRT display in front of the monkey at four different times within a single trial. Targets are represented by vertical bars, and the cursor by a closed circle. The position of the cursor reflected the angular position of the working forearm. At the start of the trial the centre target appeared and the monkey 'captured' it by making the appropriate arm movement to align the cursor with the target. Throughout the ensuing 'pre-instruction' interval, the monkey held the cursor stationary over the centre target for 1.5–3.0 s. During this time, the monkey could predict neither the location to which the target would shift nor the required direction of the next limb movement. When the centre target shifted to one of the two (randomly selected) side locations, the monkey was required to capture this new target by moving his forearm in the appropriate lateral direction. After the first side target had been captured (and the cursor held in alignment with it for 500 ms), it shifted back to the centre position and monkey was required to track the apparent target movement by returning the cursor to the centre position. During the ensuing 1.5–3.0 s 'post-instruction' interval, the monkey knew the direction of the next (second) lateral movement of the target because he was required to recapture the target at the same location as for the first lateral movement. The simultaneous appearance of both side targets was the cue for the monkey to make the second lateral movement. To dissociate the direction of preparatory set (and limb movement) from the task-related pattern of tonic (and phasic) muscular activation, some trials included the application of a constant torque load (0.1 N m) that either opposed or assisted the movements required by the behavioural paradigm. From Alexander and Crutcher (1990b). (B) Schematic representation of the two tasks used to dissociate expected target location and direction of intended limb movement. In the standard task, the cursor which the monkey used to 'capture' the targets moved in the same horizontal direction as the monkey's forelimb. When the target shifted from the centre position to the right, the monkey would move his right forearm to the right (elbow extension) in order to position the cursor over the target. In contrast, for the visuomotor inversion task, the cursor and forearm moved in opposite directions. When the target shifted from the centre position to the right side position, the monkey would move his forearm towards the left (elbow flexion) in order to move the cursor to the right and capture the target. From Alexander and Crutcher (1990a).

proportion of such cells (24%). Approximately one-quarter of the cells in the SMA (23%) and MC (26%) showed a combination of preparatory and movement-related activity, while the proportion of putamen cells showing this combination was much smaller (9%). This would seem to indicate that the processes subserving the preparation for movement may be carried out by a distributed neural system that is largely separate from that concerned with movement execution, though clearly some neurones appear to play a role in both processes.

Movement-related cells were classified as *directional* if they showed an increase in discharge rate predominantly or exclusively during movements in one direction and did not have significant static or dynamic load effects. A cell was classified as *muscle-like* if its directional movement-related activity was associated with static and/or dynamic load effects whose pattern was similar to that of flexors or extensors of the forearm. Both types of task-related discharge pattern are illustrated in Fig. 11.3. Cells with directional activity appeared to be coding for *trajectory/kinematics*-level variables while cells with muscle-like activity appeared to be coding for *dynamics/muscle*-level variables. Both directional and muscle-like cells were found in all three motor areas. The largest proportion of directional cells was located in the putamen (52%), with significantly smaller proportions in the SMA (38%) and MC (41%). Conversely, a smaller proportion of muscle-like cells was seen in the putamen (24%) than in the SMA (41%) or MC (36%).

The majority of cells with preparatory activity showed selective activation in anticipation of elbow movements in a particular direction (SMA, 86%; MC, 87%; putamen, 78%), and in most cases the preparatory activity was found to be independent of the loading conditions (80% in SMA, 83% in MC, and 84% in putamen). This type of discharge pattern is illustrated in Fig. 11.4. Thus, there was little evidence that preparatory activity coded for dynamics/muscle-level variables.

11.6. **Neural representations of the target or goal of a movement**

Regardless of whether visually guided limb movements are controlled in a predominantly serial or parallel fashion, it seems clear that the brain must contain neural representations of the target or goal of the movement, as well as of the movement itself (Bernstein, 1984). However, it is not known whether such representations occur within the motor system, or only in structures that are 'upstream' from the cortical motor areas (e.g. frontal or parietal association cortex). It is conceivable, for example, that the motor system may be concerned only with *translating* the goal or target of a movement into the appropriate muscle activation patterns, and not with the goal or target itself. One of the central issues addressed by our recent neurophysiological studies was whether there are neural repre-

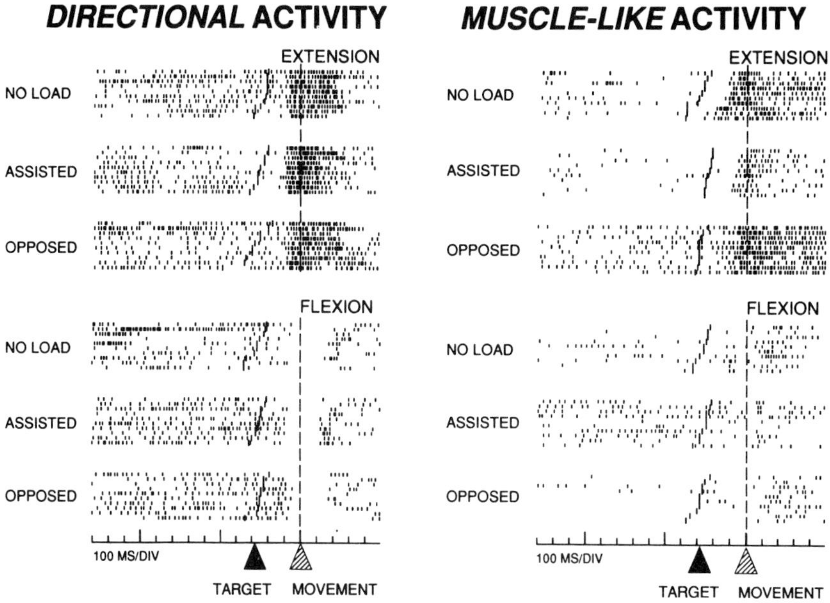

Fig. 11.3. Raster displays of the discharge patterns of two motor cortex neurones with movement-related activity. Each small tick indicates the occurrence of a single action potential, and each row represents the neuronal activity recorded during one trial. The large ticks indicate the times of occurrence of the target shift that triggered the first centre-to-side movement of each trial. Trials are aligned on the time of the onset of movement (vertical dashed line) and sorted by reaction time. On 'loaded' trials, a continuous torque was applied to the handle that either opposed or assisted the required lateral movements. One-third of the trials were 'unloaded'. Trials of all six classes (three loading conditions × two directions of movement) were presented to the monkey in a balanced but unpredictable sequence, although they are grouped here and in subsequent raster displays according to class. The 'directional' cell on the left showed increased discharge in relation to extension movements that was independent of the loading conditions. The cell also showed a reciprocal reduction in activity during flexion movements. In contrast, the 'muscle-like' cell whose activity is presented in the right panel showed increased discharge during extension movements that was characterised by a dynamic load effect (the movement-related activity was increased with loads that opposed extension and decreased with loads that assisted extension). The cell also showed a comparable static load effect during the hold period prior to the onset of movement. From Crutcher and Alexander (1990).

sentations within the motor system of the 'highest' or most abstract level of motor processing, namely the target or goal of the movement in terms of the coordinate system of external space. It has been suggested that for limb movements that involve direct reaching to targets, the primary variable controlled by the brain might be the endpoint of the hand movement within a Cartesian frame of reference that also includes the

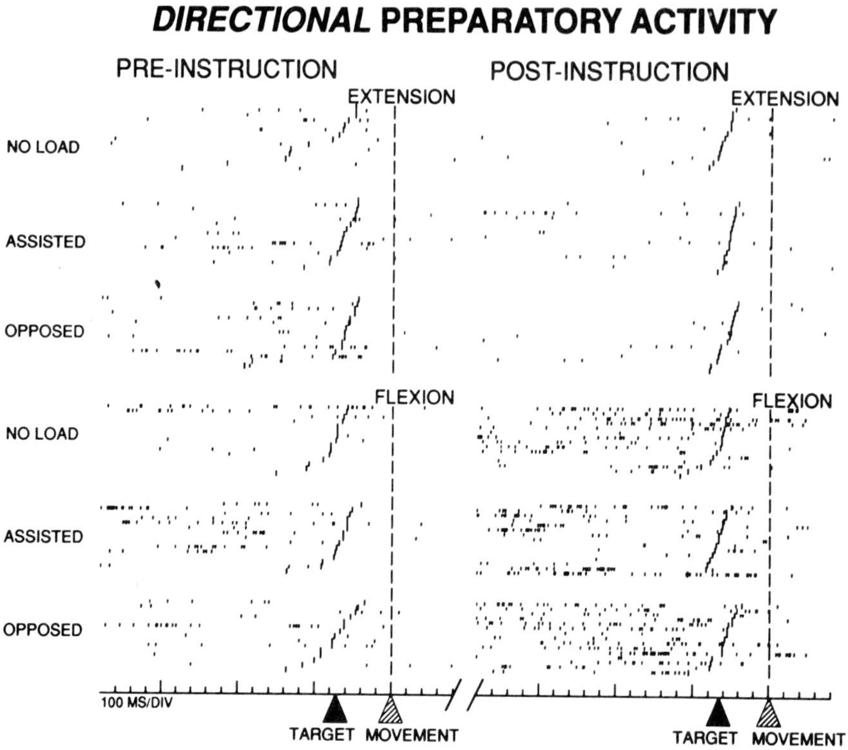

Fig. 11.4. Raster display of a putamen neurone that showed directional preparatory activity. The activity associated with the first half of each trial is aligned on the onset of the first lateral movement of the forearm, and that of the second half of the trial is aligned on the onset of the second lateral movement. The rasters are sorted according to class and reaction time, using the same conventions as in Fig. 11.3. This cell showed a sustained increase in discharge throughout the post-instruction interval on flexion trials, irrespective of the loading conditions. From Alexander and Crutcher (1990b).

target (Morasso, 1981; Soechting and Lacquaniti, 1981). This is an attractive hypothesis as it would readily explain the observation that target-directed movements tend to have straight trajectories, which are difficult to achieve by controlling joint dynamics (Morasso, 1981; Abend *et al.*, 1982), and would also help to explain the phenomenon of 'motor equivalence' (the fact that hand trajectories are easily scaled both in time and space, despite disproportionately complex changes in the associated joint dynamics: Soechting and Lacquaniti, 1981; Viviani and Terzuolo, 1982).

In the preceding section we showed that neurones may be found in the SMA, MC and putamen that discharge selectively during the planning (Alexander and Crutcher, 1990b) and/or execution (Crutcher and

Alexander, 1990) of directionally specific arm movements, irrespective of the joint dynamics (i.e. regardless of the loading conditions) associated with the movements. While it may be natural to assume that this type of activity (especially when recorded in motor structures) is specifically related to the *direction of arm movement*, there is no assurance of this unless the direction of arm movement has been clearly dissociated from the *target or goal* of the movement.

In direct reaching to targets, both the location of the target and the trajectory of the hand are readily defined within the same Cartesian coordinate system even though the hand's trajectory is fundamentally associated with a body-centred frame of reference (Saltzman, 1979; Hogan *et al.*, 1987). The optimal frame of reference for describing the hand's trajectory need not coincide with that of the target, however, as is demonstrated by situations in which reaching or tracking movements are guided by indirect visual feedback (from a mirror, for example, or a cursor). This principle can be exploited experimentally by varying the spatial features of indirect visual feedback in order to dissociate the neural events subserving arm movement from those subserving the more abstract process of target capture. As an extension of the studies described in the preceding section, monkeys were trained to perform a pair of arm movement tasks in which the spatial features of the target or goal of the movement were dissociated from those of the limb movement itself (see Fig. 11.2B).

Recordings of preparatory and movement-related activity in the same three motor areas (SMA, MC and putamen) showed that the motor system does indeed contain neural representations of the target or goal of the movement as well as representations of the movement itself. Thus, for example, two types of directionally selective *preparatory* activity were seen in each motor area. Cells with '*limb-dependent*' preparatory activity showed selective discharge prior to all pre-planned elbow movements in a particular direction (extension or flexion), irrespective of whether the target to which the cursor was moved was located on the right or left side of the display (Fig. 11.5). Cells of this sort appeared to be coding for *trajectory/kinematics*-level variables. The SMA contained a higher proportion of limb-dependent preparatory cells (40%) than either MC (15%) or putamen (9%). Cells with '*target-dependent*' preparatory activity showed selective discharge prior to all pre-planned movements of the cursor towards one of the side targets (right or left), irrespective of whether the limb movement involved extension or flexion of the elbow (Fig. 11.6). Cells of this sort appeared to be coding for *target*-level variables. Comparable proportions of target-dependent preparatory cells were seen in the SMA (36%), MC (40%) and putamen (38%).

Comparable types of directionally selective *movement-related* activity were also seen in each motor area. For cells that showed *limb-dependent*

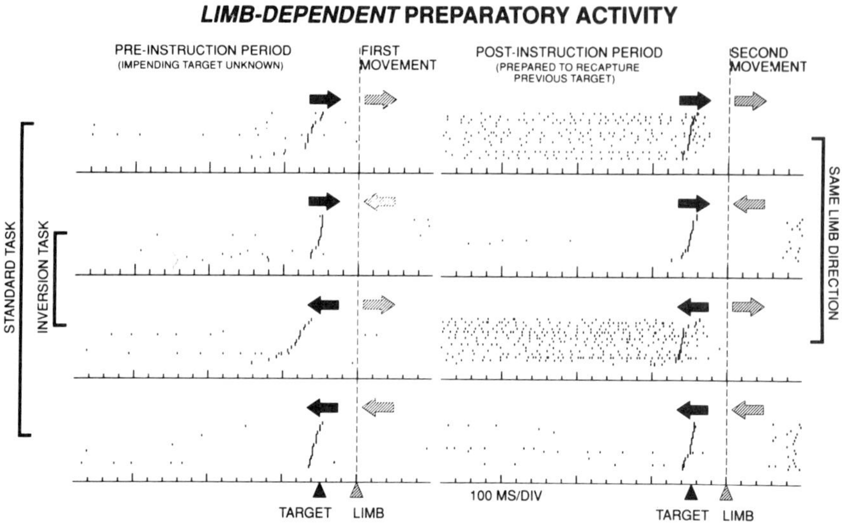

Fig. 11.5. Limb-dependent preparatory activity in the supplementary motor area. The trials from both the standard and cursor/limb inversion tasks are sorted by trial type and reaction time, and the split-plot rasters are aligned on the onsets of the first and second lateral movements. Limb direction is indicated by the shaded arrows and cursor direction by the solid arrows. This cell showed sustained activation during the post-instruction period preceding elbow extensions that moved the forearm to the right, irrespective of the direction of cursor movement or target displacement. From Alexander and Crutcher (1990a).

activity, the movement-related discharge was associated with elbow movements in a particular direction (extension or flexion) (Fig. 11.7). In contrast, for the cells that showed *target-dependent* activity, the movement-related discharge was associated with all movements of the cursor towards one of the side targets (right or left) regardless of whether the elbow was being extended or flexed (Fig. 11.8). As expected most movement-related neurones in all three motor areas were limb-dependent (SMA, 65%; MC, 71%; putamen, 63%). However, some movement-related cells showed target dependence, with the percentages of such cells in the SMA (16%) and MC (14%) being twice as high as that in the putamen (6%).

11.7. **Neurophysiological evidence for parallel, distributed motor processing**

The average onsets and offsets of directional preparatory activity in the SMA and MC were significantly earlier than those in the putamen. This is consistent with the possibility that some of the preparatory activity in the

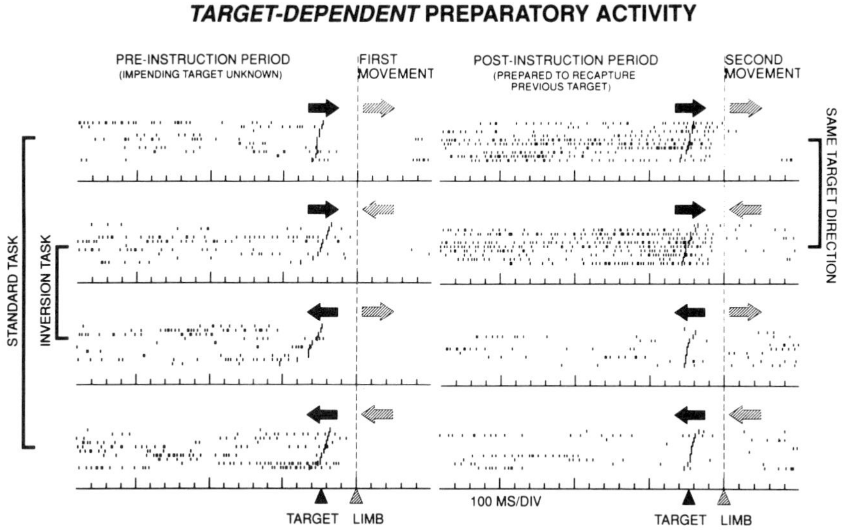

Fig. 11.6. Preparatory activity in the supplementary motor area that depended upon the expected direction of displacement of the target. This neurone showed sustained activation throughout the post-instruction period prior to all pre-planned movements of the cursor directed towards the right lateral target, regardless of whether the limb movement that was used to position the cursor involved elbow extension (as on standard trials, in which the forearm moved to the right), or elbow flexion (as on inversion trials, in which the forearm moved to the left). There was also a reciprocal reduction in activity during the post-instruction period that preceded pre-planned movements to capture the left target. From Alexander and Crutcher (1990a).

putamen may arise from corticostriatal inputs to this nucleus from the SMA and/or MC. It should be noted, however, that preparatory neurones in all three motor areas were active simultaneously throughout most of the post-instruction interval. The time of onset of movement-related discharge relative to the onset of movement ('lead time') was computed for each cell with movement-related activity. On average, SMA neurones discharged significantly earlier (SMA lead times 47 ± 8 ms; mean ±SE) than those in MC (23 ± 6 ms), which in turn were earlier than those in the putamen (−33 ± 6 ms). However, the degree of overlap of the distributions of lead times for the three areas was extensive, and population analyses revealed that movement-related neurones in these three areas were active largely concurrently, as illustrated in Fig. 11.9.

When comparisons of timing were made *within* each of the three motor areas, it appeared that different levels of motor processing were occurring simultaneously. Thus, for example, no differences were found between the latencies of either the onset or offset of target- vs limb-dependent

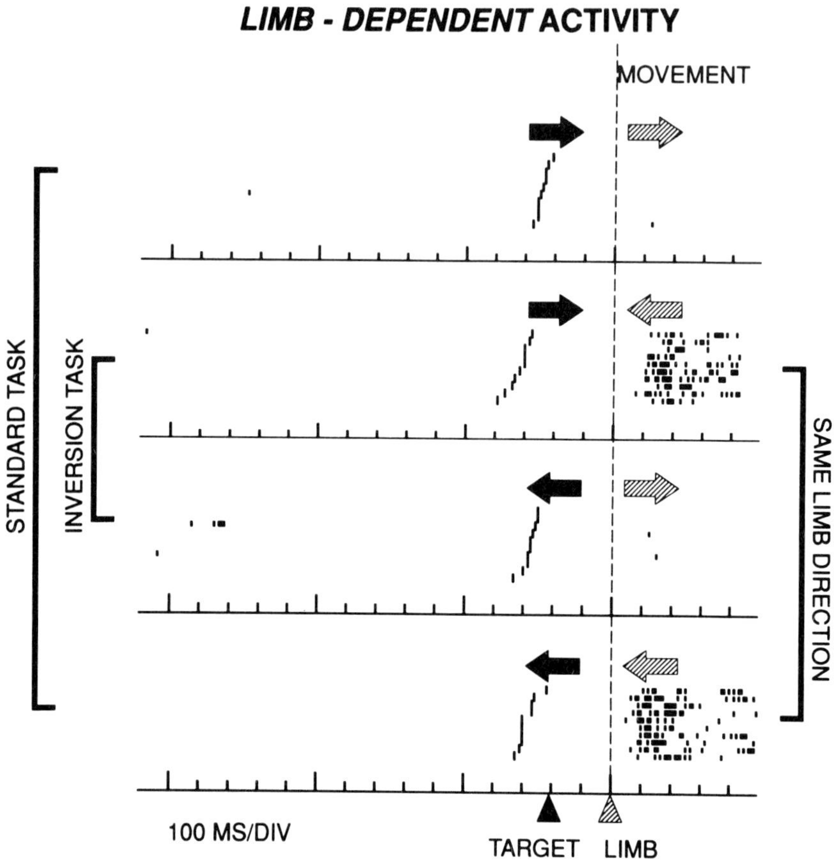

Fig. 11.7. Movement-related activity in the putamen that depended upon the direction of limb movement. This cell discharged during elbow flexions, in which the forearm moved to the left, irrespective of the direction of target displacement. From Alexander and Crutcher (1990a).

preparatory activity within the SMA, MC or putamen, nor were there differences between the onsets of target- vs limb-dependent movement-related activity within any of these structures (see Fig. 11.9).

The results of this series of studies are relevant to several aspects of the question as to whether motor processing is organised in a serial or parallel fashion. For example, it appears that *multiple levels of motor processing* are represented within the motor system. Within the arm regions of the SMA, MC and putamen there were neural representations of the target (or goal), limb direction (or trajectory), and muscle pattern (or joint torques) associated with an overlearned, visually guided elbow movement. By demonstrating the representation of at least one intermediate level of

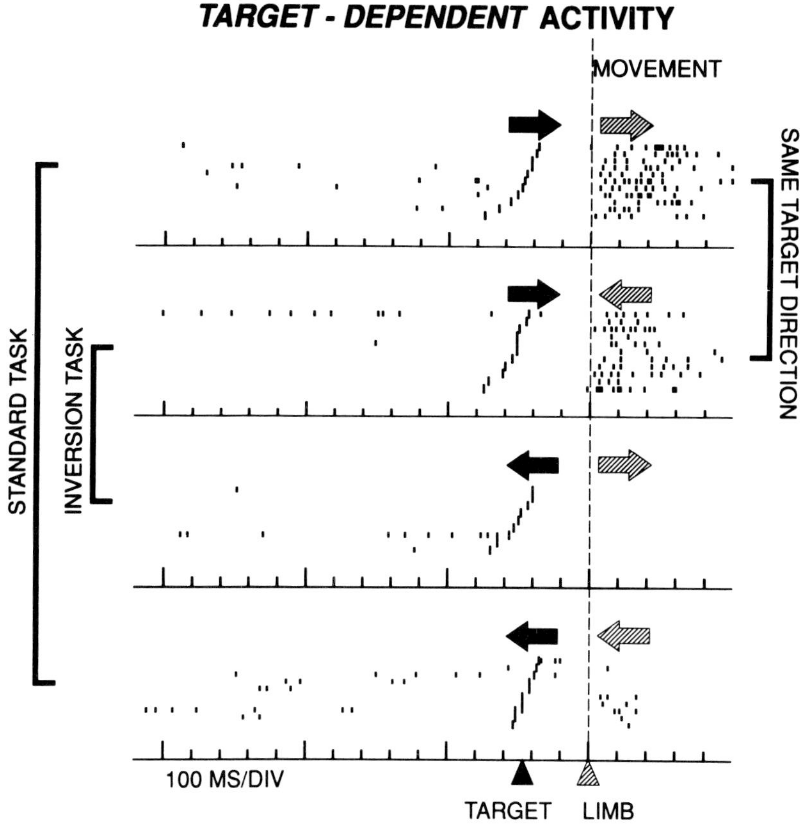

Fig. 11.8. Movement-related activity in the motor cortex that depended upon the direction of target displacement. This cell discharged during movements of the cursor to capture the right target, irrespective of whether the forearm was moving to the right (standard task) or to the left (inversion task). From Alexander and Crutcher (1990a).

processing, namely limb direction (i.e. trajectory- and/or kinematics-level processing), these results argue against the collapse of intervening levels into a direct (parallel) mapping of target coordinates on to muscle activation patterns. In other words, these findings suggest that at least three of the five analytically defined levels of motor processing (see Fig. 11.1) are actually represented within the motor system.

The studies also suggest that the functional levels of processing that are emphasised during the preparation for movement may be different from those emphasised during movement execution. Thus, within the three motor areas studied, *target-dependent* representations were more common than limb-dependent representations during the *preparation* for move-

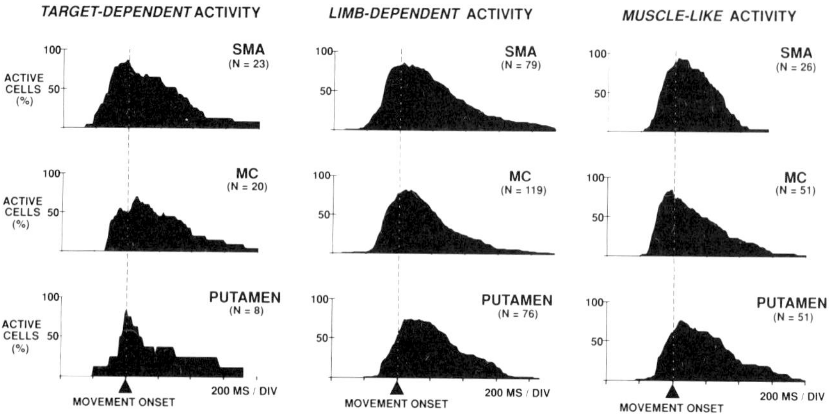

Fig. 11.9. Histograms of the times over which different categories of cells discharged relative to the movement. Each histogram shows the percentage of cells of a given type which were simultaneously active at different times relative to the onset of movement. The three histograms on the left show the duration of activity of cells in the SMA, MC and putamen whose movement-related activity was target-dependent. The centre and right-hand panels show corresponding histograms for cells whose movement-related activity was limb-dependent and muscle-like, respectively. Although there are subtle differences between histograms, it is apparent that cells related to different levels of motor processing and in different motor areas were simultaneously active.

ment (except in the SMA, where there were comparable proportions of both types of preparatory response). Conversely, during movement *execution* these same three areas (SMA, MC and putamen) showed a marked predominance of *limb-dependent* representations. Moreover, in these same three motor areas the overwhelming majority of preparatory representations reflected the *direction* of intended movement independent of force or muscle pattern, while substantial proportions of '*muscle-like*' representations were seen in each structure during movement execution. Together these findings suggest that, at least within the SMA, MC and putamen, relatively 'high' levels of motor processing are emphasised during the preparation for movement, with an incomplete shift towards 'lower' levels during movement execution.

Our comparisons of neuronal activity in the SMA, MC and putamen support the view that motor processing is *spatially distributed*, rather than being strictly localised or hierarchically segregated. Thus, for example, roughly the same proportion of target-dependent preparatory cells was seen in each of the three motor areas (SMA, 36%; MC, 40%; putamen, 38%), and the same was true for the limb-dependent movement-related cells (SMA, 65%; MC, 71%; putamen, 63%). This is not to suggest that these three motor areas should be considered functionally equivalent, of course, as there was also abundant evidence that certain types of motor processing (e.g. processing related to the preparation vs execution of movements) were more strongly represented in one or the other of these structures. In addition, it should be emphasised that these experiments involved relatively simple, stimulus-triggered movements. It is possible that the functional differences between SMA, MC and putamen would have been more striking if neuronal relations to more complex movements had been examined.

The average onsets and offsets of preparatory activity in the SMA and MC were earlier than those in the putamen. Likewise, the average onset of movement-related activity in the SMA was earlier than that in the MC, which in turn was earlier than that in the putamen. These differences suggest that there is some degree of sequential processing from the SMA to the MC and thence to the putamen. This is consistent with the commonly held view that the SMA is at the 'highest' level of the motor system (Roland *et al.*, 1980; Eccles, 1982). It is also consistent with the view that activity in the putamen arises from cortical inputs to this nucleus from either or both of these precentral motor fields. However, despite these differences in average onsets (and offsets) between the three areas studied, it is important to stress that preparatory neurones in the SMA, MC and putamen were active *simultaneously* throughout most of the post-instruction interval and, likewise, the cells with movement-related activity in each of these areas were also active largely simultaneously. In addition, when comparisons of timing were made *within* each of the three motor areas, there was little evidence of timing differences between neurones coding for different levels of motor processing (target, trajectory/kinematics, dynamics/muscle). Thus, the primary impression that emerged from this study was that multiple levels of motor processing appear to progress largely *in parallel* in each of these three motor areas during both the preparation and execution of visually guided limb movements.

11.8. **Summary and conclusions**

The control of goal-directed limb movements in three-dimensional space represents an enormously complex challenge to the motor system. Theorists have embraced the concept of the motor programme in an attempt to account for the remarkable speed and flexibility of skilled motor behaviours. However, fundamental issues regarding the origin and nature of motor programmes have yet to be addressed in any detail. PDP (parallel distributed processing) models of motor control represent an exciting new approach to the study of the neural substrates of motor behaviour. Such models provide detailed and testable hypotheses concerning the neural basis for the complex and highly adaptive motor behaviour of primates. These models are particularly appealing because they are self-organising, non-algorithmic and require only local activity-dependent processes. Features such as these effectively circumvent many of the difficulties inherent in other models of motor control, and make PDP models extremely plausible from a neurobiological perspective.

Our recent neurophysiological investigations of the motor system in awake, behaving primates have shown that each of three interconnected motor areas contains neural representations of several different 'levels' of motor processing. These include target-level, trajectory/kinematics-, and dynamics/muscle-level processing. Moreover, in all three motor areas the neurones coding for these representations were active largely simultaneously. Such results are difficult to explain from either a serial/analytical or a hierarchical motor programming perspective, but they are quite consistent with a PDP approach to motor processing.

The concept of parallel, distributed motor processing seems to represent an important theoretical advance in the study of the neural control of movement. PDP approaches to neural modelling will very likely provide the first detailed, testable and neurobiologically relevant models of motor processing in primates. The advent of such models should eventually make it possible to integrate the theoretical/behavioural and neurophysiological approaches to the study of primate motor control in a manner that previously has not been possible.

Acknowledgement

The research described in this paper was supported by grant NS-17678 from the United States Public Health Service.

References

Abend, W., Bizzi, E. and Morasso, P. (1982) Human arm trajectory formation. *Brain* **105**: 331–348.

Alexander, G. E. (1987) Selective neuronal discharge in monkey putamen reflects intended direction of planned limb movements. *Exp. Brain Res.* **67**: 623–634.

Alexander, G. E. and Crutcher, M. D. (1990a) Neural representations of the target (goal) of visually guided arm movements in three motor areas of the monkey. *J. Neurophysiol.* **64**: 164–178.

Alexander, G. E. and Crutcher, M. D. (1990b) Preparation for movement: neural representations of intended direction in three motor areas of the monkey. *J. Neurophysiol.* **64**: 133–150.

Alexander, G. E., DeLong, M. R. and Strick, P. L. (1986) Parallel organization of functionally segregated circuits linking basal ganglia and cortex. *Ann. Rev. Neurosci.* **9**: 357–381.

Alexander, G. E., Crutcher, M. D. and DeLong, M. R. (1990) Basal ganglia–thalamocortical circuits: parallel substrates for motor, oculomotor, 'prefrontal' and 'limbic' functions. *Progr. Brain. Res* **85**: 119–146.

An, C. H., Atkeson, C. G. and Hollerbach, J. M. (1988) *Model-based Control of a Robot Manipulator*, MIT Press, Cambridge, MA.

Anastasio, T. J. and Robinson, D. A. (1990) Distributed parallel processing in the vestibulo-oculomotor system. *Neural Comp.* **1**: 230–241.

Bear, M. F., Cooper, L. N. and Ebner, F. F. (1987) A physiological basis for a theory of synapse modification. *Science* **237**: 42–48.

Bernstein, N. (1984) The problem of the interrelation of co-ordination and localization (originally published in *Arch. Biol. Sci.* **38** 1935). In: *Human Motor Actions: Bernstein Reassessed*, H. T. A. Whiting, ed. North-Holland, Amsterdam, pp. 77–119.

Crutcher, M. D. and Alexander, G. E. (1990) Movement-related neuronal activity selectively coding either direction or muscle pattern in three motor areas of the monkey. *J. Neurophysiol.* **64**: 151–163.

Eccles, J. C. (1982) The initiation of voluntary movements by the supplementary motor area. *Arch. Psychiatr. Nervenkr.* **231**: 423–441.

Evarts, E. V., Shinoda, Y. and Wise, S. P. (1984) *Neurophysiological Approaches to Higher Brain Functions*, John Wiley & Sons, New York.

Gibson, J. J. (1941) A critical review of the concept of set in contemporary experimental psychology. *Psychol. Bull.* **38**: 781–817.

Grajski, K. A. and Merzenich, M. M. (1990) Hebb-type dynamics is sufficient to account for the inverse magnification rule in cortical somatotopy. *Neural Comp.* **2**: 71–84.

Grossberg, S. and Kuperstein, M. (1989) *Neural Dynamics of Adaptive-Motor Control*, Pergamon Press, New York.

Hinton, G. (1984) Parallel computations for controlling an arm. *J. Mot. Behav.* **16**: 171–194.

Hogan, N., Bizzi, E., Mussa-Ivaldi, F. A. and Flash, T. (1987) Controlling multijoint motor behavior. *Exerc. Sports Sci. Rev.* **15**: 153–190.

Hollerbach, J. M. (1982) Computers, brains and the control of movement. *Trends Neurosci.* **5**: 189–192.

Keele, S. W. (1981) Behavioral analysis of movement. In: *Handbook of Physiology, sect. 1. The Nervous System, Vol. II. Motor Control, part 2*, J. M. Brookhart, V. B. Mountcastle, V. B. Brooks and S. R. Geiger, eds. American Physiological Society, Bethesda, MD, pp. 1391–1414.

Kien, J. and Altman, J. S. (1991) Decision making in the insect nervous system: a model for selection and maintenance of motor programmes. In: *Neurobiology of Motor Programme Selection: New Approaches to the Study of Behavioural Choice*, J. Kien, C. R. McCrohan and W. Winlow, eds. Manchester University Press, Manchester, pp. 147–169.

Kuperstein, M. (1988) Neural model of adaptive hand–eye coordination for single postures. *Science* **239**: 1308–1311.

Linsker, R, (1986) From basic network principles to neural architecture: emergence of orientation columns. *Proc. Natl Acad. Sci. USA* **83**: 8779–8783.

Loeb, G. E. (1983) Finding common ground between robotics and physiology (letter to the editor). *Trends Neurosci.* **6**: 203–204.

Morasso, P. (1981) Spatial control of arm movements. *Exp. Brain Res.* **42**: 223–227.

Paul, R. P. (1981) *Robot Manipulators: Mathematics, Programming and Control*, MIT Press, Cambridge.

Pellionisz, A. and Llinas, R. (1985) Tensor network theory of the metaorganization of functional geometries in the central nervous system. *Neuroscience* **16**: 245–273.

Pew, R. W. (1989) Human perceptual–motor performance. In: *Distributed Motor Control. Cognition of Motor Processes*, W. Prinz and A. F. Sanders, eds. Springer-Verlag, Berlin, pp. 19–27.

Roland, P. E., Larsen, B., Lassen, N. A. and Skinhoj, E. (1980) Supplementary motor area and other cortical areas in organization of voluntary movements in man. *J. Neurophysiol.* **43**: 118–136.

Rosenbaum, D. A. (1985) Motor programming: a review and scheduling theory. In: *Motor Behavior: Programming, Control, and Acquisition*, H. Heuer, U. Kleinbeck and K.-H. Schmidt, eds. Springer-Verlag, Berlin, pp. 1–34.

Rumelhart D. E. and McClelland, J. L. (1986) *Parallel Distributed Processing*, MIT Press, Cambridge, MA.

Ryan, T. A. (1970) *Intentional Behavior: An Approach to Human Motivation*, The Ronald Press Company, New York.

Saltzman, E. (1979) Levels of sensorimotor representation. *J. Math. Psych.* **20**: 91–163.

Schmidt, R. A., Zelaznik, H. N., Hawkins, B., Frank, J. S. and Quinn, J. T. (1979) Motor-output variability: a theory for the accuracy of rapid motor acts. *Psychol. Rev.* **86**: 415–451.

Schwartz, E. L. (1990) *Computational Neuroscience*, MIT Press, Cambridge, MA.

Soechting, J. F. and Lacquaniti, F. (1981) Invariant characteristics of a pointing movement in man. *J. Neurosci.* **1**: 710–720.

Tanji, J. and Evarts, E. V. (1976) Anticipatory activity of motor cortex neurons in relation to direction of an intended movement. *J. Neurophysiol.* **39**: 1062–1068.

Thach, W. T. (1978) Correlation of neural discharge with pattern and force of muscular activity, joint position, and direction of intended next movement in motor cortex and cerebellum. *J. Neurophysiol.* **41**: 654–676.

Viviani, P. and Terzuolo, C. (1982) Trajectory determines movement dynamics. *Neuroscience* **7**: 431–437.

Weinrich, M. and Wise, S. P. (1982) The premotor cortex of the monkey. *J. Neurosci.* **2**: 1329–1345.

Winlow, W., Moroz, L. L. and Syed, N. I. (1991) Mechanisms of behavioural selection in *Lymnaea stagnalis*. In: *Neurobiology of Motor Programme Selection: New Approaches to the Study of Behavioural Choice*, J. Kien, C. R. McCrohan and W. Winlow, eds. Manchester University Press, Manchester, pp. 52–72.

12 *Gary Goldberg*

Premotor systems, attention to action and behavioural choice

12.1. Internal and external sources of constraint in motor performance

Adaptive biological action can be conceptualised as a goal-directed dynamic process which evolves in space and time and through which the physical configuration of the organism is altered in relation to the environment to solve a specific motor problem relevant to the organism. This process develops through the simultaneous satisfaction of emergent constraints that arise from a variety of sources, *both intrinsic and extrinsic to the organism*, on a variety of different time scales, ranging from phylogenetic to microgenetic (Yakovlev, 1948). These lawful constraints serve to select the act through a 'self-organising' process (Prigogine, 1980; Yates, 1986) which progressively reduces the degrees of freedom for an act so that the performance itself is low-dimensional and the behaviour of the system remains stable, controllable and ordered in spite of its complexity (Nicolis and Prigogine, 1989). Through the application of lawful constraints, adaptive cooperativity between autonomous elements develops with long-range correlations between functional elements appearing throughout the system. This emergent cooperativity may itself be driven through inherent variability or low-level 'microscopic' fluctuations that become selectively amplified to macroscopic dimensions by the application of a selective constraining process (Prigogine, 1980; Kugler and Turvey, 1987; Mpitsos *et al.*, 1988; Kelso, 1990). Furthermore, this fluctuation or apparently 'noisy' variability, critical to self-organising processes, may be produced through nonlinear dynamic interactions inherent in the system that convey a lawful, deterministic basis for the generation of the fluctuations (Crutchfield *et al.*, 1986; see also Skarda and Freeman, 1987).

How in the nervous system might this process of specifying action proceed through the elaboration of constraints? The function of the organism (and the assurance of its survival) is attained ultimately through its ability to control energy (and matter) flows within itself and at the interface between the organism and its surrounding environment. Interactions between an animal and its environment are of two forms: first, a low-energy 'perceptual' coupling, where external inputs are coupled to the organism through sensory receptors; and, secondly, a high-energy mechanical coupling with the surroundings and/or within the body to effect controlled changes in the relationship of body parts with respect to each other. There are two major sources of informational constraint used to control the high-energy mechanical coupling: an endogenous source related to the intentionality of the organism, and an extrinsic source related to the existing external conditions. The intrinsic source enables the organism to function in an anticipatory adaptive mode using information that is derived from intrinsic models constructed through experienced, biologically relevant historical contingency or, in the human, through prior verbal instruction. The extrinsic source enables responses that are reactive to immediate (and possibly unanticipatable) environmental conditions (see Rosen, 1985). It is hypothesised that that there are separable premotor systems in the brain associated with the endogenous or intentional versus reactive or extrinsic behavioural modes (Goldberg, 1985a, 1985b, 1987).

The premotor process is that whereby informational constraints generated within the nervous system are coupled to the effector apparatus by the transformation of intent into action. This is equivalent to the concept of '*attention to action*' (Norman, 1981, 1988; Shallice, 1982; Norman and Shallice, 1986; Posner *et al.*, 1988b), the intrinsically based selection of specific strategies or 'motor programmes' that are allowed to gain access to the output systems. This selectivity may be viewed as emerging through a process of competing interaction between different intentions that are attempting to activate executive motor areas. It is clear that the issue of behavioural choice, or how an organism selectively acts, relates very closely to the premotor process or how the organism selectively activates a particular motor pattern in a particular context so as to achieve a particular outcome. I will now turn to a more specific discussion of the premotor concept and to premotor systems of the human and non-human primate brain.

12.2. **The premotor concept**

The historical development of the premotor concept has recently been reviewed by Freund and Hummelsheim (1985). The concept of a 'premotor' area in the human frontal lobes can be traced to Fulton (1935) and his co-workers at Yale who drew a distinction between the 'motor area' of the precentral gyrus (Brodmann's area 4) and the regions anterior to it, in particular the upper part of Brodmann's area 6 (labelled 6aα and 6aβ by the Vogts in 1919). This distinction was not without controversy, however, since others (e.g. Walshe, 1935) argued that the 'motor area' was a continuum including both area 6 and area 4 as a single entity. Based on electrical stimulation studies in monkeys, Woolsey *et al.* (1952) defined the motor field of the precentral area to have a continuous epaxial structure (Kwan *et al.*, 1978) that extended rostrally into area 6, and saw no reason to distinguish anterior area 6β as a motor association area. However, they did identify an additional separate motor field on the mesial surface of the frontal lobe anterior to the leg representation of the primary motor area corresponding to the supplementary motor area (SMA) of Penfield and Welch (1949) in humans. Woolsey *et al.* (1952) labelled the precentral excitable region MI and the mesial motor field MII. For recent reviews of the SMA see Wiesendanger and Wiesendanger (1984), Goldberg (1985b), and Wiesendanger (1986). The SMA and the anterior cingulate cortex are adjacent to each other on the medial surface of the hemisphere and are strongly interconnected, and both areas project topographically to the primary motor cortex (Muakkassa and Strick, 1979). Both of these regions are critical cortical structures in the putative medial premotor system to be described.

For the purposes of the development of the hypothesis to be presented below, I will define a premotor area in terms of both function and structure. Functionally, a premotor area will be viewed as a region that conveys the conditions for action to those areas that are involved in organising the actual execution of action (see also Passingham, 1987). A premotor area may thus be viewed as a cortical region influencing the executive areas and so serving as a route to action. Following Muakkassa and Strick (1979), anatomically a premotor area will be defined as a region of the frontal cortex projecting in topographically organised fashion to MI. Several such regions have been identified in the monkey including the SMA, the anterior cingulate cortex, cortex in the region of the arcuate sulcus (which will be referred to as the arcuate premotor area, APA), as well as a region in the dorsolateral precentral sulcus (Muakkassa and Strick, 1979; see also Matsamura and Kubota, 1979; Godschalk *et al.*, 1984). Matsumura and Kubota (1979) looked specifically at those regions which projected to the arm–hand area of MI and found that the medial frontal cortex, including SMA and anterior cingulate cortex, projected most strongly into this region with a separate less dense projection from a region around the spur and inferior limb of the arcuate sulcus.

It is possible that such specifically organised corticocortical projections to MI, along with transthalamically mediated subcortical projections, may operate through context-dependent modulation of the gain of transcortical sensorimotor loops traversing MI (Hummelsheim *et al.*, 1986; Noth, 1986; Dick *et al.*, 1987). These transcortical loops may play a critical role in creating a functional, topographically organised linkage between somatic afference and corticospinal efferent activity (Murphy *et al.*, 1978; Asanuma and Arissian, 1984; Favorov *et al.*, 1988). Furthermore, electrical stimulation of cerebral cortex through subdural electrodes in humans undergoing investigation of epilepsy has shown that stimulation-produced disruption of normal function in the SMA medially and in a lateral inferior frontal region rostral to the face region of MI produces a transient 'block' of ongoing motor behaviour: a 'negative motor response' (Dinner *et al.*, 1987; Lüders *et al.*, 1987, 1988). This effective suspension of ongoing motor behaviour suggests that these two cortical regions of the frontal lobe are particularly involved in the higher order integration of motor behaviours in humans and correspond to the SMA and APA in the monkey.

12.3. **The dual premotor systems hypothesis**

A 'premotor system' is defined as a premotor cortical region identified as above together with all of its distributed influences at the cortical and subcortical level. The dual premotor systems hypothesis thus identifies two separate but interactive *premotor systems*, one organised around the SMA and anterior cingulate cortex medially (see Fig. 12.1) and the other organised around the APA laterally in monkey (probably relating to inferior area 6 in humans and corresponding to the inferior frontal 'negative motor area' identified by Lüders *et al.*, 1988). The medial premotor system organised around the SMA is proposed to be selectively involved in the control of isolated spontaneous voluntary movements based on endogenous or interoceptive signals, whilst the lateral premotor system organised around the APA is recruited when the conditions for action involve the selective incorporation of external 'sensory' context.

These two systems should typically be in a state of mutually inhibitory interactive balance to deal with and resolve conflicts between 'internally' and 'externally' driven movements. This type of conflict has been referred to as 'interference' between competing intentions (Freud, 1924), and one of the major consequences of frontal lobe damage is the inability to resolve such interference so as to accomplish intended goals. As a result, systematic errors of action or 'action slips' occur (Reason, 1979; Norman, 1981; Mayer *et al.*, 1990; Schwartz *et al.*, 1991). The occurrence of such errors can be viewed as the result of a failure of constraints being applied to an emergent act in the transformation of intent into action (Goldberg, 1991). The medial system would be expected to be particularly active in suppressing pre-potent tendencies to respond in circumstances of high stimulus–response compatibility where an external stimulus 'primes' or selectively activates a habitual response that conflicts with the internal context for action (e.g. as in the Stroop paradigm, see below).

The putative medial and lateral premotor systems can be linked to two evolutionary trends in frontal cortical differentiation that gave rise to separable cortical moieties within the hemisphere (Sanides, 1964; Goldberg, 1985b; Pandya and Barnes, 1987). The lateral parainsular system is phylogenetically older than the medial system. It originates from the piriform cortex (i.e. primitive sensory cortex) and includes the peripaleocortical proisocortex of the insular region (Sanides, 1964; Pandya and Barnes, 1987). This lateral parainsular trend is more developed in the parietal lobe particularly in the non-language hemisphere (Eidelberg and Galaburda, 1984). It receives visual and somatic inputs from the inferior parietal lobule and tends to be interconnected with other cortical zones in the lower cortical moiety (Godschalk *et al.*, 1984; Petrides and Pandya, 1984; Matelli *et al.*, 1986) including the auditory association cortex of the temporal lobe (Galaburda and Pandya, 1983). Furthermore, there appears to be a modular segregated organisation of parieto-frontal networks in which parallel connections are established between different regions of the 'sensory' parietal and 'limbic' frontal cortex (Cavada and Goldman-Rakic, 1989). The APA, which has access to a wealth of processed polymodal sensory information, is thus linked to registration and decoding of external sensory information (Goldberg, 1987) and thereby postulated to be involved in the control of action that is directed or guided by sensory input conveying external context (Rizzolatti *et al.*, 1983; Godshalk *et al.*, 1984; Matelli *et al.*, 1987).

The arcuate premotor area and the lateral premotor system (LPS) (Fig. 12.1) including the lateral parietal cortex which projects to the APA (Godshalk *et al.*, 1984; Petrides and Pandya, 1984) may be particularly important for differential activation of actions that are organised as conditional stimulus–response linkages rather than as response–reward linkages (Dabrowska, 1971; Konorski, 1972; Petrides, 1986). It becomes active when action must be generated in a 'responsive' mode rather than in an anticipatory predictive one (Rosen, 1985). If this system is 'released' or 'disinhibited' through damage to its medial counterpart, compulsive automatisms that are directed towards object use (i.e. 'utilization behaviour' — Lhermitte, 1983) or which are imitative (Lhermitte *et al.*, 1986) may appear. With large bilateral frontal lesions, patients may become totally dependent upon environmental cueing in the control of behaviour in a social context, thus effectively losing the ability to act autonomously (Lhermitte, 1986).

The medial paralimbic system (Fig. 12.1) is phylogenetically newer than the lateral system but older than the primary motor cortex (Sanides, 1964, 1969). It takes on special significance in mammalian evolution with the development and elaboration of cortical re-entrance of outflow from the basal ganglia (Reiner *et al.*, 1984). It originates from the hippocampal cortex and includes the periarchicortical proisocortex of the anterior cingulate region (Sanides, 1964). This derivation from the hippocampal source suggests a functional relationship between hippocampus and the related, interconnected structures on the medial surface of the frontal lobe: the anterior cingulate and the supplementary motor area (SMA). The SMA is the cortical premotor centre in this bilaterally organised

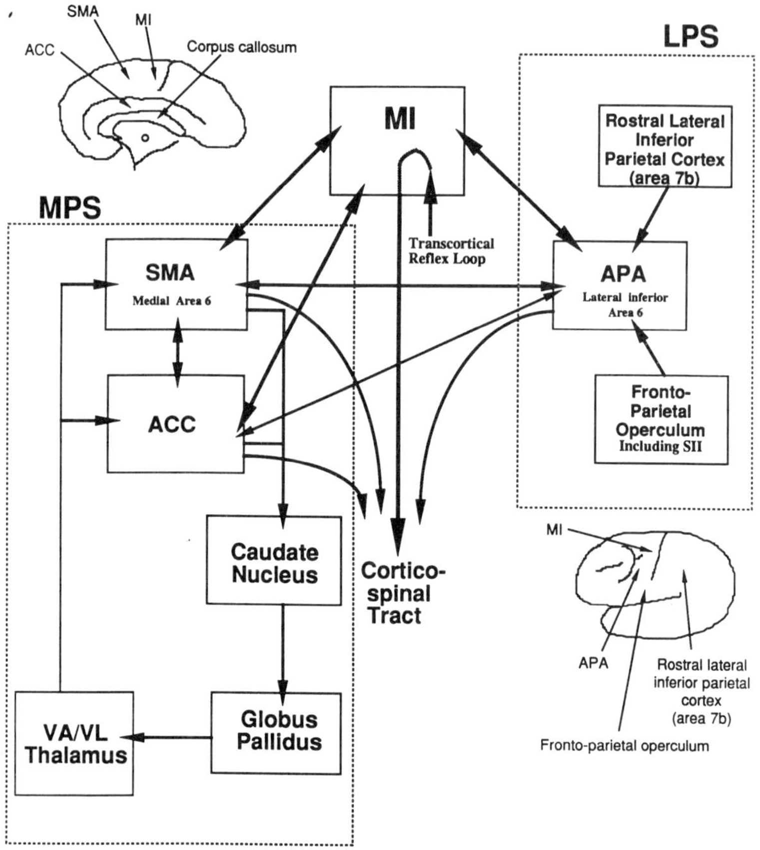

Fig. 12.1. Some of the major interconnections between movement-related regions in the medial premotor system (MPS) and the lateral premotor system (LPS). The supplementary motor area (SMA) and anterior cingulate cortex (ACC) on the medial surface of the hemisphere project with topographic ordering to the primary motor cortex (MI). The arcuate premotor area (APA) located posterior and inferior to the spur and inferior limb of the arcuate sulcus on the lateral surface of the cerebral hemisphere of the monkey also projects to the MI. Also shown is the medial premotor loop through the basal ganglia as well as major cortical zones on the lateral surface of the hemisphere which project to the APA within the LPS. Convergent influences on the MI may modulate trans-cortical reflex gains, and all premotor regions project themselves directly into the corticospinal tract in addition to the MI itself.

system (Wiesendanger, 1986). The medial premotor system (MPS) is connected to both primary motor cortices with stronger connections ipsilaterally. It also projects directly into the corticospinal tract (Hutchins *et al.*, 1988). Regional cerebral blood flow (rCBF) studies have suggested bilateral activation of the SMA and the contralateral precentral motor area when subjects performed the 'motor-sequence' test — learned unilateral sequences of ballistic movements of the fingers (Roland *et al.*, 1980, 1982). Only bilateral SMA activation occurred, without activation of the contralateral primary cortex, when the subjects were asked to internally simulate the performance of such movements without actually executing them (Orgogozo and Larsen, 1979; Roland *et al.*, 1980). The medial paralimbic trend, possibly through its relation to the hippocampal source, is more directly linked to the endogenous or interoceptive and internal mnemonic sources of information as well as to ordered temporal structure (Ingvar, 1985). It is a system in which hippocampal and basal ganglia outflow converge.

Internal information may be integrated in the medial system through subcortical re-entrant pathways. The basal ganglia outflow through the internal segment of the globus pallidus projects by way of the ventral-anterior and ventrolateral thalamus to the SMA, anterior cingulate and medial prefrontal cortex (Kalil, 1978; Schell and Strick, 1984; Strick, 1985). The medial prefrontal cortex also projects heavily to the head of the caudate nucleus through the medial subcallosal fasciculus (Yakovlev and Locke, 1961; Jürgens, 1984) forming a pathway through the basal ganglia that begins and ends at the medial frontal cortex. The connections of the basal ganglia with widespread cortical input to the striatum and delimited projections back to the cortex suggest that this loop may act as a selective filter to allow specific aspects of cortical activity to reach medial frontal cortex and the SMA in the process of selecting motor strategies (Goldberg, 1985b). The medial trend would therefore be used to control action when temporal structure is present, and memory (i.e. some form of internal model) can be used in conjunction with selective perception to allow prospective control of behaviour (Goldberg, 1987). Intentionality implies the capacity to follow an internal plan to control action prospectively using anticipation (Brand, 1984; Rosen, 1985). Thus, it is mainly through participation of the medial system (including the basal ganglia) that the feeling of volitional control over action emerges and the capacity to act autonomously — with selective rather than obligatory dependence on external information — develops.

12.4. **Recent evidence supporting the dual premotor systems hypothesis**

Recently, a number of lines of investigation have produced evidence supportive of the dual premotor systems hypothesis. Recording cortical unit activity from the SMA, the APA and MI in monkeys, Okano and Tanji (1987) found a unique population of units in the SMA which fired selectively with a pre-movement latency of several hundred milliseconds prior to self-paced upper limb movements. Units which fired usually with a shorter pre-movement latency in advance of movements made in response to a visual signal were found in the SMA and APA as well as in the MI. These authors thus suggested a unique role for the SMA in self-paced movement but in addition the SMA was found to participate together with the APA in visually triggered movement.

Schlag and Schlag-Rey (1987) have reported the discovery of an eye movement area anterior to the SMA representation in the monkey frontal cortex, which they have labelled the supplementary eye field (SEF). Microstimulation of this region of cortex produced saccades. Mapping of this region by electrical microstimulation suggests that it may encode for eye position in a head-based reference system so that gaze is biased towards a particular part of space, independent of initial gaze direction. Thus the saccades produced by microstimulation at various points in the SEF converge to a region of space, whereas saccades produced by microstimulation at a point in the frontal eye field (FEF) on the lateral surface of the frontal lobe generally produced movements in a fixed direction. Unit behaviour in this region is similar to that in the SMA (Okano and Tanji, 1987) in that a population of units can be found which fire with long pre-movement latencies prior to exploratory eye movements performed during visual search in the dark — that is, in the absence of visual targets. This form of activity is quite different from that found in the FEF. In the FEF, activation of units appears to relate to the presence of an external visual target to which the saccade is directed (Bruce *et al.*, 1985). Moster and Goldberg (1990) have reported a study of normal human subjects using topographic mapping of pre-saccadic brain electrical activity. When subjects choose to make saccadic eye movements either to the left or right, at the time at which the subject makes a choice a large, vertex-centred, slowly increasing negativity appears over the scalp at least 500 ms prior to the onset of the saccade. On the other hand, when saccades are directed to lateral visual targets which suddenly appear at random to either the left or the right of a central fixation point, this pre-saccadic potential is either greatly attenuated or absent (Goldberg and Moster, 1988). While it is not possible to determine definitively the origin of this vertex-centred pre-saccadic negative potential from topographic mapping alone, correlation with other data obtained, for example, by Schlag and Schlag-Rey (1987) and by Fox *et al.* (1985) in positron-emission tomographic studies in human subjects suggests that activation of medial frontal cortex could be contributing to its generation. It thus appears possible to extend the differentiation of a medial and lateral premotor cortical system for limb movement to the volitional control and context dependence of voluntary saccadic eye movements.

Monkeys with SMA lesions have difficulty with the endogenous generation of movements in a task in which they needed to move the arm in a self-paced manner to a specific point in space in order to break an infrared beam for a food reward (Thaler and Passingham, 1989).

Monkeys with SMA lesions (Thaler, 1988) or cingulate lesions (Stern, 1987) made far fewer movements in the self-initiated movement condition than monkeys with premotor lesions in which the movements were also more accurate. However, when an external auditory cue was added to the paradigm to help 'externally instruct' the animal to move the limb in exactly the same task, animals with SMA lesions performed almost at their pre-operative level of performance. Passingham *et al.* (1989) reported that monkeys with premotor lesions that spare the SMA are only minimally impaired with self-initiated conditioned movements. They concluded that the SMA is necessary whenever the animal must make an internal decision about what to do, whereas animals can utilise the premotor cortex to facilitate performance when an external cue provides an instruction about what to do. Thus monkeys with SMA lesions 'are better at reacting than they are at acting of their own accord' (Passingham *et al.*, 1989, p. 22). These findings are clearly consistent with the dual premotor systems hypothesis.

Recent studies in human subjects, using positron emission tomography to examine regional blood flow patterns during cognitive activation, have demonstrated that the medial frontal lobe, and particularly the anterior cingulate cortex, may participate in a network responsible for 'attention to action', a construct which may be closely related to the premotor concept (Posner *et al.*, 1988b). In tasks that involve internally based selection with overt motor output, the pattern of activation tends to be more superior and posterior in the region of the SMA, whereas when overt motor activity is not present (as with imagined movement), or when the effects of movement are deleted using data from a control condition (e.g. Petersen *et al.*, 1988), the focus of activity tends to be more anterior and inferior in the area of the anterior cingulate cortex.

In the Stroop interference task (Stroop, 1935), the subject is asked to identify and name as quickly as possible the colour of the lettering of words presented visually. In the 'compatible' condition, the words are all colour names presented in the named colour (e.g. the word 'red' presented in red lettering), while in the 'incompatible' condition, the colour names are presented in lettering of a colour different from that named (e.g. the word 'red' presented in green lettering). This latter condition presents a conflict between the colour primed by the habitual response of reading the presented word (interference) and the instructed task of naming the colour of the letters (target). When patterns of cerebral activation were studied in human subjects performing the Stroop interference task, the right anterior cingulate cortex was found to be the region with the largest significantly increased activation in the incompatible condition with reference to the compatible condition (Pardo *et al.*, 1990; see also Brown and Marsden 1988, 1990). Similarly, strong anterior cingulate activation was observed in a reading task when subjects were instructed to generate verbs associated with visually presented nouns, thus requiring the intentional suppression of the habitual response of reading the nouns (Petersen *et al.*, 1988). Pardo *et al.*, (1990, p. 259, emphasis added) concluded that the 'anterior cingulate is involved in the selection process between competing processing alternatives on the basis of some pre-existing *internal, conscious plan*'. This 'attentional' formulation of medial frontal cortex function as providing an endogenous means of overcoming strong, exogenously indicated, habitual stimulus–response associations is completely compatible with the dual premotor systems hypothesis as outlined here. An impairment of this endogenously dependent selective process could also explain the appearance of the alien hand sign (see below) with medial frontal cortex damage.

It is also possible that this same selective mechanism, functioning short of the point at which overt action develops, may be employed to manipulate covert attention to bring specific information into conscious focus, suggesting a 'premotor' theory of selective attention (Rizzolatti *et al.*, 1987; see also Edelman, 1989). Overlap between the 'premotor' concept, 'attention to action' and intentionality is suggested because these concepts all address the processes whereby selective access to output mechanisms is determined. Patients with medial frontal damage and with Parkinson's disease are impaired on this and similar tasks. Patients with schizophrenia may also show specific deficits on attentional conflict tasks (Posner *et al.*, 1988a). This suggests that these concepts could be usefully applied to the understanding of psychiatric illness (Frith, 1987; Early *et al.*, 1989).

12.5. **The alien hand sign as a consequence of medial premotor system dysfunction**

Damage to the medial aspects of the frontal lobe, including the anterior cingulate cortex and the SMA, may thus be responsible for a peculiar disturbance of intentionality, the alien hand sign (Goldberg, *et al.*, 1981; Goldberg and Bloom, 1990). These patients, who most commonly have suffered damage to the medial frontal lobe as a result of a thrombosis of the anterior cerebral artery, display the very curious phenomenon of purposeful organised movements of the affected limb contralateral to the damaged hemisphere, for which the patient has no accompanying conscious feeling of volition. The movements are not perceived as self-generated and the patients interpret them as being produced by an external force. They do not feel that they are initiating or controlling the behaviour of the hand and often express dismay at the ability of the hand to act purposefully, independent of conscious volition. While the patients perceive and recognise the hand as their own, thus distinguishing this condition from anosognosia in which the hand is not recognised as part of the self, they do not recognise the intentionality associated with the behaviour of the hand as being self-generated.

I have observed and studied several individuals with 'alien hand sign' associated with medial frontal cortex infarction demonstrated by X-ray computed tomographic (CT) imaging and/or magnetic resonance imaging (MRI). In all cases, the 'alien' behaviour was observed in the hand contralateral to the damaged hemisphere and the patient attempted to control this aberrant behaviour using the opposite 'obedient' hand. One case history is given as an example.

Alien hand sign: Example case history

A 75-year-old right-handed woman was admitted to the hospital because of a right hemiparesis and non-fluent aphasia characterised by preserved comprehension but difficulty with initiation of spontaneous speech. The patient was admitted to a rehabilitation hospital nine days after the onset of hemiparesis. On physical examination, the patient was alert and oriented. She had a right central facial weakness and her right side was hyperreflexic with decreased sensation to pinprick. Her ability to repeat words and complex phrases was normal. Her ability to name objects to confrontation was accurate. However, her volitional word production was slow and laboured in propositional speech. She was non-fluent, with very limited spontaneous language production. When addressed, the patient spoke with hesitancy but without language impairment. She was able to read without difficulty and could write names appropriately with her right hand. Her ability to produce serial numbers was intact. There was no spontaneous movement in the right lower extremity, which displayed profound weakness (grade 0/5) throughout. The right arm had isolated finger, wrist and elbow activity. There was profound weakness at the shoulder (grade 1/5) but distally the patient was stronger (elbow and wrist grade 3/5). Bimanual coordination was severely impaired with a tendency to produce mirror-symmetric movement. A tonic grasp reflex was present in her right hand, which would reach out uncontrollably and grasp things around her. Once her hand clasped something, the patient was unable to voluntarily release the object. To effect release, she forcibly extended the fingers of the right hand using the left hand, effectively 'peeling' the right fingers away from the grasped object. The patient expressed dismay and perplexity with regard to these 'involuntary' actions. She indicated that she did not feel as if she herself was moving the limb but that some unknown force was directing these actions. When asked who or what, then, was moving her right arm if she was not, the patient, a religious woman, replied that 'It must be God'. The right hand would occasionally interfere with tasks she was performing with her left hand. She would hold on to her right hand with her left hand to prevent it from performing involuntary movements. Purposeful movement of the right hand was also highly responsive to verbal suggestion. When asked if she ever had problems with her right hand touching her face, she answered no as the right hand then began touching her face. When asked if she ever had difficulty with her right hand reaching over to grasp on to her left forearm she indicated no as the right hand then reached across to grasp onto her left arm. She then had to struggle to free the left arm from the grasp of the right arm. This type of behaviour was readily reproduced. Yet she continued to have some voluntary control over the right hand although the movements produced were effortful, slow and hypometric. Thus two modes of action could be distinguished in the limb: an 'alien' mode and a voluntary control mode. Some improvement in voluntary control over the limb was noted when the hand was brought across the midline into left hemispace. The troublesome extravolitional activities of the right hand operating in 'alien mode' gradually diminished as time passed so that, after about eight months, the alien action occurred only rarely.

The described behaviours were extensively documented in videotaped examinations and interviews after informed consent was obtained to videotape for educational and research purposes. A photographic sequence obtained from the video tape is shown in Fig. 12.2.

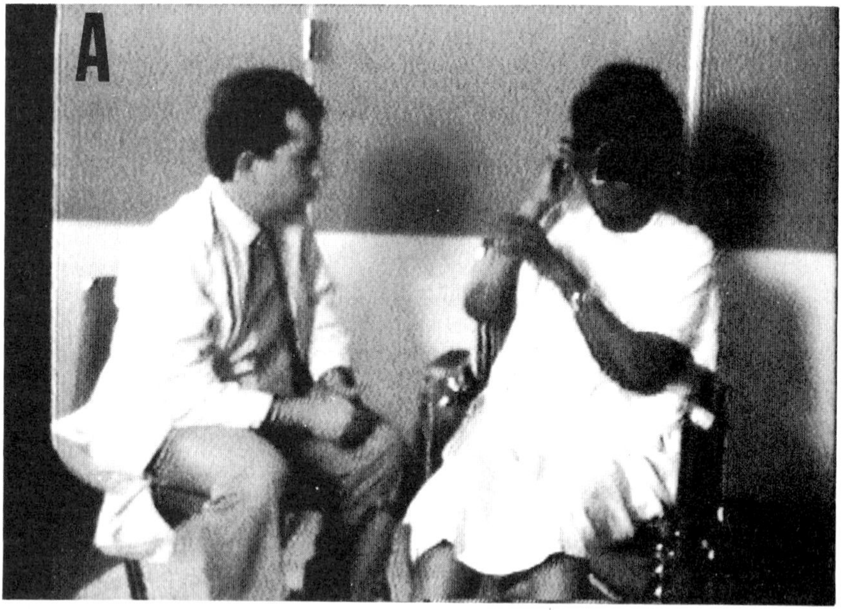

Fig. 12.2. Right alien hand behaviour documented on video tape. (A) Moments after being asked whether her right arm ever disturbs her by pulling her hair, to which she indicated to the negative, the right hand reaches up to the right side of the head and grasps on to her hair with the left hand in pursuit.

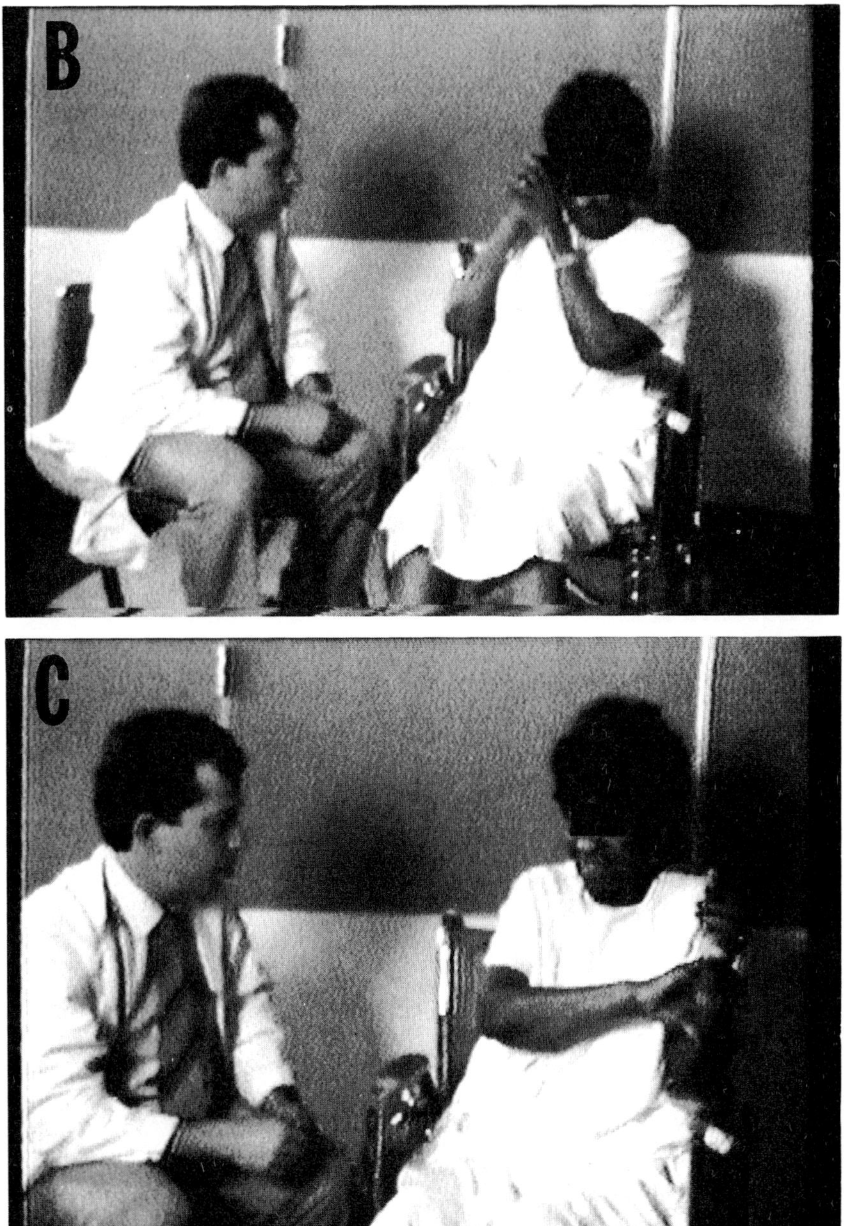

Fig. 12.2. cont. (B) Left hand reaches up and grasps the right hand to pull it back down to her lap. (C) Patient is asked if the right hand ever reaches over to grasp on to the left arm. After indicating 'No', the right hand crosses over to grasp on to the left forearm. The patient is then unable to voluntarily release her grasp of the forearm . . .

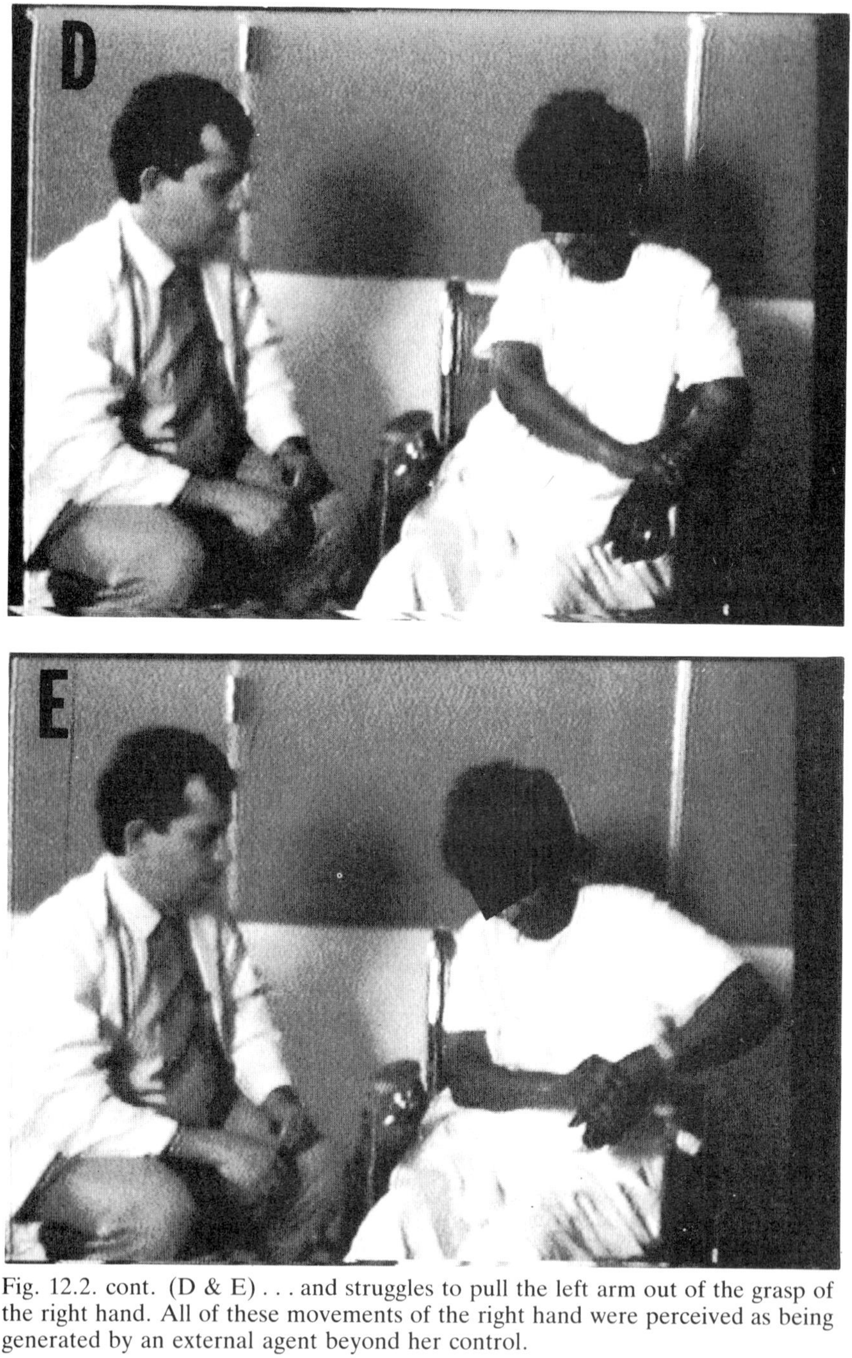

Fig. 12.2. cont. (D & E) . . . and struggles to pull the left arm out of the grasp of the right hand. All of these movements of the right hand were perceived as being generated by an external agent beyond her control.

A cranial CT revealed a large left medial posterior frontal infarct (see Fig. 12.3). Cranial MRI revealed a corpus callosum infarct and involvement of the medial surface of the left frontal lobe including the anterior cingulate cortex and supplementary motor area with increased signal compatible with ischaemic damage in these areas (see Fig. 12.4).

While the alien hand sign is observed in the hand contralateral to the damaged hemisphere, there also appears to be a hemispatial modulation of this effect suggesting that the influence of the hemispheric systems may be spatially bounded to some extent. The hand is observed to display most of its alien behaviour when it is in the hemispace contralateral to the damaged hemisphere, and voluntary motor performance in the hand improves when it crosses the midline into the hemispace opposite the intact hemisphere (Meador *et al.*, 1986). Furthermore, we have observed one case in which medial frontal hemisphere and lateral parietal hemisphere lesions were combined within the same left hemisphere (Coslett *et al.*, 1990). In this patient, dramatic hemispatial effects were observed in which the right hand was totally without voluntary control when extended laterally into right hemispace but could be used purposefully when brought over into left hemispace.

How can the 'alien hand' dissociative phenomena be conceptualised in the context of the foregoing discussion? It is possible that all action is initiated through a subconscious process that eventually penetrates into consciousness as the process reaches a particular threshold (Libet *et al.*, 1983; Libet, 1985). One may then postulate that the 'alien' acts are not accompanied by the normally acquired attribute of conscious volition and self-reference (see also Brown, 1987). Alternatively, one can view the emergence of the alien hand sign with medial frontal cortex damage as a manifestation of dysfunction in the wilful initiation of voluntary movement (Eccles, 1982, 1986). Another way of viewing this phenomenon is by invoking the concept of the 'corollary discharge' (Sperry, 1950; Teuber, 1964, 1972; Amassian *et al.*, 1989). The corollary discharge is a proposed flow of anticipatory information from effector regions to sensory regions that provides advanced information to the sensory system about the self-generated act so that the sensory consequences of the emerging action are interpreted as being internally as opposed to externally produced. Teuber (1972) proposed that the corollary discharge can be viewed as a 'physiologic marker for the "voluntariness" of self-initiated movement . . . different degrees of disruption of the corollary mechanism would become manifest as different degrees of *compulsiveness*, or of abnormally *stimulus-bound behaviour*' (p. 640, emphasis added). One may thus propose that one of the functions of the medial premotor system may be the anticipatory elaboration of the corollary discharge that

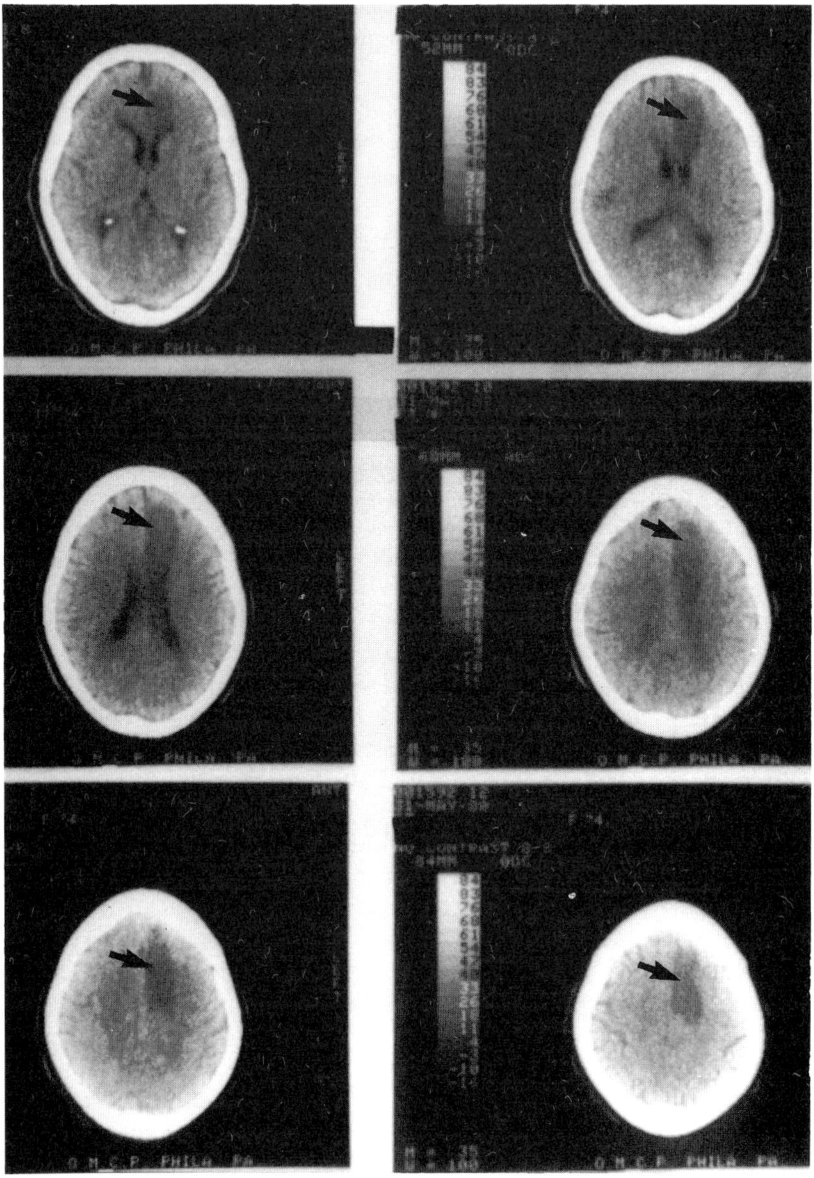

Fig. 12.3. Axial computer tomography (CT) scan of the brain documenting low signal intensity in the left medial frontal lobe consistent with cerebral infarction involving anterior cingulate, medial prefrontal and supplementary motor contex (SMA).

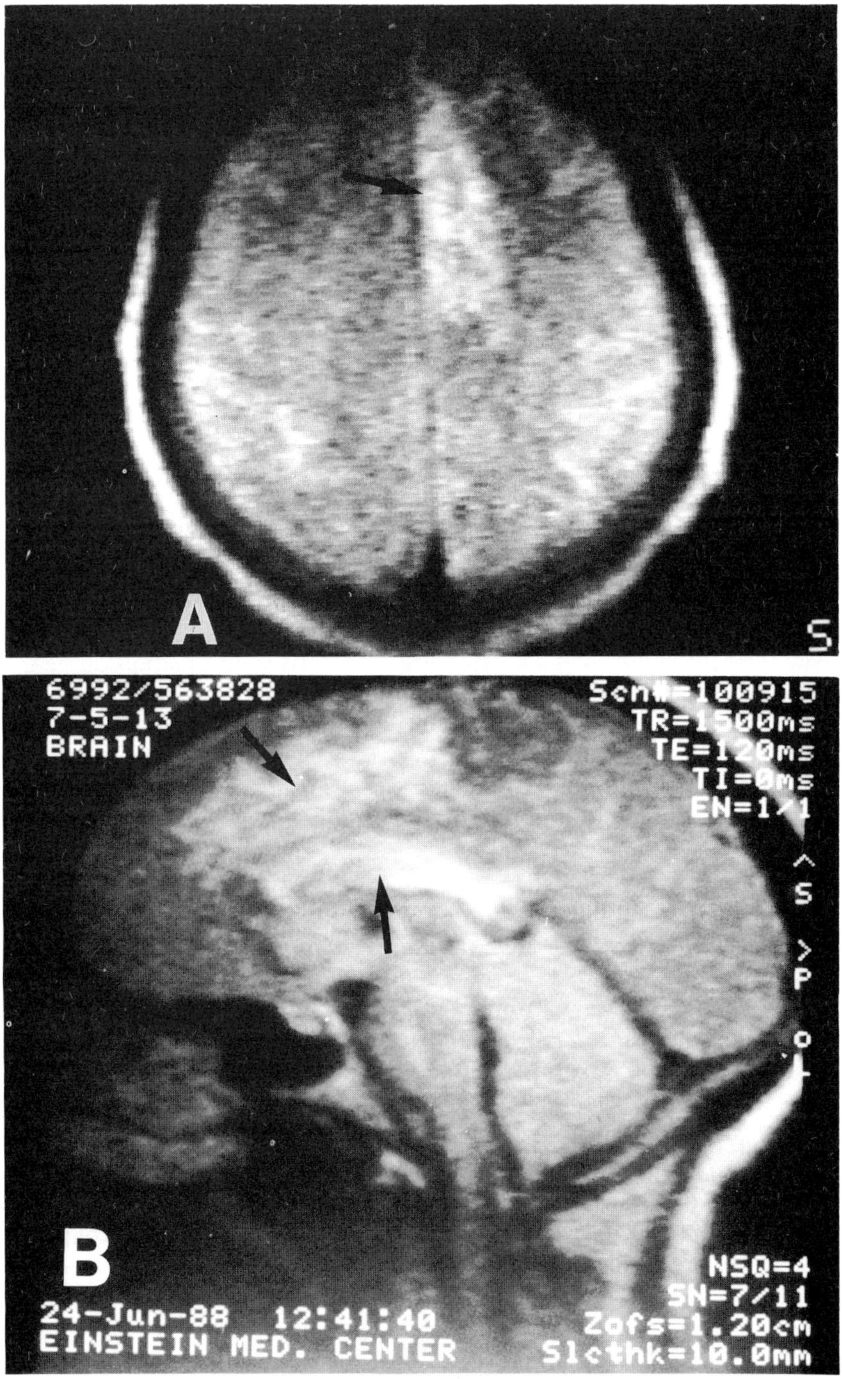

Fig. 12.4. (A) Axial magnetic resonance (MR) image demonstrates increased signal intensity along the left medial frontal lobe as the result of prior ischaemic infarction. T2-weighted (TR = 1500 ms; TE = 120 ms). (B) T2-weighted mid-sagittal MR image demonstrates increased signal intensity in the corpus callosum and cingulate gyrus as well as more dorsal and anterior regions on the medial surface of the frontal lobe.

is associated with the 'voluntariness' of an act and enables the act to be interpreted as self-generated.

In terms of the dual premotor systems hypothesis, the alien hand sign is proposed to result from an imbalance between the influence of the medial as opposed to the lateral premotor system of the damaged hemisphere. Damage to the medial surface of the frontal lobe effectively 'releases' the lateral premotor system to facilitate action that is directly responsive to the sensory context. The hand is thus placed in a 'free-running' mode in which action is dissociated from internal context, and the biasing tends to be determined by external conditions and influences. This is dramatically indicated in the case study presented here by the verbal suggestibility of action and the dissociation between what the patient said about her own actions and what was subsequently observed to happen. We have noted this responsiveness to verbal suggestion only in patients with medial frontal infarction of the dominant hemisphere.

12.6. **Conclusion**

The dual premotor systems hypothesis synthesises information on neural function and behaviour from a wide variety of sources. It suggests that different brain systems have evolved to facilitate and control action, and that the degree to which these systems are activated depends upon whether an act is generated endogenously or linked to immediate environmental conditions. The control of action emerges through a global process of progressive constraint application and self-organisation within the nervous system and neuromuscular apparatus. The sources of constraint may be generated endogenously through uncued recall or 'internal instruction' (Thaler and Passingham, 1989) or may be tied directly to environmental context. It is proposed that endogenous generation selectively activates the medial premotor system, while the incorporation of environmental context recruits the lateral premotor system. While it is efficient and necessary to habitualise useful subroutines and stimulus–response linkages to improve and stabilise motor performance as well as to 'free up' cortical systems, there is also a need to be able, where necessary, to monitor, supervise and intervene in ongoing habitualised behaviours with deliberate conscious control (Norman and Shallice, 1986). However, interference generated by the participation of conscious volition may actually impede the performance of highly skilled, precisely timed motor acts. It is proposed that the medial premotor system is involved in this process of voluntary supervision of motor behaviours through a process of selective facilitation and inhibition. The medial frontal cortex, through its position at the point where the outflow from the basal ganglia reaches the cortex; the anterior cingulate cortex, in its relationship to the outflow from the ventral striato-pallidal complex; and the medial prefrontal cortex and SMA, in their relationship to the outflow from the dorsal striato-pallidal complex, may all serve as critical network nodes controlling cortical access to brainstem and spinal motor mechanisms. Under typical conditions, both medial and lateral premotor systems would be involved together in the control of movements that are both voluntary and also dependent upon salient external events. The dual premotor hypothesis predicts that those acts which involve a greater degree of voluntary effort and attention to action, for example in setting conflictual activations between triggered habitual tendencies and conscious volition mediating internal instruction (as in the Stroop paradigm), would involve a greater degree of activation of the medial premotor system. It is the medial system's participation which conveys the conscious 'voluntariness' of the action. Damage to this system may lead to a variety of different types of impairment of voluntary action, including difficulty with initiation of internally instructed behaviour and the appearance of 'alien' behaviours, or compulsive automatisms dissociated from a sense of conscious volition. It is argued that through its activity, including the capacity to suppress pre-potent tendencies based on environmental factors, the medial system bestows behavioural autonomy and supports the capacity to act selectively, in anticipation of future possibility rather than in an automated habitual response to immediate circumstance.

Application of this broad conceptual framework for understanding the neurophysiology of premotor mechanisms may provide useful insights into the neural correlates of voluntary behaviour under conditions of normal functional performance. It may also help to develop heuristic models for understanding the wide variety of neurological and psychiatric disorders which manifest as impairments of voluntary behaviour and everyday functioning. Such models may then be applied to the development of rational methods for the treatment and rehabilitation of the dysfunctional states produced by these conditions.

Acknowledgements

The author gratefully acknowledges grant support from the Moss Rehabilitation Hospital Research Fund and from the Whitaker Foundation. Karen K. Bloom MD provided valuable assistance in the preparation of the clinical case materials.

References

Amassian, V. E., Cracco, R. Q. and Maccabee, P. J. (1989) A sense of movement elicited in paralyzed distal arm by focal magnetic coil stimulation of human motor cortex. *Brain Res.* **479**: 355–360.

Asanuma, H. and Arissian, K. (1984) Experiments on functional role of peripheral inputs to motor cortex during voluntary movements in the monkey. *J. Neurophysiol.* **52**: 212–227.

Brand, M. (1984) *Intending and Acting: Toward a Naturalized Action Theory*, MIT Press, Cambridge, MA.

Brown J. W. (1987) The microstructure of action. In: E. Perecman ed., *The Frontal Lobes Revisited.* IRBN Press, New York, pp. 251–272.

Brown, R. G. and Marsden, C. D. (1988) Internal vs external cues and the control of attention in Parkinson's disease. *Brain* **111**: 323–345.

Brown, R. G. and Marsden, C. D. (1990) Cognitive function in Parkinson's disease. From description to theory. *Trends Neurosci.* **13**: 21–29.

Bruce, C. J., Goldberg, M. E., Buschnell, M. C. and Stanton, G. B. (1985) Primate frontal eye fields. II. Physiological and anatomical correlates of electrically evoked eye movements. *J. Neurophysiol.* **54**: 714–734.

Cavada, C. and Goldman-Rakic, P. S. (1989) Posterior parietal cortex in rhesus monkey: II. Evidence for segregated corticocortical networks linking sensory and limbic areas with the frontal lobe. *J. Comp. Neurol.* **287**: 422–445.

Coslett, H. B., Schwartz, M., Goldberg, G., Haas, D. and Perkins, J. (1990) Hemispatial motor, sensory, and language deficits after parietal and mesial frontal infarcts. *Neurology* **40** (Suppl 1): 302.

Crutchfield, J. P., Farmer, J. D., Packard, N. H. and Shaw, R. S. (1986) Chaos. *Sci. Am.* (Dec): 46–57.

Dabrowska, J. (1971) Dissociation of impairment after lateral and medial prefrontal lesions in dogs. *Science* **171**: 1037–1038.

Dick, J. P. R., Rothwell, J. C., Day, B. L., Wise, R. J. S., Benecke, R. and Marsden, C. D. (1987) Modulation of the long-latency reflex to stretch by the supplementary motor area in humans. *Neurosci. Lett.* **75**: 349–354.

Dinner, D. S., Lüders, H., Morris, H. H., Wyllie, E. and Kramer, R. (1987) Human supplementary motor area (SMA) electrical stimulation study. *Epilepsia* **28**: 619.

Early, T. S., Posner, M. I., Reiman, E. M. and Raichle, M. E. (1989) Hyperactivity of the left striato-pallidal projection. Part II: Phenomenology and thought disorder. *Psychiatr Dev* **7**: 109–121.

Eccles, J. C. (1982) The initiation of voluntary movements by the supplementary motor area. *Arch. Psychiatr. Nervenkr.* **231**: 423–441.

Eccles, J. C. (1986) Do mental events cause neural events analogously to the probability fields of quantum mechanics? *Proc. R. Soc. London Ser. B* **227**: 411–428.

Edelman, G. M. (1989) *The Remembered Present. A Biological Theory of Consciousness*, Basic Books, New York.

Eidelberg, D. and Galaburda, A. M. (1984) Inferior parietal lobule: divergent architectonic asymmetries of the human brain. *Arch. Neurol.* **41**: 843–852.

Favorov, O., Sakamoto, T. and Asanuma, H. (1988) Functional role of corticoperipheral loop circuits during voluntary movements in the monkey: a preferential bias theory. *J. Neurosci.* **8**: 3266–3277.

Fox, P. T., Fox, J. M., Raichle, M. E. and Burde, R. M. (1985) The role of cerebral cortex in the generation of voluntary saccades: a positron emission tomographic study. *J. Neurophysiol.* **54**: 348–369.

Freud, S. (1924) *A General Introduction to Psychoanalysis*, Washington Square Press, New York, 1952 (reprinting of original edition published in 1924).

Freund, H-J. and Hummelsheim, H. (1985) Lesions of the premotor cortex in man. *Brain* **108**: 697–733.

Frith, C. D. (1987) The positive and negative symptoms of schizophrenia reflect impairments in the perception and initiation of action. *Psychol. Med.* **17**: 631–648.

Fulton, J. F. (1935) A note on the definition of the 'motor' and 'premotor' areas. *Brain* **58**: 311–316.

Galaburda, A. M. and Pandya, D. N. (1983) The intrinsic, architectonic and connectional organization of the superior temporal region of the rhesus monkey. *J. Comp. Neurol.* **221**: 169–184.

Godschalk, M., Lemon, R. N., Kuypers, H. G. J. M. and Ronday, H. K. (1984) Cortical afferents and efferents of monkey postarcuate area: an anatomical and electrophysiological study. *Exp. Brain Res.* **56**: 410–424.

Goldberg, G. (1985a) Response and projection. A reinterpretation of the premotor concept. In: *Neuropsychological Studies of Apraxia and Related Disorders*, E. A. Roy, ed., North-Holland, Amsterdam, pp. 251–266.

Goldberg, G. (1985b) Supplementary motor area structure and function: review and hypotheses. *Behav. Brain Sci.* **8**: 567–616.

Goldberg, G. (1987) From intent to action. Evolution and function of the premotor systems of the frontal lobe. In: *The Frontal Lobes Revisited*, E. Perecman, ed., IRBN Press, New York, pp. 273–306.

Goldberg, G. (1991) Microgenetic theory and the dual premotor systems hypothesis. Implications for the rehabilitation of the brain-damaged subject. In: *Cognitive Microgenesis. A Neuropsychological Perspective*, R. Hanlon, ed., Springer, New York, in press.

Goldberg, G. and Bloom, K. K. (1990) The alien hand sign. Localization, lateralization and recovery. *Amer. J. Phys. Med. Rehab.* **69**: 228–238.

Goldberg, G., Mayer, N. H. and Toglia, J. U. (1981) Medial frontal cortex infarction and the alien hand sign. *Arch. Neurol.* **38**: 683–686.

Goldberg, G. and Moster, M. (1988) Topographic mapping of presaccadic brain electrical activity in humans: support for the dual premotor systems hypothesis. *Soc. Neurosci. Abstr.* **14**: 1236.

Hummelsheim, H., Wiesendanger, M. and Bianchetti, M. (1986) The supplementary motor area modulates perturbation-evoked discharges of neurones in the precentral motor cortex. *Neurosci. Lett.* **67**: 119–122.

Hutchins, K. D., Martino, A. M. and Strick, P. L. (1988) Corticospinal projections from the medial wall of the hemisphere. *Exp. Brain Res.* **71**: 667–672.

Ingvar, D. H. (1985) 'Memory of the future'. An essay on the temporal organization of conscious awareness. *Hum. Neurobiol.* **4**: 127–136.

Jürgens, U. (1984) The efferent and afferent connections of the supplementary motor area. *Brain Res.* **300**: 63–81.

Kalil, K. (1978) Neuroanatomical organization of the primate motor system. Afferent and efferent connections of the ventral thalamic nuclei. In: *Multidisciplinary Perspectives in Event-related Brain Potential Research. Proc Fourth Int. Cong. Event-related Slow Potentials of the Brain (EPIC IV)*. D. A. Otto, ed., US Government Printing Office, Washington, DC, pp. 112–123.

Kelso, J. A. S. (1990) Phase transitions: foundations of behavior. In: *Synergetics of Cognition*, H. Haken and M. Stadler, eds, Springer-Verlag, Berlin, pp. 249–268.

Konorski, J. (1972) Some hypotheses concerning the functional organization of prefrontal cortex. *Acta Neurobiol. Exp.* **32**: 595–613.

Kugler, P. N. and Turvey, M. T. (1987) *Information, Natural Law, and the Self-assembly of Rhythmic Movement*. Lawrence Erlbaum Publishers, Hillsdale.

Kwan, H. C., Mackay, W. A., Murphy, J. T. and Wong, Y. C. (1978) Spatial

organization of precentral cortex in awake primates. II. Motor outputs. *J. Neurophysiol.* **41**: 1120–1131.
Lhermitte, F. (1983) Utilization behavior and its relation to lesions of the frontal lobes. *Brain* **106**: 237–255.
Lhermitte, F. (1986) Human autonomy and the frontal lobes. Part II. Patient behavior in complex and social situations: The 'Environmental Dependency Syndrome'. *Ann Neurol.* **19**: 335–343.
Lhermitte, F., Pillon, B. and Serdaru, M. (1986) Human autonomy and the frontal lobes. Part I. Imitation and utilization behavior: a neuropsychological study of 75 patients. *Ann. Neurol.* **19**: 326–334.
Libet, B. (1985) Unconscious cerebral initiative and the role of conscious will in voluntary action. *Behav. Brain Sci.* **8**: 529–566.
Libet, B., Gleason, C. A., Wright, E. W. and Pearl, D. K. (1983) Time of conscious intention to act in relation to onset of cerebral activity (readiness-potential): the unconscious initiation of a freely voluntary act. *Brain* **106**: 623–642.
Lüders, H., Lesser, R. P., Dinner, D. S., Morris, H. H., Wyllie, E., Godoy, J. and Hahn, J. (1987) A negative motor response (NMR) elicited by electrical stimulation of the human frontal cortex. In: *Advances in Epileptology*, vol. 16, P. Chauvel, E. Halgren, J. Bancaud, A. Escueta, eds. Raven Press, New York, pp. 229–231.
Lüders, H., Lesser, R. P., Dinner, D. S., Morris, H. H., Wyllie, E. and Godoy, J. (1988) Localization of cortical function: new information from extraoperative monitoring of patients with epilepsy. *Epilepsia* **29** (Suppl 2): S56–S65.
Matelli, M., Camarda, R., Glickstein, M. and Rizzolatti, G. (1986) Afferent and efferent projections of the inferior area 6 in the macaque monkey. *J. Comp. Neurol.* **251**: 281–298.
Matsamura, M. and Kubota, K. (1979) Cortical projection to hand-arm motor area from post-arcuate in macaque monkey: a histological study of retrograde transport of horseradish peroxidase. *Neurosci. Lett.* **11**: 241–246.
Mayer, N. H., Reed, E., Schwartz, M. F., Montgomery, M. and Palmer, C. (1990) Buttering a hot cup of coffee. An approach to the study of errors of action in patients with brain damage. In: *The Neuropsychology of Everyday Life. Vol. 2: Assessment and Basic Competencies*, D. E. Tupper and K. D. Cicerone, eds. Kluwer Academic Publishing, Boston, pp. 259–284.
Meador, K. J., Heilman, K. M. and Bowers, D. (1986) Hypometria with hemispatial and limb motor neglect. *Brain* **109**: 293–305.
Moster, M. L. and Goldberg, G. (1990) Topography of scalp potentials preceding self-initiated saccades. *Neurology* **40**: 644–648.
Mpitsos, G. J., Creech, H. C., Cohan, C. S. and Mendelson, M. (1988) Variability and chaos. Neurointegrative principles in self-organization of motor patterns. In: *Dynamic Patterns in Complex Systems*, J. A. S. Kelso, A. J. Mandell and M. F. Shlesinger, eds, World Scientific, New Jersey, pp. 162–190.
Muakkassa, K. F. and Strick, P. L. (1979) Frontal lobe inputs to primate motor cortex: evidence for four somatotopically organized 'premotor' areas. *Brain Res.* **177**: 176–182.
Murphy, J. T., Kwan, H. C., Mackay, W. A. and Wong, Y. C. (1978) Spatial organization of precentral cortex in awake primates. III. Input–output coupling. *J. Neurophysiol.* **41**: 1132–1139.
Nicolis, G. and Prigogine, I. (1989) *Exploring Complexity. An Introduction*, W. H. Freeman, New York.
Norman, D. A. (1981) Categorization of action slips. *Psychol. Rev.* **88**: 1–15.

Norman, D. A. (1988) *The Psychology of Everyday Things*, Basic Books, New York.

Norman, D. A. and Shallice, T. (1986) Attention to action. Willed and automatic control of behavior. In: *Consciousness and Self-regulation. Advances in Research and Theory, Vol. 4*, R. J. Davidson, G. E. Schwartz and D. Shapiro, eds, Plenum Press, New York, pp. 1–18.

Noth, J. (1986) Long-loop reflexes: concepts and consequences. *Appl. Neurophysiol.* **49**: 262–268.

Okano, K. and Tanji, J. (1987) Neuronal activities in the primate motor fields of the agranular frontal cortex preceding visually triggered and self-paced movement. *Exp. Brain Res.* **66**: 155–166.

Orgogozo, J. M. and Larsen, B. (1979) Activation of the supplementary motor area during voluntary movement suggests it works as a supramotor area. *Science* **206**: 847–850.

Pandya, D. N. and Barnes, C. L. (1987) Architecture and connections of the frontal lobe. In: *The Frontal Lobes Revisited*, E. Perecman, ed., IRBN Press, New York, pp. 41–72.

Pardo, J. V., Pardo, P. J., Janer, K. W. and Raichle, M. E. (1990) The anterior cingulate cortex mediates processing selection in the Stroop attentional conflict paradigm. *Proc. Natl Acad. Sci. USA* **87**: 256–259.

Passingham, R. E. (1987) Two cortical systems for directing movement. *CIBA Foundation Symposium* **132**: 151–164.

Passingham, R. E., Chen, Y. C. and Thaler, D. (1989) Supplementary motor cortex and self-initiated movement. In: *Neural Programming*, M. Itoh, ed. Karger, Basel, pp 13–24.

Penfield, W. and Welch, K. (1949) The supplementary motor area in the cerebral cortex of man. *Trans. Am. Neurol. Assoc.* **74**: 179–184.

Petersen, S. E., Fox, P. T., Posner, M. I., Mintun, M. and Raichle, M. E. (1988) Positron emission tomographic studies of the cortical anatomy of single word processing. *Nature* **331**: 585–589.

Petrides, M. (1986) The effect of periarcuate lesions in the monkey on the performance of symmetrically and asymmetrically reinforced visual and auditory go, no-go tasks. *J. Neurosci.* **6**: 2054–2063.

Petrides, M. and Pandya, D. N. (1984) Projection to the frontal cortex from the posterior parietal region in the rhesus monkey. *J. Comp. Neurol.* **228**: 105–116.

Posner, M. I., Early, T. S., Reiman, E., Pardo, P. J., Dhawan, M. (1988a) Asymmetries in hemispheric control of attention in schizophrenia. *Arch. Gen. Psychiat.* **45**: 814–821.

Posner, M. I., Petersen, S. E., Fox, P. T. and Raichle, M. E. (1988b) Localization of cognitive operations in the human brain. *Science* **240**: 1627–1631.

Prigogine, I. (1980) *From Being to Becoming. Time and Complexity in the Physical Sciences*. W. H. Freeman, New York.

Reason, J. T. (1979) Actions not as planned. The price of automatization. In: *Aspects of Consciousness, Vol. 1*, G., Underwood and R., Stevens eds, Academic Press, London, pp. 67–89.

Reiner, A., Brauth, S. E. and Karten, H. J. (1984) Evolution of the amniote basal ganglia. *Trends Neurosci.* **7**: 320–325.

Rizzolatti, G., Matelli, M. and Pavesi, G. (1983) Deficits in attention and movement following removal of postarcuate (area 6) and prearcuate (area 8) cortex in macaque monkeys. *Brain* **106**: 655–673.

Rizzolatti, G., Riggio, L., Dascola, I. and Umiltá, C. (1987) Reorienting attention across the horizontal and vertical meridians. Evidence in favor of a pre-

motor theory of attention. *Neuropsychologia* **25**: 31–40.
Roland, P. E., Larsen, B., Lassen, N. A. and Skinhøj, E. (1980) Supplementary motor area and other cortical areas in organization of voluntary movements in man. *J. Neurophysiol.* **43**: 118–136.
Roland, P. E., Meyer, E., Shibasaki, T., Yamamoto, Y. L. and Thompson, C. J. (1982) Regional cerebral blood flow changes in cortex and basal ganglia during voluntary movements in normal human volunteers. *J. Neurophysiol.* **48**: 467–480.
Rosen, R. (1985) *Anticipatory Systems. Philosophical, Mathematical and Methodological Foundations*. Pergamon Press, New York.
Sanides, F. (1964) The cyto-myeloarchitecture of the human frontal lobe and its relation to phylogenetic differentiation of the cerebral cortex. *J. Hirnforsch.* **6**: 269–282.
Sanides, F. (1969) Comparative architectonics of the neocortex of mammals and their evolutionary significance. *Ann. NY Acad. Sci.* **167**: 404–423.
Schell, G. R. and Strick, P. L. (1984) The origin of thalamic input to the arcuate premotor and supplementary motor areas. *J. Neurosci.* **4**: 539–560.
Schlag, J. and Schlag-Rey, M. (1987) Evidence for a supplementary eye field. *J. Neurophysiol.* **57**: 179–200.
Schwartz, M. F., Reed, E. S., Montgomery, M., Palmer, C. and Mayer, N. H. (1991) The quantitative description of action disorganization after brain damage: a case study. *Cognitive Neuropsychol.* in press.
Shallice, T. (1982) Specific impairments of planning. *Phil. Trans. R. Soc. London Ser. B* **298**: 199–209.
Skarda, C. A. and Freeman, W. J. (1987) How brains make chaos in order to make sense of the world. *Behav. Brain Sci.* **10**: 161–195.
Sperry, R. W. (1950) Neural basis of the spontaneous optokinetic response produced by visual inversion. *J. Comp. Physiol. Psychol.* **43**: 482–489.
Stern, C. E. (1987) Functions of the ventral striatum. PhD Thesis, Oxford University.
Strick, P. L. (1985) How do the basal ganglia and cerebellum gain access to the cortical motor areas? *Behav. Brain Res.* **18**: 107–123.
Stroop, J. R. (1935) Studies of interference in serial verbal reactions. *J. Exp. Psychol.* **18**: 643–662.
Teuber, H-L. (1964) The riddle of frontal lobe function in man. In: *The Frontal Granular Cortex in Man*, J. M. Warren and K. Akert, eds, McGraw-Hill, New York, pp. 410–444.
Teuber, H-L. (1972) Unity and diversity of frontal lobe functions. *Acta Neurobiol. Exp.* **32**: 615–656.
Thaler, D. E. (1988) Supplementary motor cortex and the control of action. PhD Thesis, Oxford University.
Thaler, D. E. and Passingham, R. E. (1989) The supplementary motor cortex and internally directed movement. In: *Neural Mechanisms in Disorders of Movement*, A. R. Crossman and M. A. Sambrook, eds, Libbey, London, pp. 175–182.
Walshe, F. M. R. (1935) On the 'syndrome of the premotor cortex' (Fulton) and the definition of the terms 'premotor' and 'motor': with a consideration of Jackson's views on the cortical representation of movements. *Brain* **58**: 49–80.
Wiesendanger, M. (1986) Recent developments in studies of the supplementary motor area of primates. *Rev. Physiol. Biochem. Pharmacol.* **103**: 1–59.
Wiesendanger, M. and Wiesendanger, R. (1984) The supplementary motor area in the light of recent investigations. *Exp. Brain Res. Suppl.* **9**: 382–392.

Woolsey, C. N., Settlage, P. H., Meyer, D. R., Sencer, W., Hamuy, T. P. and Travis, A. M. (1952) Patterns of localization in precentral and 'supplementary' motor areas and their relation to the concept of a premotor area. *Res. Pub. Assoc. Res. Nerv. Ment. Dis.* **30**: 238–264.

Yakovlev, P. (1948) Motility, behavior and the brain. *J. Nerv. Ment. Dis.* **107**: 313–335.

Yakovlev, P. I. and Locke, S. (1961) Limbic nuclei of thalamus and connections of limbic cortex. III. Corticocortical connections of the anterior cingulate gyrus, the cingulum and the subcallosal bundle in the monkey. *Arch. Neurol.* **5**: 364–400.

Yates, F. E. (1986) *Self-organizing Systems. The Emergence of Order.* Plenum Press, New York.

13 *Paul Grobstein*

Directed movement in the frog: motor choice, spatial representation, free will?

The sooner we recognize the fact that the complex higher functional Gestalts which leave the reflex physiologist dumbfounded in fact send roots down to the simplest basal functions of the CNS, the sooner we shall see that the previously terminologically insurmountable barrier between the lower levels of neurophysiology and higher behavioral theory simply dissolves away.

von Holst and Mittelstaedt (1950)

13.1. **Introduction**

There might, on the face of it, seem to be nothing in the way of 'complex higher functional Gestalts' in the directed movements made by frogs in response to prey items. Indeed, until quite recently, a large part of the appeal of anuran orienting responses for experimental neuroscientists was the apparent simplicity of the behaviour, and a belief that the underlying neuronal organisation was clearly understandable in reflex terms with only the anatomical and physiological details remaining to be filled in. However, as has happened with notable frequency in a variety of other contexts over the past ten years or so (Grobstein, 1990a), efforts to fill in the details of frog orienting behaviour have revealed an unexpected complexity in neuronal organisation, and forced a recognition that the behaviour itself is far more sophisticated than one might have thought. Frogs respond rapidly and accurately to prey items, but the processes involved in doing so have proven to have so little in common with stereotyped stimulus/response relations that explorations of alternative ways of thinking about how the nervous system works are necessary (Grobstein, 1988a,b, 1989). Frogs get where they need to go, but how they get there, I will suggest in this chapter, is a matter not of reflexes but instead of choices, of gestalts, and perhaps of a rudimentary form of free will.

The appropriateness of using, in a discussion of nervous system organisation, terms whose origins reflect observations of a quite different sort may not be obvious, particularly when the meaning and even the putative significance of the terms being used is a matter of no little disagreement. What I hope to show is that if terms such as 'choice', 'gestalt', and 'free will' did not exist, something very like them would have had to be invented to make sense of observations on nervous system organisation, and, further, that by at least temporarily borrowing the existing terms for this purpose one can contribute to a clearer understanding of what they actually mean in the context from which they arose. The existence of varying degrees of unpredictability of output in response to external perturbation, and that of distinctive overall forms which persist despite variations in the elements making them up, are characteristics of most biological systems at all levels of organisation (Grobstein, 1988c). While both the material substrates giving rise to these characteristics and the roles they play in biological function almost certainly differ at different levels of organisation (Grobstein, 1988c, 1990a), the admonition of von Holst and Mittelstaedt suggests that a recognition of similar and related phenomena at different levels of organisation should not only enrich studies of 'the lower levels of neurophysiology' but also contribute to those of 'higher behavioral theory' as well. One may ultimately want to acknowledge differences between phenomena at different levels of organisation with appropriate terminological distinctions, but this is best done with full awareness of how they relate to one another rather than across an 'insurmountable barrier'.

13.2. **From the reflex frog to activity-gated divergence: of choices and choice**

Classical work on anuran orienting behaviour led to working models which were clearly reflex in character (cf. Ewert *et al.*, 1983; Grobstein *et al.*, 1983; Ingle, 1983; see Fig. 13.1). A prey item at a given location in space is imaged by the optics of the eye at a given location on the retina. This in turn, by way of the topographic retinotectal projection, causes activity at a given location in superficial layers of the optic tectum. Different locations in the latter, it was supposed, are individually and distinctively linked to motor pattern generating circuitry in such a way as to cause the particular directed movement appropriate for the corresponding target location. The essence of such models was the presumption that what one was dealing with was an ensemble of parallel reflex pathways, each associating in a one-to-one fashion a given retinal and superficial tectal locus with a given movement.

Models of this sort were appealing to experimental neurobiologists because they seemed to provide a full explanation of behaviour in terms requiring no more than two already well understood neurobiological entities: sensory maps and motor pattern generating circuitry (Grobstein, 1988b). Such models were additionally appealing in that they reduced the seemingly vague and operationally undefined problem of accounting for directed movement to the more obvious and technically feasible problem of characterising the presumably different patterns of projection of differ-

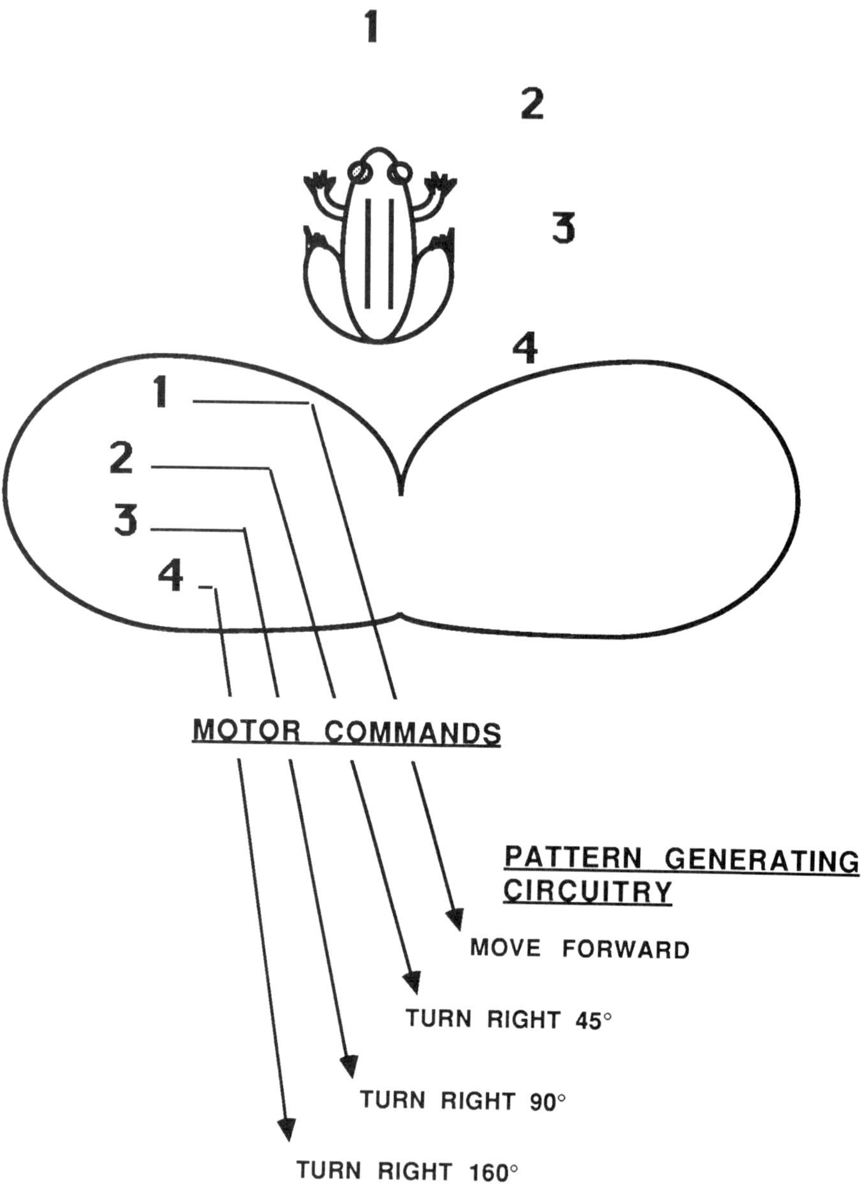

Fig. 13.1. Schematic illustration of early models of the neuronal organisation underlying frog prey orienting behaviour. Locations in visual space are represented by numbers around the frog. Below is shown the representation of these locations in the left lobe of the optic tectum. The descending parallel paths ('motor commands') link each of these tectal locations with 'pattern-generating circuitry' in such a way that activity at each causes a distinct and appropriate motor pattern, as diagrammed.

ent tectal locations. There was thus a certain amount of both scepticism and frustration when it began to be reported that neither lesion analyses (Grobstein *et al.*, 1983) nor anatomical studies (Masino and Grobstein, 1990, and references therein) were consistent with a simple parallel coupling between the optic tectum and directed movements. I have reviewed elsewhere the kinds of evidence that encouraged a re-evaluation of the information processing requirements involved in anuran orienting (Grobstein, 1988b, 1989, 1990b), and discussed some of the methodological implications of this and similar stories for strategy and tactics in integrative neurobiology generally (Grobstein, 1988a, 1989, 1990a). Here I want to focus on the new theoretical perspective, and on the significance of finding that the neuronal circuitry involved in orienting proved to be more complicated than it was expected to be.

In essence, what was recognised was a fundamental inconsistency between neural model and behavioural reality: the latter requires that each retinal and superficial tectal locus must be related to elements of the movement repertoire, not in an obligatory one-to-one fashion, but rather in a way which makes possible the association of a given input with any of an array of outputs. The most straightforward evidence for this divergent organisation comes from the observation that the prey-orienting movements of a frog vary with stimulus location in all three dimensions of space (Grobstein *et al.*, 1985; Grobstein, 1988b). While not in itself particularly surprising, and indeed earlier documented (Ingle, 1970, 1972; Comer and Grobstein, 1981), the implications of this variation are profound but were largely unrecognised until recently. The retina, and the superficial retinotectal projection, are both two-dimensional surfaces, and there is simply no way to associate, in a one-to-one fashion, locations in a two-dimensional surface with elements of a motor repertoire which itself varies continuously in three dimensions. Instead what one should be looking for is another kind of neuronal organisation which I have termed 'activity-gated divergent' (Grobstein, 1988b; see Fig. 13.2). Each point in the superficial tectum must be linked by intervening circuitry not in an invariant fashion with a particular movement but rather in a labile fashion with an array of possible movements, including at a minimum all of those appropriate for all of the stimulus locations which are imaged at the same point on the retina but differ in distance. The particular movement which occurs at any given time must reflect not only the pattern of connections from a particular superficial tectal locus but also other activity in the nervous system at that time, relaying information related to stimulus distance. Other important parameters not represented in the location of superficial tectal activity must also be involved (Grobstein, 1988b, 1989).

That an activity-gated divergent network, rather than parallel paths, intervenes between superficial tectum and anuran orienting movements very much changes one's expectations of the efferent anatomical organ-

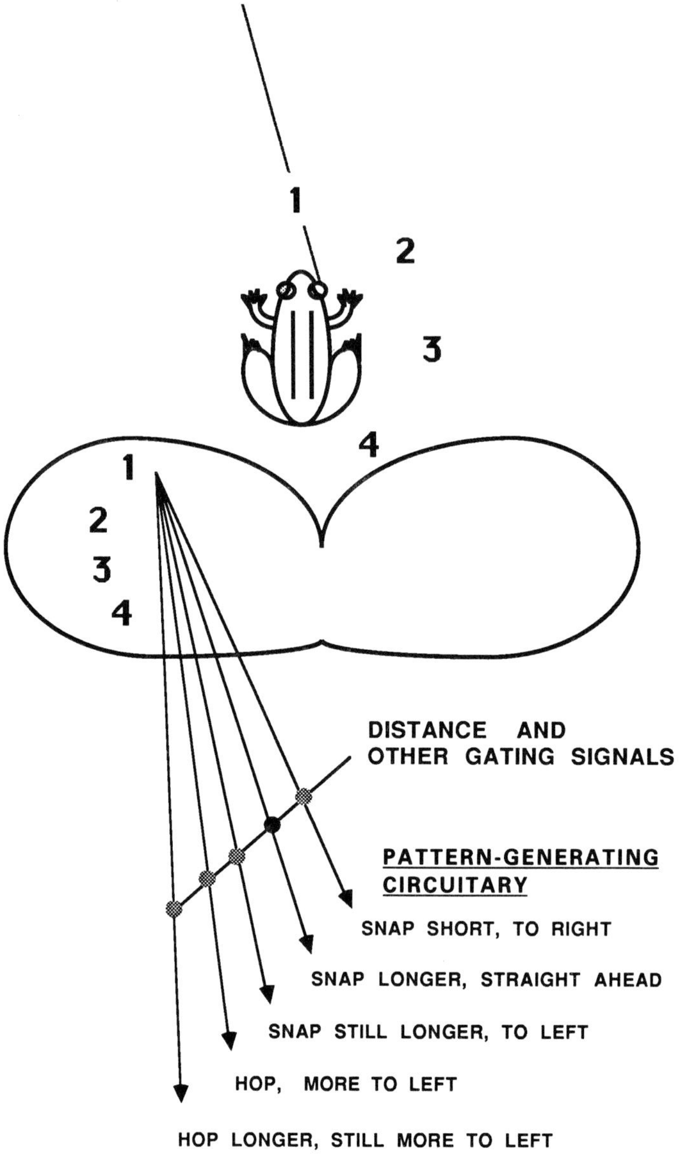

Fig. 13.2. Schematic illustration to show the more realistic, activity-gated divergent character of the linkage between tectum and motor output in the case of frog orienting behaviour. A given retinal and tectal locus receives input not from a particular point in three-dimensional space but rather from all points along a given line of sight from the eye. One such line of sight, passing through location '1' in space, is diagrammed. Stimuli at different points along this line of sight must trigger different movements. Hence, the pathways linking each tectal locus with movement pattern must be divergent, as diagrammed, rather than parallel, with the particular pathway effective at a particular time (black dot in contrast to shaded dots, presuming a stimulus at the distance of '1') reflecting other signals, including those related to stimulus distance ('distance and other gating signals').

isation of superficial tectum in anurans, and has in the present context a quite specific additional significance as well. Inherent in the concept of an activity-gated divergent network is the notion that a fundamental design requirement of many neural networks is not that of associating a given input with a particular output, but rather that of supporting 'choice', in the sense of making possible the association of a given input with any of several outputs. I do not expect all readers to instantly accept that my use of the term 'choice' is appropriate in the present context, but I do think that the unexpected complexity in tectal efferent organisation clearly requires acknowledgement by some term more weighty than 'reflex pathways'. Moreover, I will try to show that 'choice', at least in an important rudimentary sense, is real in the situation described and, further, that the tentative use of the term in the present context helps us to understand its meanings in more general usage.

An initial concern is that many readers are likely to be most comfortable using the term 'choice' for situations in which, for a given input, there is variation in the motor output of an *organism*, and no variation in the motor output of a frog in response to a stimulus at a given location has been asserted so far (although it will be in the following). All that has been claimed is that there is a degree of indeterminacy in the output associated with activation of a given superficial tectal locus. Frogs, of course, have two eyes and overlapping visual fields (cf. Grobstein *et al.*, 1980), and one might reasonably expect the indeterminacy to disappear if one considers the input from both eyes rather than from only one. It does not in fact, but, even putting this temporarily aside, an important point is being missed if one considers the matter from the perspective of binocular convergence (or the whole organism) rather than from the perspective of one eye. Each retina is two-dimensional, as is its projection to the tectum, in so far as this is organized to represent information about stimulus location. Hence, each location in the plane of the two-dimensional superficial tectum must project in such a way as to trigger any of a variety of movements. Convergence of a signal from the other eye may provide a way to gate this divergent circuitry, but the possibility does not vitiate the argument for its existence.

What all this suggests as an important generalisation is that 'choice' is frequently a matter of perspective. From the vantage point of a location on the retina or in superficial tectum, one is dealing with an activity-gated divergent circuit. If one stands outside the animal, however, then one might be dealing not with activity-gated divergence but simply with two input channels whose combined activity determines the output. 'Choice' seems to have disappeared. This raises the issue of whether 'activity-gated divergence' is really a distinctive concept as opposed to being a fancy new name for what most neurophysiologists presume about the nervous system anyhow: that outputs typically reflect the interaction of signals on several input channels. The answer is a little bit of both, with no apologies for the latter; it all depends on perspective. From the point of view of an observer sitting at a particular point on the retina, or at the corresponding terminals in the optic tectum, the output resulting from activity at that location differs from occasion to occasion and clearly reflects a process of choice. Neurobiologists often work from local vantage points of this kind, only to find subsequent circuitry more complex than they expected it to be. Only later do they realise that the information processing going on from that point necessarily involves the kind of convergence of additional sensory information which constitutes the choice process (Grobstein, 1988a, 1990a). That hypothetical mechanisms underlying such a capability exist within the imagination of neurobiologists ought not to be allowed to detract from the recognition that particular circuits are indeed structured for this purpose.

Indeed the presence of 'choice' at one level of organisation, and its disappearance at a second, seems to me not so much an embarrassment as an instructive realisation. If one is looking at any given system or subsystem and it displays variation in outputs for a given known input, then that system or subsystem is organised to support the capability for choice. That a larger system, of which the system or subsystem being analysed is a component, is more predictable in terms of *its* output relationships this may be significant for understanding the larger system, but may equally be irrelevant for characterising the smaller. Components with variable response characteristics are not infrequently combined so as to yield apparently more stereotyped input/output relations at a higher level of complexity (Oyama, 1985; Grobstein, 1990a). What this highlights is that 'choice' (or something very like it) is not a phenomenon which appears only with increasing complexity. It is a property of components at a variety of levels of organisation, a phenomenon which sends roots 'down to the simplest basal functions of the CNS'.

The concept of 'activity-gated divergence' not only recognises a known capability of neural nets with regard to 'choice', but also distinguishes one of several different general ways in which the nervous system is organised so as to yield the phenomenon of different outputs associated with a given input. Use-dependent synaptic lability has been well established as a mechanism of this sort (cf. Carew and Sahley, 1986), and recently it has become clear that relatively diffusely acting pharmacological agents can influence both synaptic events and intrinsic cell properties in such a way as to produce substantial changes in the responses of networks to particular inputs (cf. Getting, 1989; Harris-Warwick *et al.*, 1990). Activity-gated divergence differs from the former in that it provides a kind of variation which neither reflects nor depends on cellular changes produced by prior patterns of activation of the network. Perhaps more fundamentally, activity-gated divergence differs from both use-dependent synaptic lability and the effects of diffusely acting pharmacological agents in that different input/output states occur with no changes whatsoever in either synaptic or intrinsic cellular properties; they reflect instead transient activity patterns. This has interesting methodological implications, in that potential variability is unobservable by studies aimed at detecting changes of intrinsic cellular and synaptic properties. It also probably means that activity-gated divergence represents the most rapidly achieved and labile form of input/output variation, since the need for a coupling of neuronal activity to metabolic processes is obviated. Indeed, it seems likely that this rather simple form of functional lability may be the most common (cf. Forssberg *et al.*, 1975; Fukson *et al.*, 1980; Hoy and Nolen, 1987). One need only reflect on how frequently one's own behavior is altered by a given experience (or a given thought) so rapidly and specifically that neither changes in synaptic strengths nor diffusely acting pharmacological agents are likely to be involved. As noted by Lashley many years ago, 'input is never into a quiescent or static system, but always into a system which is already excited and organized . . . behavior is the result of interaction of this background of excitation with input from any designated stimulus' (Lashley, 1951). This background of excitation is a significant information store for the functioning nervous system (Grobstein, 1990a), and its variations almost certainly provide a major source of input/output variability.

My hope is that the discussion so far not only contributes to an understanding of the potential for 'choice' in stable networks with stable neuronal properties, but also illustrates what seems to me a natural and useful logic for further analysing some of the multiple meanings inherent in this term as it is generally used. I have argued that a minimal criterion of 'choice' is that a system exhibits several possible outputs for a particular input, and that this same criterion may be applied whether one is speaking of an organism or a part of an organism (a particular neural circuit). What this calls attention to is the need for careful definition of both system and input, together with a willingness to let 'choice' come and go depending on one's perspective and the outcome of particular lines of investigation.

My definition of choice, albeit rudimentary, is actually more demanding than one might prefer in some contexts. Organisms generally exhibit a variety of outputs and one can reasonably ask how they 'choose' among them with no preconceptions whatsoever about the determinants of the choice. The parallel reflex model with which I began this chapter would be a perfectly reasonable outcome of such an investigation. Prey items at different locations in space represent different inputs to the nervous system (because of the optics of the eye), and the act of 'choice' involved in selecting the right output for each stimulus presentation is materialised as a series of parallel reflex paths associating each retinal locus with a distinctive movement. Examples of choice in this sense are quite real and behaviourally significant (for both some examples and some subsequently recognised complexities see Wine, 1984; Stein, 1986; and Newland and Neil, 1990). At the same time they do not involve an underlying circuitry which associates several different outputs with a particular input; they instead associate each of a set of different outputs with each of a set of different inputs. For this reason, they lack something of the drama of what is meant by 'choice' in general usage.

The activity-gated divergent network linking a retinal and tectal locus with movement clearly does have the property of associating several different outputs with a particular input, so long as one keeps in mind that by input one means the signal originating in one eye, rather than in two. Such circuits are thus not only neurobiologically interesting, but represent something closer to what comes to most people's minds when one uses the word 'choice'. However, if one thinks of the input as occurring not on one channel but on two, as in the organism as a whole, then, as already noted, one could be dealing not with a system which associates a given input with several outputs, but rather with one that combines several inputs and associates with each combination a particular output. That the output associated with a given input is influenced by other inputs seems to me to reflect more of what is generally meant by 'choice', but an important something seems still to be missing and I expect readers will agree. The missing something, I suggest, is an answer to the question of whether a knowledge of all input signals would suffice to predict the movement, or whether something else is involved. If the latter, then one is still closer to what is generally meant by 'choice'.

At this point it becomes relevant that binocular convergence is not necessary for a frog's variation of movement with variation in stimulus distance. Frogs show such variations for stimuli in areas of the visual field seen by only one eye, and continue to vary their movements appropriately with variations in stimulus distance for stimuli in the binocular visual field even after interruption of one optic nerve (Ingle, 1972; Grobstein *et al.*, 1985). Frogs can use binocular convergence as a distance cue (Collett and Harkness, 1982) but, like most animals, have a number of other cues at their disposal (a point to which I will return in the following section). Among these cues, one which is particularly germane to the present discussion is accommodative state (Jordan *et al.*, 1980; Douglas *et al.*, 1986). Stimuli at different distances require different lens settings for their images to be sharply focused on the retina. Definitive experiments have not as yet been done, but it seems likely that what gates the divergent network from the tectum in this situation is not an afferent signal but rather a corollary discharge signal related to the motor output to the ciliary muscles (Douglas *et al.*, 1986). If this is so, then a knowledge of activity in all of the sensory pathways of the nervous system would still not suffice to predict a frog's movement in response to a stimulus at a given location. For different object distances a feedback mechanism will adjust the power of the lens, but once this has been accomplished all afferent signals to the nervous system will be the same, and still the resulting output will differ. The gating signal would thus not be an afferent signal at all, but rather one internal to the nervous system.

While still hypothetical in the case of the frog, the reality of different responses to the same input signal resulting from differences in ongoing activity in the nervous system rather than to other afferent signals is well documented in other systems (Sperry, 1950; von Holst and Mittelstaedt, 1950; Kovac and Davis, 1980, and more recently Moore-Ede *et al.*, 1982; Sparks, 1986; Huang and Satterlie, 1990; and Posner and Petterson, 1990). This, it seems to me, adds an important further level of sophistication to the potential display of choice by neuronal networks which are stable anatomically, synaptically, and in terms of intrinsic cell properties. Parallel systems can account for different outputs resulting from different inputs. Activity-gated divergent systems can account for different outputs resulting from a particular input and can do so by two important mechanisms: one involving additional significant afferent signals and the other involving internal changes in neuronal activity.

This breakdown, in neural terms, of the concept 'choice' seems to me useful not only in specifying some of the multiple meanings of the general concept, but also in identifying what additional features remain to be accounted for. Two are likely to occur to most readers: the sense of oneself having picked between alternatives, and the feeling that one is free to choose, that is that there is something more going on than simply responding to the environment. The latter is to some extent dealt with by the recognition that internally generated signals may determine the responses of activity-gated divergent circuitry, but the situation remains more deterministic than one might like. I will return to this issue, and briefly to the sense of oneself as an agent of choice, in the last two sections of this chapter. A useful breakdown of 'choice' in neural terms ought also to be helpful in providing a defining paradigm for posing novel experimental questions about neuronal organisation itself. Here too an aspect of frog orienting briefly mentioned earlier is illustrative. Even for a fixed stimulus location, the movement exhibited by a frog is not stereotyped, but rather displays significant variation. Two fairly obvious forms of this variation are relevant here, and I will describe a third later.

The visual field of each eye in the frog includes in excess of 225 degrees on the horizontal plane, meaning that a frog can not only see behind itself but can do so with both eyes (Grobstein *et al.*, 1980). Stimuli directly behind a frog trigger an orienting turn of 180°. This turn could in principal be either clockwise or counterclockwise. It is, in fact, either, with roughly equal probability (Kostyk and Grobstein, 1987a). Since the target is seen by both eyes, one might entertain the possibility that the clockwise/counterclockwise 'choice' relates to which of the two eyes is more effectively activated, and thus has a parallel pathway character to it. This hypothesis was tested by optic nerve section and it was found that both clockwise and counterclockwise turns persist (Kostyk and Grobstein, 1987a). The implication is that one is dealing with an activity-gated divergent circuit, and that the gating, or 'choice', must be occurring at some location central to the retina. Corresponding findings of persistent variation following unilateral tectal lobe lesions, and unilateral interruption of descending tectofugal pathways, imply an activity-gated divergent circuit subsequent to both of these locations, suggesting that the 'choice' is actually being made in the spinal cord. Some additional evidence identifying spinal cord as an important locus of motor choices in the frog will be given in the following section. My major point here is that one can fairly readily design experiments which not only help to specify the nature of a particular 'choice' along the lines discussed, but also yield information about its anatomical location.

A second documented variability in the movements made by frogs in response to stimuli at particular locations relates to the fact that frogs use different motor patterns for stimuli at different distances (Ingle, 1970; Comer and Grobstein, 1981; Grobstein *et al.*, 1985; Grobstein, 1988b). For nearer stimuli, what is involved is a 'snap' involving a lunge and a tongue flick. Over the range where it is used, the amplitude of this movement increases linearly with increasing stimulus distance. For more distant stimuli, the response is a 'hop', a reorienting movement which differs in a number of ways, including the absence of both lungeing and the tongue flicking. Frogs switch rather abruptly from one motor pattern to another at a characteristic distance, termed the 'snap/hop border' (the basis of this motor 'choice' will be discussed in the following section). Around this distance, however, there is a small zone of ambiguity where stimuli may elicit on any given trial either a snap or a hop. Since the snap/hop choice is in general related to stimulus distance, one might well entertain the hypothesis that the variation in movement within the zone of ambiguity reflects variation in the distance signal. An analysis of snap amplitudes, however, shows this not to be the case (Grobstein *et al.*, 1985). Snap amplitudes increase with stimulus distance throughout the zone of ambiguity, indicating that the distance signal does not in fact vary in such a way as to account for particular snap/hop choices. The distance signal must therefore be the input to a subsequent activity-gated divergent circuit, with some additional signal responsible for the actual motor choice. While the logic here is similar to that in the previous case of clockwise and counterclockwise turning, the situation is interestingly different in representing a distinctly more abstract analysis. For reasons discussed in the following section, the distance signal is unlikely to have any simple relation to a particular sensory input pathway, and to be instead an abstraction constructed at some as yet unknown central nervous system location. In short, what the present form of analysis allows is not only a characterisation of choice with regard to inputs and known pathways, but also a formal characterisation which can be used to predict the existence of abstract forms of organisation for use in locating and identifying particular central nervous system structures and signals.

I have not discussed the nature of the gating signal for either the snap/hop choice within the zone of ambiguity or the clockwise/counterclockwise choice. We do not in fact have any information on this matter in either case, but it is not difficult to imagine appropriate experiments to provide such information. The gating signal could be a second sensory signal, one related to posture, for example. It could also be a deterministic internal signal of the kind mentioned already, perhaps a corollary discharge signal or one related to motivational state. However, both examples, particularly that of clockwise/counterclockwise turning, have an air of 'what difference does it make?' This raises the possibility that one may be dealing with a non-deterministic internal gating signal, something equivalent to a coin toss (Grobstein, 1988c). The existence and possible usefulness of such a signal will be considered further in the final sections of this chapter.

13.3. **An intermediate spatial representation: gestalts and choices**

In the preceding sections I have been concerned with the overall character of the neuronal organisation which links sensory maps with output pattern generating circuitry in the case of anuran orienting behaviour, and with the contributions that an appreciation of this organisation can make to an understanding of the concept of 'choice'. An additional perspective on phenomena of 'choice' and, in particular, their relation to other 'higher order gestalts' comes from studies that reveal something of the details of the individual information processing steps which together yield the overall organisation of the anuran sensorimotor interface. Of particular importance in the present context are a series of observations suggesting that there exists in the interface a generalised, non-sensory, non-motor, three-dimensional representation of stimulus location.

The relevant studies (Kostyk and Grobstein, 1982, 1987a,b,c; Masino and Grobstein, 1989a,b; Grobstein *et al.*, 1988; Grobstein and Staradub, 1989) largely involve observations of behaviour following selected central nervous system lesions, a form of analysis whose validity, logic, and distinctive appropriateness for investigations of central nervous organisation in relation to behaviour have been discussed elsewhere (Grobstein, 1990a,b). The particular form of spatial representation observed appears not to be idiosyncratic to the frog but instead to be common to a wide

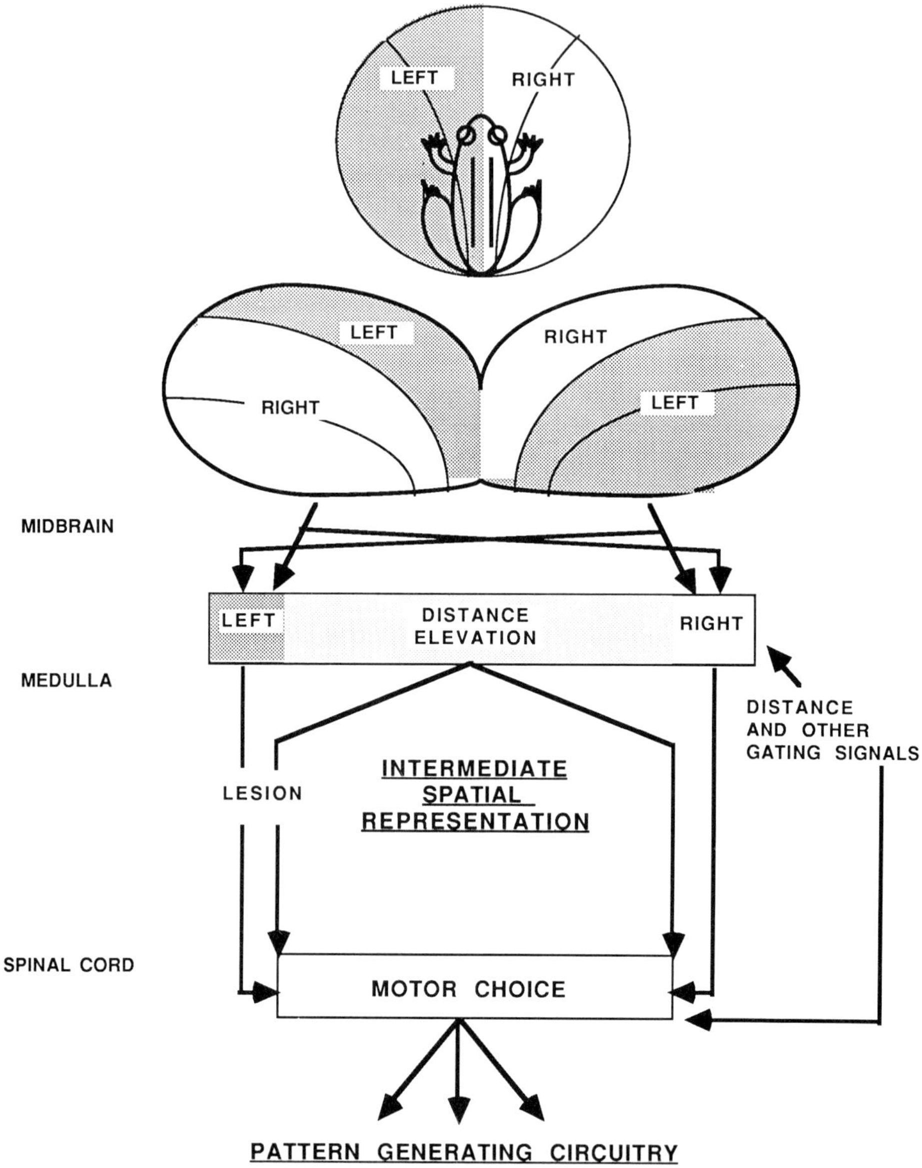

Fig. 13.3. Schematic illustration of a still more realistic model of the linkage between tectum and motor output in the case of frog orienting behaviour. The visual world of a frog and its representation on the two lobes of the optic tectum is shown in the top diagram. The frog should be imagined as centred in a hemispherical dome, on which the left visual hemifield is shaded and the edges of the visual fields of the two eyes (LFL, RFL) are indicated with curved lines. As illustrated, the visual field of each eye includes a full hemifield and extends across

array of vertebrate and perhaps invertebrate organisms (Grobstein, 1988b, 1989, 1990b), suggesting the existence of some important generalities about the neuronal organisation underlying directed movement and raising interesting questions as to why a rather diverse set of problems are solved in such similar ways (Grobstein, 1988c, 1990b). The observed spatial representation also appears to be distributed and fundamentally dependent on activity coding (Grobstein and Staradub, 1989; Masino and Grobstein, 1989a,b, 1990; Grobstein, 1991), suggesting that the map-like representations of spatial information seen near the nervous system periphery may not be a good model for more central representations of such information. While of interest in their own right, neither of these matters is directly germane to the present discussion, and I will confine myself to a description of those observations which bear most immediately on 'gestalts' and 'choices' (see Ewert, 1987, for a discussion in a different context of 'gestalt' phenomena in anuran prey orienting behaviour).

Figure 13.3 provides a schematic summary of our current understanding of the sensorimotor interface with an emphasis on the intermediate representation of stimulus location and its relation to the determination of motor output patterns. The key observation on which this figure is based is the finding that unilateral lesions of a defined fibre tract at locations between the midbrain and the spinal cord (Kostyk and Grobstein, 1987b; Masino and Grobstein, 1989a) produce a deficit whose character implies a disturbance in neuronal processing at some level between initial sensory analysis and motor choice. Frogs with such lesions do not fail to respond to prey items. Orienting movements are triggered by stimuli

---

the midsagittal plane to include a portion of the opposite hemifield. Each eye projects in its entirety to the opposite tectal lobe, so that, as illustrated, the midsagittal plane (upper curved line ML) and portions of visual space seen by both eyes (to lower curved line LFL, RFL) are represented in both tectal lobes.

Within the midbrain, but after the tectum, information about target location is transformed (arrows from tectum and box immediately below it) into an 'intermediate spatial representation,' as discussed in the text. Information about the horizontal eccentricity of a stimulus in the left and right hemifields is carried through the medulla along descending axons on the left and right sides of the brain respectively (arrows descending from left and right sides of box). Information about the distance and elevation of such stimuli is carried on a separate bilaterally descending pathway or set of pathways (arrows descending from middle of box). In consequence, a left sided 'lesion' prevents leftward turning, but spares other components of responses to stimuli on the left and has no effect on responses to stimuli on the right. Horizontal eccentricity and distance signals remain separate until the spinal cord. It is not until they converge that a 'motor choice' occurs and pattern generating circuitry can be appropriately activated. 'Distance and other gating signals' may be involved both in creating the central spatial representation, and in motor choice, as shown to the right.

presented at all locations in visual space. For stimuli in the hemifield contralateral to the lesion (clear area in Fig. 13.3), the orienting movements themselves are normal, and vary normally with variations in stimulus location in all three spatial dimensions. For stimuli in the hemifield ipsilateral to the lesion (shaded area in Fig. 13.3), however, the movements are abnormal, in a way which relates quite specifically to one of the three dimensions of stimulus location. The lesioned frogs show no variation in their orienting movements associated with variations in stimulus angle on the horizontal plane (horizontal eccentricity). The orienting movements are forwardly directed, regardless of the angle of stimulus presentation. Even more significantly, these orienting movements, while not varying with horizontal eccentricity, continue to vary appropriately with the elevation and distance of stimuli in the affected hemifield.

In addition to their continued responsiveness, two other features of the behaviour of the lesioned frogs indicate a disturbance which is subsequent to the initial stages of sensory processing. Unilateral lesions of the optic tectum, the locus of initial sensory processing, produce deficits in response to stimuli in regions of space defined by the borders of the visual fields of the eyes, instead of by the midsagittal plane as in the hemifield deficit described previously (Kostyk and Grobstein, 1982, 1987a). The difference is about 45° of visual angle (see Fig. 13.3), easily measured and quite significant. The hemifield deficit indicates disturbance of a process in which spatial information is represented in a head or body coordinate frame, rather than in the retinocentric coordinate frame of the eye and its projection to the tectum. The other aspect of the deficit in lesioned frogs which distinguishes it from a primary sensory disturbance is that the deficit is related specifically to one dimension of the three dimensions of stimulus location. In the retinotectal projection, each location codes simultaneously two dimensions of stimulus location, one of which is horizontal eccentricity (or something related to it). There is no way in which a tectal lesion could selectively affect only one dimension of stimulus location. One is dealing with a process in which components of a three-dimensional representation of spatial location exist independently of one another (lines carrying separate signals in the 'intermediate spatial representation' of Fig. 13.3), rather than being corepresented as in the initial retinotectal projection. Consistent with these two lines of evidence for a post-tectal process is a related series of findings suggesting that what is damaged by the lesions is a descending, indirect tectofugal projection containing axons of tegmental neurones which themselves receive tectal input, and it is presumably in the tecto-tegmental relay (upper box in Fig. 13.3) that the non-retinotopic representation of spatial information is created (Masino and Grobstein, 1989b, 1990).

That the deficit in lesioned frogs reflects a disturbance of neural processing at some level prior to motor choice is less obvious. Our initial hypothesis was that the deficit reflected the existence of distinct sets of pattern-generating circuitry, and of distinct pathways linking them with the tectum (Kostyk and Grobstein, 1982, 1987a). The failure of lesioned frogs to turn towards stimuli in one hemifield seemed straightforwardly understandable as a selective interruption of a pathway responsible specifically for activation of pattern-generating circuitry for horizontal turning in one direction, and the persistence of other movements and movement components was presumed to reflect sparing of additional, distinct pathways responsible for activation of turns in the opposite direction, as well as of movement components related to stimulus elevation and distance. The organisation of the descending pathways in Fig. 13.3 is consistent with this interpretation, except for the fact that they converge at a motor choice step rather than on distinct sets of pattern-generating circuitry. The reason for this additional feature is that the simpler motor hypothesis predicts that the movements of lesioned frogs in response to stimuli in the affected hemifield should be identical to those of normal frogs except for the missing horizontal turn component. More careful examination of the behaviour of lesioned frogs showed that this was not in fact the case: the lesions affected not movement *per se*, but rather motor pattern choice.

In the preceding section, I briefly described a snap/hop motor pattern choice which occurs with increasing stimulus distance, and it is this choice which is most dramatically altered, in a quite reliable way, by the lesions under discussion here. The switch between one motor pattern and the other normally occurs at a distance of two body lengths for stimuli in front of the frog on the midsagittal plane, and at one body length for stimuli at 90° of horizontal eccentricity. If the misdirected movements of lesioned frogs were normal except for the missing turn component, they should have consisted of forwardly directed snaps in response to lateral stimuli at distances up to one body length and of forwardly directed hops for stimuli at greater distances. What was actually observed was a snap/hop motor choice occurring at a distance of two body lengths, the same distance at which it occurs for frontal stimuli (Kostyk and Grobstein, 1987a; Masino and Grobstein, 1989a). In short, the abnormality in lesioned animals consists not only of responding with forwardly directed movements rather than turns, but also in making qualitative motor pattern choices which are appropriate for frontally rather than for laterally located stimuli.

The fact that lesioned animals snap for stimuli at locations where normal animals hop clearly indicates that the disturbance has altered the process of motor choice, and hence that the lesion must be disturbing something at this stage or prior to it, rather than something involved in the elaboration of particular motor output patterns. This something is most reasonably interpreted as a signal representing the horizontal eccentricity of the stimulus. Such a signal must be combined with one related to stimulus distance before the decision as to what movement to make is accomplished. The absence of a signal in the horizontal eccentricity pathway is apparently interpreted as indicating zero degrees of horizontal eccentricity (a frontal stimulus) and so the motor choices are those appropriate for frontal stimuli. This has several implications, one of which was briefly alluded to in an earlier section. The tegmental axons of the critical tract run all the way to the spinal cord, and the same behavioural deficits result from interrupting them at the junction of midbrain and medulla, or at the junction of medulla and spinal cord. The latter implies that the process of motor choice has not in fact occurred by the time signals leave the medulla, and hence that this process (and the associated box of Fig. 13.3) must be in the spinal cord.

The other, and perhaps more general, implication of the findings described is that this body-centred, distributed, three-dimensional form of spatial representation is, in reality, an intermediate spatial representation in the sense that its character is intrinsic to itself, reflecting organisational features of neither the sensory nor the motor sides of the nervous system. This is true not only in detail but also in a broad sense. As I have briefly discussed elsewhere (Grobstein, 1988b, 1989, 1990b), there is no obvious reason why a specifically three-dimensional representation of target location should exist at all. Associated with a stimulus at a given location in space are a large number of potentially independent input signals, certainly more than three, and similarly a large number of independent output signals, again certainly greater than three. I will here elaborate on this, since I think it not only sharpens the question of why the intermediate spatial framework exists but also sheds some further light on 'choice' and 'higher functional gestalts', which in turn helps to define some interesting neurobiological problems. What I want to suggest is that sensory 'gestalts' emerge as a consequence of going from neuronal signals of higher dimensionality to signals of lower dimensionality, that 'choice' and motor 'gestalts' are phenomena associated with going from lower dimensionality to higher dimensionality, and that both sorts of transition may require rather elaborate neuronal circuitry.

While the three-dimensional description of a given stimulus location may be obvious to the experimental observer, it is not apparent in the sensory pathways to the central nervous system. In the simple case of binocular vision, the input signal is not actually three-dimensional but rather four-dimensional, corresponding to a two-dimensional input signal from each eye. The situation is, however, even more complicated than this. As alluded to earlier, frogs may use binocular vision to extract a distance signal, but have additional cues available as well: monitoring of accommodative state and at least one or two others (Jordan *et al.*, 1980;

Collett and Harkness, 1982; Grobstein *et al.*, 1985; Douglas *et al.*, 1986; Collett and Udin, 1988). There is also reason to believe that a frog's visual localisation abilities take account of its own body size as well as its posture (Reyes and Grobstein, 1984; Grobstein *et al.*, 1985), so that the number of input variables involved in the production of a directed movement must be nine or greater. In addition, frogs can localise, in three dimensions (Comer and Grobstein, 1981) using tactile stimuli, suggesting that cutaneous input (in at least two dimensions) and some distance variable related to limb position (Grobstein *et al.*, 1988) may also need to be included in thinking about input dimensionality.

My point in all this is that stimulus location in three-dimensional space is not a sensory reality but rather an abstraction, something very close to a 'higher order functional Gestalt'. A given signal in the intermediate spatial representation is presumably not associated with any particular set of input signals but rather with any of a wide array of sets of input signals all corresponding to the same three-dimensional spatial location (what I have called elsewhere a 'bounded variance'; Grobstein, 1988c). Furthermore, these bounded variant sets are not continuous in any mathematical sense. Nearby distances may reflect similar binocular signals, or instead a given binocular signal in comparison with a particular accommodative signal. Nearby horizontal angles may reflect similar retinal eccentricities, or a given retinal eccentricity in comparison with a particular somatosensory signal. Consistent with this gestalt interpretation of the intermediate spatial representation in the frog is that lesions altering visually triggered turn angles produce qualitatively and quantitatively similar effects on turn angles triggered tactually (Grobstein *et al.*, 1988). Recently we have also found that damage to tracts carrying distance information produces similar effects on responses to stimuli in binocular and monocular visual fields (Grobstein and Staradub, 1989).

Two points follow from this, one having to do with the notion of gestalts, the other with its implications for studies of the nervous system. The first is that sensory 'gestalts' can be thought of as a consequence of reducing input dimensionality, of creating an internal signal of low dimensionality which corresponds to a bounded-variant set of a large number of input signals that exist in a higher dimensional space. The second is that describing the neuronal circuitry which underlies such a process is a non-trivial problem. Simple convergence of inputs will not do. What is needed is a form of neuronal organisation that will yield a given output for an array of quite different inputs, and which is further capable of using or ignoring signals along one input pathway depending on the presence or absence of input signals along another. The characterisation of such circuitry, even at a hypothetical level, is a significant challenge for neurobiology.

The dimensionality of neuronal signalling is higher not only on the input side of the intermediate spatial representation but also on its output side. Here too the actual number of independent variables is unknown but certainly greater than three. Frog orienting movements involve, at a minimum, displacement in three dimensions together with three degrees of rotational freedom, yielding at least six dimensions. Depending on the organisation of the underlying pattern-generating circuitry, the output dimensionality may be much higher (Grobstein, 1990b). What this suggests is that just as a bounded variant set of input combinations is associated with a given three-dimensional localisation signal, so too may it be the case that a given three-dimensional localisation signal is associated with a bounded variant set of outputs. We have recently begun to acquire evidence that this is indeed the case not only for turn direction, as described earlier, but quite generally (Grobstein *et al.*, 1990). A close analysis of frog orienting movements shows that an accurate turn towards a stimulus at a fixed location involves not only a rotation but also a two-dimensional displacement of the body, with the two components covarying so as to yield in a variety of different ways an accurate turn (Fig. 13.4; see Collett and Land, 1974, for a related phenomenon in a quite different organism). In short, an 'accurate' turn is itself an abstraction, a motor 'gestalt' whose origin is the association of a given central spatial representation signal with a bounded variant set of possible movements. Here, too, one encounters some quite challenging neurobiological problems: for example, what kind of neuronal circuitry produces an association between a given spatial signal and a bounded-variant set of movements? In addition, one re-encounters the problem of 'choice', of what is responsible for the particular movement exhibited on a particular occasion. I will return to this question in the next section of this chapter.

The central spatial representation not only provides a second example of 'higher behavioral theory' with roots deep in the 'simplest basal functions of the CNS', but also raises an intriguing question about 'functional Gestalts' in general: why do they exist? I have argued that there is no obvious explanation on either the sensory or motor sides of the nervous system for a specifically three-dimensional central spatial representation. The frog might in principle have been organised in some parallel reflex fashion, with each of the possible ensembles of inputs corresponding to a given location in space linked independently and appropriately to pattern-generating circuitry. Instead, one has a convergence from many input dimensions to three dimensions and then a divergence from three to many output dimensions, a bottleneck so to speak. It is not impossible that a key to the question of why this bottleneck exists will be found in general rules of information acquisition and processing (Grobstein, 1988c), that it is, in some sense, the optimal solution to the problem of relating large numbers of different kinds of information defined in dif-

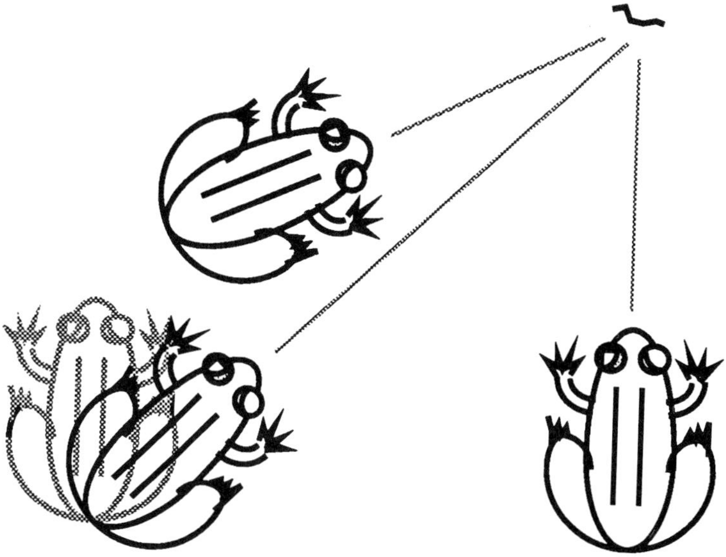

Fig. 13.4. Schematic diagram to illustrate one kind of variability in the responses of frogs to stimuli at a fixed target location. The end points of three different movements from a single starting position (lighter frog) are shown. The movement represented to the lower left involves a pure rotation of the body, that to the lower right a pure translation of the body with no rotation, and that above a combination of body rotation and translation. Note that, in all three cases, the frog achieves the objective of aligning its midsagittal plane with the stimulus (worm to upper right). An infinite number of other combinations of rotation and translation would similarly satisfy the objective. The variability normally observed for a target at the location shown is within this set of correlated rotations and translations. Pure translation is uncommon under the circumstances studied.

ferent coordinate frames (Grobstein, 1989, 1990b). It would be interesting if this were in some formal sense true of 'gestalts' in general.

13.4. **Motor equivalence and internal feedback: on the road to free will?**

The models illustrated Figs 13.2 and 13.3 represent a degree of sophistication in neuronal circuitry substantially greater than that of the model in Fig. 13.1, reflecting, as I have argued is necessary, aspects of both choice and higher order gestalt representations. At the same time they share with the model of Fig. 13.1, and indeed with many — perhaps most — models of neuronal organisation, a peculiar flavour of being somehow less than what most people feel behaviour to be. To call a nervous system organised like that in Fig. 13.3 a reflex machine is to miss the point of the intermediate spatial representation and of activity-gated divergence; but, at the same time, a nervous system like that in Fig. 13.3 seems to lack both purpose and playfulness. Given a knowledge of all inputs, and the value of some internal signals, one can reasonably expect to be able to predict the output of the system, and this output can reasonably be expected to occur whether or not it achieves any desired objective in a given context. In this section I want to review two quite new and still somewhat preliminary findings which suggest that the model of Fig. 13.3 will need further supplementation to approximate the actual neuronal organisation underlying anuran orienting, precisely because of its lack of elements which would yield a sense of purpose and playfulness. As in previous sections, my hope is both to use these general terms to help illuminate nervous system organisation and to use the nervous system to better understand the general terms, together with the related concept of free will.

One of the two sets of relevant new observations has already been mentioned: the finding that frogs can and do achieve accurate orienting turns in a variety of different ways. This is a novel finding in the context of frog orienting behaviour, but not in the context of motor control generally. A variety of workers in a variety of systems have noticed a similar variance in the actual motor output patterns underlying a series of apparently equally successful movements (Lashley, 1930; Bernstein, 1967; see also Möhl, 1989, Stein, 1989, for recent examples). The phenomenon has been termed 'motor equivalence', and variously perceived as a more or less biologically irrelevant but computationally interesting problem associated with going from low-dimensional to high-dimensional signals (Hogan, 1985; Pellionisz, 1985), or as evidence for successive translations from higher level to lower level motor commands (Lacquaniti and Soechting, 1982; Soechting, 1984), or as an indication of a desirable capability to adjust movements during their execution based on afferent or reafferent signals (Berkinblitt *et al.*, 1986; Abbs and Cole, 1987). In the frog, elements of the first two of these views of motor equivalence are clearly present. The intermediate spatial representation can certainly be regarded as a higher level motor command signal. Going from this three-dimensional signal to a motor output signal of greater dimensionality does mean that any of a number of different movements are potentially associated with a given localisation signal, and hence does raise what might be regarded as computational problems. My own inclination though, consistent with the entire direction of this chapter, is to regard such matters as problems of 'choice', and so to focus on the nature of the additional signals which gate the divergent circuitry linking the intermediate spatial signal with movement.

The possibility remains that the motor equivalence phenomenon in the frog is a reflection of afferent or reafferent-based adaptive adjustment of movements during their occurrence. However, for several reasons, I am inclined to suspect a different explanation for the observed motor variation, and a different biologically meaningful significance. We have as yet been unable to correlate the variation with other likely phenomena which might require movement adjustment, and frog movements are in general rather ballistic, with relatively little influence of afferent and reafferent signals during their execution. Moreover, the variation described, like those mentioned earlier, has, under the circumstances in which it was observed, a 'what difference does it make?' character. The movements of a given frog differ from trial to trial in a way that seems to have no pattern and appear almost random, but this makes no difference to their achieving their object.

I have noted elsewhere that an element of randomness may be a general feature of biological systems, and that such elements can, without difficulty, be conceived as concrete, material realities (Grobstein, 1988c). In the present case, a neurone or group of neurones with a truly random or quasi-random firing pattern could provide the relevant gating signal. For reasons I will come to in a moment, an element of this kind could play a quite significant biological role. It also adds a sense of playfulness to models of the neuronal organisation underlying orienting, bringing them more in accord with one's impression of frogs and of nervous systems generally. The model of Fig. 13.5 incorporates such an element as a testable hypothesis. In so doing, it also raises some intriguing new neurobiological questions. What kind of circuitry, for example, would allow a gating signal to act to generate a choice which is random but at the same time constrained to some outcome within a defined set of movements (those which successfully point the frog towards the target; the phenomenon, previously alluded to, of bounded variance, but with the added twist of an element of randomness).

The model of Fig. 13.5 differs from that of Fig. 13.3 in two additional ways, reflecting the second set of recent observations mentioned at the outset of this section. For years we have restricted our observations on anuran orienting behaviour to the initial movement triggered by a stimulus, since we were interested in the coupling between sensory maps and pattern-generating circuitry and did not want our interpretations confused by the possibility of afferent-dependent corrective movements. Over the past year we have begun to look not only at the first movement but at subsequent ones as well, particularly in those lesioned animals, described earlier, which fail to turn towards stimuli in one visual hemifield and whose initial movement is forwardly directed. As in normal frogs, however, the forwardly directed movements are not invariant, and may include small components of body rotation. What we have recently found is that such frogs may, by a sequence of small movements, end up pointing towards the target location even under conditions where the

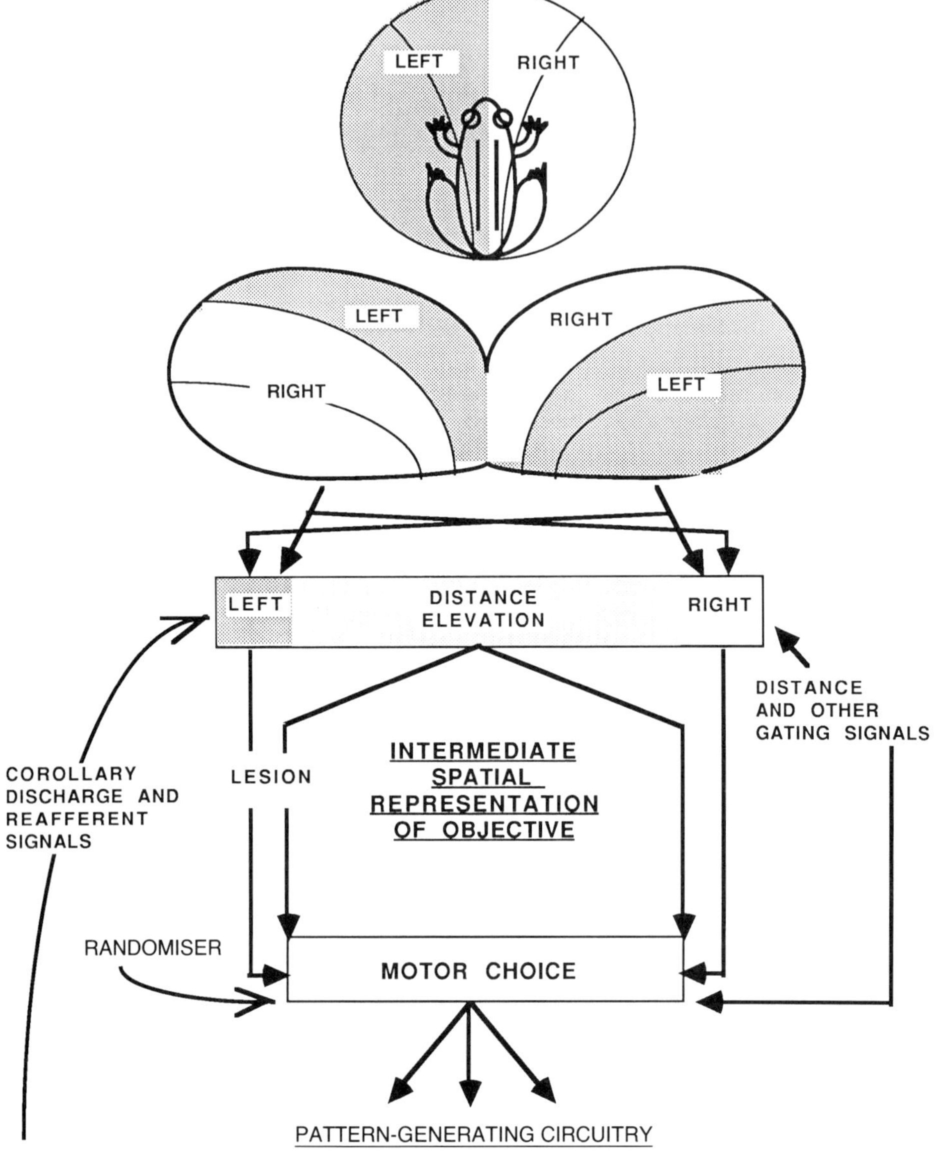

Fig. 13.5. Schematic illustration to show new directions in understanding the neuronal organisation underlying prey orienting movements in the frog. The model shown is identical to that of Fig. 13.3, with one slight modification ('intermediate spatial representation *of objective*') and two additions (to left). First, the output of the motor choice box is not determined by the central spatial representation but involves in addition a 'randomiser' input. Secondly, 'corollary discharge and reafferent signals' are used to determine whether a particular objective defined by the central spatial representation was in fact achieved. If not, a new movement (again with an element of randomness) is generated. Such a system, as discussed in the text, has the capacity to create novel, successful, motor pattern sequences.

target was removed subsequent to their first movement (Grobstein *et al.*, 1990).

These observations suggest that the spatial localisation signal, rather than being conceived simply as an intermediate signal generated transiently and then disappearing once transmitted to the spinal cord, might instead be better understood as a signal defining an objective to be accomplished, a signal which potentially can remain present for an extended period of time ('intermediate spatial representation of objective' in Fig. 13.5). They also suggest the existence of an internal feedback loop, making possible a comparison between the objective signal and signals indicating the actual movement accomplished. Once a motor choice is made, based on the objective signal, corollary discharge signals indicating the particular choice and reafferent signals indicating the actual effects of the motor pattern selected (to left in Fig. 13.5) could be compared with the objective signal itself. If the two match (as in much of normal behaviour), the objective signal is cancelled. If they fail to match, an objective signal persists and a new motor choice and movement occur, with the process potentially generating an extended sequence of movements that ultimately achieve the defined objective.

Feedback loops provide a well understood basis for endowing biological (or other) systems with properties that correspond to purpose, in the sense of displaying capabilities to adjust for unpredictable perturbing events. Internal feedbacks loops, of the kind illustrated in Fig. 13.5, were in fact a prominent characteristic of models developed by von Holst and Mittelstaedt forty years ago to account for directed movement (von Holst and Mittelstaedt, 1950). These models, however, differed in two important respects from the present one. First, the internal feedbacks and associated comparators were developed to account for the organisation of a given movement pattern, rather than for successive movements. More importantly, the earlier models contained no randomising element, and the combination of purposiveness and playfulness in the present model gives it a special flavour. The deterministic model of Fig. 13.3 will yield a particular motor output for each input, but will display no capability to correct with successive movements for unpredictable variations either in the organism or in the environment. The model of Fig. 13.5 will be capable not only of corrective movements but also of the generation of novel sequences of corrective movements, enabling the frog to explore and find new effective sequences regardless of the particular challenge. In short, it is capable of behaviour which is not only purposeful but also creative. An observer would be totally incapable of predicting the effective solution which will be generated in response to a problem. The combination of internal feedback and a randomising element seems to me to take one a long way along the road to what is generally meant by 'free will'. Exactly how far along that road I will consider in the final section of this chapter.

13.5. Epilogue: the frog and beyond

I began this chapter with an excerpt from a classic paper by von Holst and Mittelstaedt. There are substantial and historically interesting coincidences between that paper and a quite different but equally important paper published in the following year by Karl Lashley (Lashley, 1951), who wrote 'I am coming more and more to the conviction that the rudiments of every behavioral mechanism will be found far down in the evolutionary scale and also represented even in primitive activities of the nervous system.' What is common to both papers is a sense of connectedness among phenomena at quite different levels of organisation, a concern that the reflex perspective was standing in the way of recognising that connectedness, and a conviction that analysis of the nervous system should both contribute to and take account of a concern for higher order behavioural phenomena. My objective in this paper has been to try to show that recent research on the frog is consistent with this prospectus. The problem of motor choice in the frog is not reducible to a parallel series of reflex pathways (see Altman and Kien, 1987, 1989, and Kien and Altman, Chapter 9 of this volume, for similar conclusions with regard to other systems). Efforts to generate more realistic models of the underlying neuronal organisation have successfully made use of such higher order behavioural concepts as 'choice' and 'gestalt' and at the same time are useful in dissecting the multiple meanings inherent in such terms. Moreover, what has emerged is that much of the richness of behavioural phenomenology which many presume to be the exclusive province of the most elaborate nervous systems is present in the less elaborate nervous system of the frog, as both Lashley, and von Holst and Mittelstaedt, suspected.

There remain, however, some deferred issues with regard to both 'choice' and 'free will' which might be brought into the framework of the present discussion. I believe that they can, and that while the frog may well prove not to be an organism in which they can be further explored experimentally, it is none the less useful for the conceptual analysis. I have argued that the frog displays a fairly sophisticated level of 'choice', with mechanisms to associate a given input with one of a number of possible outputs in a way which cannot be predicted from a knowledge of the activity in all input pathways. I have also argued that it displays elements of 'free will', at least to the extent of being capable of generating truly creative solutions to problems. What is noteworthy is that it does this in the absence of any semblance of a developed neocortex, the structure so prominent in more elaborate brains. This suggests that an identity may exist between the basic functions of neocortex, and those last troubling bits of the concepts of both choice and of free will.

What was put aside for latter consideration in the case of 'choice' was 'the sense of oneself having picked between alternatives, and the feeling that one is free to choose', and I think most readers would include these characteristics, in addition to the ability to generate novel solutions, as ingredients of 'free will'. Significantly, there are various reasons to believe these may indeed be cortical functions. Cortical activity in humans, at least as measured by encephalography, is quite similar in the waking state and during periods of dreaming, both of which are characterised by a fairly active sense of oneself, and differs in states such as quiet sleep in which a sense of self is absent. Furthermore, it has become increasingly clear in recent years that a not infrequent sequela to cortical damage is persistence of behaviour in the absence of a sense of things happening to oneself (cf. Weiskrantz, 1986). Finally, it has long been known that the cortex exerts substantial inhibitory control over subcortical circuitry.

What the studies on the frog imply is that there is a quite sophisticated information processing capability in subcortical circuitry. Cortex, and its associated afferent and efferent projections, represent a parallel input/output system with respect to the external world, but it is also capable of functioning relatively independently of that world (as during dreaming). A second 'world' with which the cortical system has input/output relations is subcortical circuitry. It seems to me not unreasonable to suggest that a sense of self emerges from the monitoring of subcortical circuitry by the cortex, and that a sense of choice arises from the capability of the cortex to exercise an executive control on the potential outputs of that subcortical circuitry. If there is indeed a capability to generate novel outputs, and an oversight function capable of approving or vetoing particular outputs that are generated, then one is indeed picking among alternatives and free to choose, in a quite full and meaningful sense. There is something satisfying in that, whether or not it is a capability which the frog has as well.

Acknowledgements

A number of collaborators have contributed in the most meaningful ways to research discussed in this chapter, as indicated in the references cited. What is not possible to acknowledge in this way is the enormously significant contribution that large numbers of undergraduates with whom I have been involved at Bryn Mawr College have made to the development of the ideas and conceptual frameworks on which this chapter is based. It would not exist but for their combination of fresh curiosity and strong critical sense, displayed both in classrooms and as colleagues in my laboratory. Among these was Roian Egnor who, stubbornly and against my advice, pursued observations which proved of great significance for this chapter. I appreciate helpful comments on drafts of this manuscript from students and colleagues working in my laboratory during the summer of 1990. My research program is supported by NIH Grant R15 NS 24968 and a grant from the Whitehall Foundation.

References

Abbs, J. H. and Cole, K. J. (1987) Neural mechanisms of motor equivalence and goal achievement. In: *Higher Brain Function*, S. P. Wise, ed. Wiley, New York, pp. 15–44.

Altman, J. S. and Kien, J. (1987) A model for decision making in the insect nervous system. In: *Nervous Systems in Invertebrates*, M. A. Ali, ed. Plenum, New York, pp. 621–643.

Altman, J. S. and Kien, J. (1989) New models of motor control. *Neural Computation* **1**: 173–183.

Berkinblitt, M. B., Feldman, A. G. and Fukson, O. I. (1986) Adaptability of innate motor patterns and motor control mechanisms. *Behav. Brain Sci.* **9**: 585–638.

Bernstein, N. A. (1967) *The Coordination and Regulation of Movement*, Pergamon, Oxford.

Carew, T. J. and Sahley, C. L. (1986) Invertebrate learning and memory: from behavior to molecules. *Ann. Rev. Neurosci.* **9**: 435–488.

Collett, T. S. and Harkness, L. I. K (1982) Depth vision in animals. In: *Analysis of Visual Behavior*, D. J. Ingle, M. A. Goodale and R. J. W. Mansfield, eds. MIT Press, Cambridge, MA, pp. 111–176.

Collett, T. S. and Land, M. F. (1974) Visual control of flight behavior in the hoverfly, *Syritta pipiens*, L. *J. Comp. Physiol.* **99**: 1–66.

Collett, T. S. and Udin, S. B. (1988) Frogs use retinal elevation as a cue to distance. *J. Comp. Physiol.* **163**: 677–683.

Comer, C. and Grobstein, P. (1981) Tactually elicited prey acquisition behavior in the frog, *Rana pipiens*, and a comparison with visually elicited behavior. *J. Comp. Physiol.* **142**: 141–150.

Douglas, R. H., Collett, T. S. and Wagner, H.-J. (1986) Accommodation in anuran Amphibia and its role in depth vision. *J. Comp. Physiol.* **158**: 133–143.

Ewert, J.-P. (1987) Neuroethology of releasing mechanisms: prey-catching in toads. *Behav. Brain Sci.* **10**: 337–387.

Ewert, J.-P., Burghagen, H. and Schurg-Pfeiffer, E. (1983) Neuroethological analysis of innate releasing mechanisms for prey-catching behavior in toads. In: *Advances in Vertebrate Neuroethology*, J.-P. Ewert, R. R. Capranica and D. J. Ingle, eds. Plenum, New York, pp. 413–475.

Forssberg, H., Grillner, S. and Rossignol, S. (1975) Phase dependent reflex reversal during walking in chronic spinal cats. *Brain Res.* **85**: 103–107.

Fukson, O. I., Berkinblitt, M. B. and Feldman, A. G. (1980) The spinal frog takes into account the scheme of its body during the wiping reflex. *Science* **209**: 1261–1263.

Getting, P. A. (1989) Emerging principles governing the operation of neural networks. *Ann. Rev. Neurosci.* **12**: 185–204.

Grobstein, P. (1988a) On beyond neuronal specificity: problems in going from cells to networks and from networks to behavior. In: *Advances in Neural and Behavioral Development, Vol. III* P. G. Shinkman, ed. Ablex, Norwood, NJ, pp. 1–58.

Grobstein, P. (1988b) Between the retinotectal projection and directed movement: topography of a sensorimotor interface. *Brain Behav. Evol.* **31**: 34–48.

Grobstein, P. (1988c) From the head to the heart: some thoughts on similarities between brain function and morphogenesis, and on their significance for research methodology and biological theory. *Experientia* **44**: 961–971.

Grobstein, P. (1989) Organization in the sensorimotor interface: a case study with increased resolution. In: *Visuomotor Coordination: Amphibians, Comparisons, Models, and Robots*, J.-P. Ewert and M. A. Arbib, eds. Plenum, New York, pp. 537–568.

Grobstein, P. (1990a) Strategies for analyzing complex organization in the nervous system. I. Lesion experiments, the old rediscovered. In: *Computational Neuroscience*, E. Schwartz, ed. MIT Press, Cambridge, MA, pp. 19–37.

Grobstein, P. (1990b) Strategies for analyzing complex organization in the nervous system. II. A case study: directed movement and spatial representation in the frog. In: *Computational Neuroscience*, E. Schwartz, ed. MIT Press, Cambridge, MA, pp. 242–255.

Grobstein, P. (1991) Directed movement in the frog: a closer look at a central representation of spatial location. In: *Visual Structures and Integrated Functions, Research Notes in Neural Computing*, M. A. Arbib and J.-P. Ewert, eds. Springer-Verlag, New York, pp. 125–138.

Grobstein, P., Comer, C. and Kostyk, S. K. (1980) The potential binocular field and its tectal representation in *Rana pipiens*. *J. Comp. Neurol.* **190**: 175–185.

Grobstein, P., Comer, C. and Kostyk, S. K. (1983) Frog prey capture behavior: between sensory maps and directed motor output. In: *Advances in Vertebrate Neuoroethology*, J.-P. Ewert, R. R. Capranica and D. J. Ingle, eds. Plenum, New York, pp. 331–347.

Grobstein, P., Crowley, K. and Spiro, J. (1988) Neuronal organization for directed movement in the frog: similarities in visual and tactile prey orienting. *Soc. Neurosci. Abstr.* **14**: 1236.

Grobstein, P., Meyer, J. and Egnor, R. (1990) Directed movement in the frog: motor equivalence, multi-dimensionality, internal feedback? *Abstr. Soc. Neurosci.* **16**: 117.

Grobstein, P., Reyes, A., Zwanziger, L. and Kostyk, S. K. (1985) Prey orienting in the frog: accounting for variations in output with stimulus distance. *J. Comp. Physiol.* **156**: 775–785.

Grobstein, P. and Staradub, V. (1989) Frog orienting behavior: the descending distance signal. *Soc. Neurosci. Abstr.* **15**: 54.

Harris-Warwick, R. M., Flamm, R. E., Johnson, B. R. and Katz, P. S. (1990) Modulation of neural circuits in Crustacea. *Amer. Zool.* **29**: 1305–1320.

Hogan, N. (1985) The mechanics of multi-joint posture and movement control. *Biol. Cyber.* **52**: 315–331.

Hoy, R. R. and Nolen, T. G. (1987) The role of behavioral context in decision making by an identified interneuron in the cricket. In: *Higher Brain Function*, S. P. Wise, ed. Wiley, New York, pp. 133–151.

Huang, Z. and Satterlie, F. A. (1990) Neuronal mechanisms underlying behavioral switching in a pteropod mollusc. *J. Comp. Physiol.* **166**: 875–887.

Ingle, D. J. (1970) Visuomotor functions of the frog optic tectum. *Brain Behav. Evol.* **3**: 57–71.

Ingle, D. J. (1972) Depth vision in monocular toads. *Psychonom. Sci.* **29**: 37–38.

Ingle, D. J. (1983) Brain mechanisms of visual localization by frogs and toads. In: *Advances in Vertebrate Neuoroethology*, J.-P. Ewert, R. R. Capranica and D. J. Ingle, eds. Plenum, New York, pp. 177–226.

Jordan, J., Luthart, G., Meyer-Naujoks, C. and Roth, G. (1980) The role of eye accommodation in the depth perception of common toads. *Z. Naturforsch.* **35**: 851–851.

Kien, J. and Altman, J. S. (1991) Decision-making in the insect nervous system: a model for selection and maintenance of motor programmes. In: *Neurobiology of Motor Programme Selection: New approaches to the study of behavioural choice*, J. Kien, C. R. McCrohan and W. Winlow, eds. Manchester University Press, Manchester, pp. 147–169.

Kostyk, S. K. and Grobstein, P. (1982) Visual orienting deficits in frogs with various unilateral lesions. *Behav. Brain Res.* **6**: 379–388.

Kostyk, S. K. and Grobstein, P. (1987a) Neuronal organization underlying visually elicited prey orienting in the frog. I. Effects of various unilateral lesions. *Neuroscience* **21**: 41–55.

Kostyk, S. K. and Grobstein, P. (1987b) Neuronal organization underlying visually elicited prey orienting in the frog. II. Anatomical studies on the laterality of central projections. *Neuroscience* **21**: 57–82.

Kostyk, S. K. and Grobstein, P. (1987c) Neuronal organization underlying visually elicited prey orienting in the frog. III. Evidence for the existence of an uncrossed descending tectofugal projection. *Neuroscience* **21**: 83–96.

Kovac, M. P. and Davis, W. J. (1980) Neuronal mechanisms underlying behavioral choice in *Pleurobranchea. Science* **198**: 469–487.

Lacquaniti, F. and Soechting, J. F. (1982) Coordination of arm and wrist motion during a reaching task. *J. Neurosci.* **2**: 399–408.

Lashley, K. S. (1930) Basic neural mechanisms in behavior. *Psych. Rev.* **37**: 1–24.

Lashley, K. S. (1951) The problem of serial order in behavior. In: *Cerebral Mechanisms in Behavior*, L. A. Jeffries ed. Wiley, New York, pp. 112–136.

Masino, T. and Grobstein, P. (1989a) The organization of descending tectofugal pathways underlying orienting in the frog, *Rana pipiens*. I. Lateralization, parcellation, and an intermediate spatial representation. *Exp. Brain Res.* **75**: 227–244.

Masino, T. and Grobstein, P. (1989b) The organization of descending tectofugal pathways underlying orienting in the frog, *Rana pipiens*. II. Evidence for the involvement of a tecto-tegmento-spinal pathway. *Exp. Brain Res.* **75**: 245–264.

Masino, T. and Grobstein, P. (1990) Tectal connectivity in the frog, *Rana pipiens*: tecto-tegmental projections and a general analysis of topographic organization. *J. Comp. Neurol.* **291**: 103–127.

Möhl, B. (1989) 'Biological noise' and plasticity of sensorimotor pathways in the locust flight system. *J. Comp. Physiol.* **166**: 75–82.

Moore-Ede, M. C., Sulzman, F. M. and Fuller, C. A. (1982) *The Clocks That Time Us*. Harvard University Press, Cambridge, MA.

Newland, P. L. and Neil, D. M. (1990) The tail flip of the Norway lobster, *Nephrops norvegicus*. II. Dynamic righting reactions induced by body tilt. *J. Comp. Physiol.* **166**: 529–536.

Oyama, S. (1985) *The Ontogeny of Information*, Cambridge University Press, Cambridge.

Pellionisz, A. J. (1985) Tensorial brain theory in cerebellar modelling. In: *Cerebellar Functions*, J. R. Bloedel, J. Dichgans and W. Precht, eds. Springer-Verlag, Berlin, pp. 201–229.

Posner, M. I. and Peterson, S. E. (1990) The attention system of the human brain. *Ann. Rev. Neurosci.* **13**: 25–42.

Reyes, A. and Grobstein, P. (1984) Frog prey orienting: effects of cerebellectomy. *Soc. Neurosci. Abstr.* **10**: 61.

Soechting, J. F. (1984) Effect of target size on spatial and temporal characteristics of a pointing movement in man. *Exp. Brain Res.* **54**: 121–132.

Sparks, D. L. (1986) Translation of sensory signals into commands for control of saccadic eye movement: role of primate superior colliculus. *Physiol. Rev.* **66**: 118–171.

Sperry, R. W. (1950) Neural basis of the spontaneous optokinetic response produced by visual inversion. *J. Comp. Physiol. Psych.* **43**:482–489.

Stein, P. S. G. (1986) The forms of a task and their blends. In: *Neurobiology of Vertebrate Locomotion*, S. Grillner, P. S. G. Stein, H. Forssberg, D. G. Stuart and R. M. Herman, eds. Macmillan, London, pp. 201–216.

Stein, P. S. G. (1989) Spinal cord circuits for motor pattern selection in the turtle. *Ann. NY Acad. Sci.* **563**: 1–10.

Holst, E. Von and Mittelstaedt, H (1950) Das Reafferenzprinzip. *Naturwiss.* **37**: 464–476. (English translation in Holst, E. von, *The Behavioral Physiology of Animals and Man*. Vol. I, University of Miami Press, Miami, Fla, 1973, pp. 139–173).

Weiskrantz, L. (1986) *Blindsight*. Oxford University Press, Oxford.

Wine, J. J. (1984) The structural basis of an innate behavioral pattern. *J. exp. Biol.* **22**: 283–319.

Part IV

Epilogue

14 *Jenny Kien, Catherine R. McCrohan and William Winlow*

Epilogue
Deciding what to do next

14.1. What have we learned about the selection of behaviour?

Functional similarities are apparent in all the systems considered here, from molluscs to humans. These should allow us to extrapolate organisational principles from one system to another. Even obvious morphological differences between invertebrate and vertebrate nervous systems, such as the far greater number of neurones in vertebrates, do not appear to result in radically different principles of functional organisation. Thus, one 'neurone' in a model system (be it a real or a simulated system) can represent a single invertebrate neurone or a group of neurones in a vertebrate. Also, the constancy of individual neurones in invertebrates may be echoed in vertebrates by a constancy at the level of groups of cells.

14.2. Principles of organisation

We have extracted the following general points from the contributions presented in this book. This list is by no means exhaustive.

(1) *Small networks may be multifunctional.* Many factors shape the final output of small networks so that their full range of functions is achieved only when the network remains embedded in the rest of the CNS. Small multifunctional or multitask networks are thus subunits of the whole system (Marder and Weimann, Chapter 1).

(2) *Flexibility is achieved by polymorphic neuronal ensembles.* The reports in this book all suggest that changes in motor patterns or different motor outputs can be produced by altering the ensemble of neurones active within a network either in terms of the relative activities of different neurones or in terms of the numbers of active units. This indicates to us that polymorphic networks may be the optimal means of achieving the enormous range of outputs required for highly flexible behaviour. The ensembles may be selected by:

(a) different inputs; for example, in humans the lateral premotor system responds to external or environmental inputs while the medial system responds more to endogenous inputs (Goldberg, Chapter 12);

(b) different neuromodulators, as in the crustacean stomatogastric system (Marder and Weimann, Chapter 1);

(c) the self-organising properties of the network itself; in the frog (Grobstein, Chapter 13) the same visual input can produce two entirely different motor responses, i.e. a snap or a hop.

(3) *Each behaviour involves large numbers of neurones*. Every behaviour, including 'simple' invertebrate behaviours, involves large numbers of neurones. Even simple withdrawal or bending responses (Falk *et al.*, Chapter 6; Kristan *et al.*, Chapter 10) involve many neurones, and more complex invertebrate behaviours, such as egg-laying, respiration, feeding, stridulation and locomotion may involve hundreds (or even thousands) of neurones (Ter Maat *et al.*, Chapter 2; McCrohan and Kyriakides, Chapter 3, Winlow *et al.*, Chapter 4, Hennig, Chapter 7, Kien and Altman, Chapter 9). Large numbers of neurones are also involved in apparently simple changes in ongoing motor pattern, e.g. the recruitment of new systems with increase in swimming speed in fish (Roberts and Mos, Chapter 8), or changes in swimming activity in tadpoles (Soffe, Chapter 5).

(4) *Motor output is achieved by parallel distributed processing*. Functions cannot be localised to a single cell or nucleus but rather to the networks in which these are embedded. The interconnections between the neurones structure the networks into multilayered architectures where the different layers function at different levels of computational complexity. In some cases, an extra layer is incorporated when a new output is needed, e.g. the white muscle system to increase swimming speed in fish (Roberts and Mos, Chapter 8).

These distributed systems, found in both invertebrates (McCrohan and Kyriakides, Chapter 3; Winlow *et al.*, Chapter 4; Kien and Altman, Chapter 9; Kristan *et al.*, Chapter 10) and vertebrates (Alexander and Crutcher, Chapter 11; Goldberg, Chapter 12; Grobstein, Chapter 13) argue for parallel distributed processing (PDP) models and against sequential or linear hierarchical command models. Behavioural choice in PDP models depends on the total input to the system which moves it to a new state, representing the choice made. Importantly, the decision is not computed analytically or serially; rather, parallel and distributed processing achieves a rapid change across the whole system.

This PDP approach can be extended to the analysis of the neuronal basis of volition or free will. The 'feeling of making a decision' (Goldberg, Chapter 12), and even free will (Grobstein, Chapter 13), may arise from the same sort of mechanism as proposed in the other papers for selection of motor programmes or for their modification.

14.3. **Where does this lead?**

We hope that the contributions in this book will further promote the systems approach to the neuronal analysis of selection of motor programmes. Neurocomputational models may play a crucial role for our understanding of behavioural choice. Not only do they provide a precise tool: their use also teaches us a new way of thinking and a new approach to experimental and theoretical analysis of the nervous system (e.g. Kristan *et al.*, Chapter 10; see also Kien and Altman, Chapter 9; Alexander and Crutcher, Chapter 11). However, neurocomputation alone will not solve all our problems; detailed simulations are only useful for small, well characterised networks whereas large and/or poorly characterised networks can only be modelled theoretically. New techniques for multi-unit recording such as the use of voltage-sensitive dyes (Falk *et al.*, Chapter 6) will also prove useful.

We predict that future research will include more analysis of the relationships between groups of neurones embedded in networks. If we are to progress, neuronal systems must be studied under more natural conditions than at present, and during changes of conditions. This will require the use of less invasive recording techniques and minimally dissected or chronic preparations. We will also need to know more about the properties of large systems, e.g. effects of changes in the input parameters on whole networks. This may require formal mathematical description.

Index